Artificial Intelligence in Medical Software

Ajit Pandey • Pramod Gupta
Naresh Kumar Sehgal

Artificial Intelligence in Medical Software

Springer

Ajit Pandey
San Juan Capistrano, CA, USA

Pramod Gupta
Brentwood, CA, USA

Naresh Kumar Sehgal
Santa Clara, CA, USA

ISBN 978-3-032-10807-4 ISBN 978-3-032-10808-1 (eBook)
https://doi.org/10.1007/978-3-032-10808-1

This Springer imprint is published by the registered company Springer Nature Switzerland AG
The registered company address is: Gewerbestrasse 11, 6330 Cham, Switzerland

Ajit dedicates this book to his late parents, whose stories live on in his heart, and to his children, whose dreams and wonders inspire him every day.

Pramod dedicates this book to his late parents, who have always inspired him throughout his life and career, and to his wife and children for their unwavering support.

Naresh dedicates this book to his mentor, Prof. Pramod Chandra P. Bhatt, for his ongoing encouragement and support, which allowed Naresh to pursue the art of writing a book. Prof. Bhatt taught Naresh the importance of continuously seeking new knowledge and striving for greater perfection.

Disclaimer

The views, thoughts, and opinions expressed in this book are solely those of the authors and do not necessarily reflect the views, policies, or positions of any employer, organization, or institution with which any author may be affiliated. This work was created independently by the authors and without the use of any employer resources.

Foreword

Artificial intelligence (AI) is no longer a distant frontier—it is here, and it is transforming healthcare in ways we could not have imagined a decade ago. For those of us working in the medical device industry, this technological shift is unprecedented. This book, *Artificial Intelligence in Medical Device Software*, arrives at a pivotal moment, offering a comprehensive and practical guide for AI beginners and professionals alike, navigating this complex and rapidly evolving landscape.

As someone who has spent the better part of two decades at the intersection of data science, healthcare, and regulated product development, I have seen firsthand how AI/ML is reshaping the medical and healthcare industry. I have seen the full spectrum of AI/ML application development from early-stage prototyping to enterprise-scale deployment. When applied thoughtfully, AI/ML can unlock new opportunities, improve patient outcomes, and reduce patient risk. But I have also seen how easily it can go wrong, when promising models fail in deployment due to bias in the training data, and regulatory misalignment derails even the most technically sound solutions.

That is why this book matters.

The authors, Ajit Kumar Pandey, Pramod Gupta, and Naresh Kumar Sehgal, bring together a wealth of experience from academia, industry, and regulatory domains. Their combined expertise is evident in the depth and clarity with which they address the full spectrum of AI/ML-enabled application development process for medical devices.

One of the strengths of this book is its life cycle perspective. AI in healthcare is not a one-time event—it is a continuous process. From foundational concepts to advanced deployment strategies, the book provides a clear road map for engineers, data scientists, and healthcare professionals. For AI beginners, it is a clear and structured introduction to the life cycle of AI/ML in medical devices. For advanced AI practitioners, this book goes deeper, offering nuanced insights grounded in real-world experience.

The inclusion of real-world case studies, such as Apple's Sleep Apnea Notification and Notal Vision's Home OCT system, adds a layer of practical relevance. These examples illustrate how AI/ML technologies are already reshaping disease diagnosis and treatment and equip the readers with knowledge to build safe, effective, and compliant AI/ML-enabled medical devices.

This book also speaks to the future. It explores the role of generative AI in healthcare, the potential of federated learning, and the challenges of deploying models at the edge. It discusses how cloud-native architectures, containerization, and microservices are reshaping the way we think about scalability and latency. It emphasizes the importance of human factors and how to design systems that are not just intelligent, but usable, interpretable, and safe.

If you are just starting out in AI/ML, this book will give you the confidence and clarity to move forward. If you are an experienced AI practitioner, this book will help you refine your approach and build AI systems that scale, adapt, and endure. Whatever your AI journey is, this book will serve as a trusted companion.

In short, this book is a bridge. It connects the technical with the clinical, the theoretical with the operational, and the beginner with the expert. It is a resource I wish I had when I started, and one I am glad to have now.

Data Science
Alcon R&D,
Vernier, Geneva, Switzerland

Ramesh Sarangapani

Preface

Artificial Intelligence (AI) and Machine Learning (ML) have transformed the development and deployment of medical device software, bringing new risks and challenges compared to traditional methods. This book starts with the history of medical software development. It then covers the current state of AI/ML-enabled software, including on-premise and public cloud deployments. The latter involves security issues, and we will suggest some possible solutions. Readers will learn how to create strategies and implement development processes that comply with regulatory and quality system requirements, ensuring security, performance, and flexible deployments. The book presents a new approach to AI/ML software development and deployment, different from traditional methods.

The idea for this book came from a spontaneous conversation between Ajit and Naresh, both of whom had previously worked at a medical device startup that used AI and ML to diagnose brain disorders. We noticed there was no comprehensive guide exploring the practical use of these emerging technologies in medicine. We also faced significant regulatory hurdles, which we successfully overcame. To share our product development journey, we consulted Professor Pramod Gupta, who had briefly contributed to our R&D team. Through their collaboration, we have put together this book for your reading enjoyment.

This book offers a comprehensive guide to AI/ML-enabled software development for medical devices, including practical projects. It builds on core software development methods used in the medical device field, emphasizing the unique features of AI/ML-enabled devices and introducing an innovative development approach. The book details each phase of development for both Software in Medical Devices (SiMD) and Software as a Medical Device (SaMD), including cloud-enabled solutions. It also explains how companies can meet current industry standards for security, privacy, risk management, and compliance with global medical device regulations, all while supporting efficient and timely new product launches.

The medical device industry is subject to stringent global regulations. Compliance with US FDA Quality System Regulations, EU Medical Device Regulations, and IEC 62304 (Medical Device Software Lifecycle Processes) is essential for typical medical device development. Regulatory agencies worldwide are issuing or planning to issue guidance documents to ensure that AI/ML-enabled medical devices meet safety and effectiveness standards while adhering to the current regulatory requirements for patient safety. This book will provide guidance on each stage of software development for AI/ML-enabled medical devices, introducing AI/ML product development processes and lifecycle management, which differ from traditional medical device software development processes.

Readers will learn the fundamentals of machine learning algorithms and data handling. The book explores core machine learning algorithms, training structures, and choosing suitable performance metrics. It covers various deployment environments, including on-premise devices on local networks and cloud-based systems. The book also offers insights into clinical validation methods and the implementation of machine learning operations. Additionally, it discusses AI-enabled medical devices, federated learning, and examines the US FDA's approach to marketing approval of these devices.

Writing this book has been a labor of love, and we are grateful for the opportunity to share it with you. If you have any feedback, please get in touch with the authors at aimds2025@gmail.com.

San Juan Capistrano, CA, USA — Ajit Pandey
Brentwood, CA, USA — Pramod Gupta
Santa Clara, CA, USA — Naresh Kumar Sehgal

Acknowledgments

Ajit sincerely thanks Naresh for his unwavering love and support in helping him write this book. He also recognizes Diane Bryant and Raaj Venkatesan for the opportunity to work briefly at a medical startup. During that time, he was introduced to Naresh and collaborated with Gary Poole, a medical software expert. Ajit appreciates Ramesh Sarangapani and Kavin Baker for their valuable guidance.

Pramod would like to acknowledge his students for their constant feedback and encouragement in the area of artificial intelligence and machine learning during his teaching at various universities. Their feedback helped him to improve the content in this book.

Naresh wants to thank Prof. PCP Bhatt for his ongoing guidance and support. Prof. Bhatt has extensive knowledge of computer hardware and software systems, which he generously shares without expecting anything in return. Naresh also appreciates Diane Bryant, a wonderful person and former CEO of NovaSignal, who hired him as the VP of Cloud Engineering at her medical startup. This role introduced Naresh to many applications of Cloud Technology in medical data and healthcare. Additionally, this book incorporates several security ideas from Prof. John M. Acken at Portland State University, Oregon. Naresh has learned a lot of hardware and software techniques while working under John at Intel.

We thank Charles Glaser and Brian Halm for their excellent support, as well as the staff at Springer Publications, for helping to bring this book to you. Various trademarks mentioned in this book are owned by their respective companies.

Competing Interests The authors have no competing interests to declare that are relevant to the content of this manuscript.

Contents

1 Introduction 1
- 1.1 Introduction 2
- 1.2 Types of Medical Device Software 3
  - 1.2.1 Software as Part of a Medical Device, also Known as Software in Medical Device (SiMD) 3
  - 1.2.2 Software-Only Medical Device, also Known as Software as a Medical Device (SaMD) 4
  - 1.2.3 Software-Only Product for Other Health Use 4
  - 1.2.4 Software as Part of Specific Health Hardware 4
- 1.3 Medical Device Software Development Process 5
- 1.4 Regulatory Framework for Medical Device Software 6
  - 1.4.1 United States (US) 6
  - 1.4.2 European Union (EU) 7
  - 1.4.3 Global Harmonization 7
- 1.5 Medical Device Software Lifecycle Management 8
  - 1.5.1 Overview of IEC 62304 8
  - 1.5.2 Critical Components of IEC 62304 9
- 1.6 US FDA Premarket Submission Guidance for Medical Device Software Functions 11
- 1.7 AAMI TIR45: Agile Process in Medical Device Software Development 13
  - 1.7.1 Project Success with Agile Software Development 15
- 1.8 Cloud Software Development for Medical Devices 16
  - 1.8.1 Benefits of Cloud-Based Medical Device Software 16
  - 1.8.2 Challenges in Cloud-Based Medical Device Software Development 17
- 1.9 Embedded and Application Software Development 18
  - 1.9.1 Challenges in Embedded Software Development 18

1.10 AI/ML Software in Medical Devices . . . 19
1.10.1 Cloud-Based AI/ML Medical Device Software . . . 20
1.10.2 On-Prem and Embedded AI/ML Medical Device Software . . . 20
1.11 Artificial Intelligence Lifecycle Framework . . . 20
1.11.1 US FDA's Good Machine Learning Practices (GMLP) . . . 23
1.11.2 Machine Learning Operation (MLOps) . . . 26
1.12 AI/ML-Enabled Medical Device Software Clinical Evaluations . . . 27
1.13 Cybersecurity in Medical Device Software . . . 28
1.13.1 Privacy in AI/ML-Enabled Medical Device Software . . . 29
1.13.2 Consent and Data Subject Rights . . . 30
1.13.3 Data Protection Measures . . . 30
1.14 The EU AI Act and AI/ML-Enabled Medical Device Software . . . 31
1.15 This Book's Organization . . . 32
1.16 Summary . . . 33
1.17 Points to Ponder . . . 34
1.18 Answers . . . 34
References . . . 35

2 Data Management for AI/ML-Enabled Medical Devices . . . 37
2.1 Introduction . . . 37
2.2 Clinical Data Needs . . . 38
2.2.1 Creating Infrastructure for Targeted Data Collection . . . 39
2.2.2 Adding Value for Providers and Patients . . . 39
2.3 Types of Healthcare Data . . . 40
2.3.1 Electronic Health Records and Electronic Medical Records . . . 40
2.3.2 Administrative Data . . . 41
2.3.3 Claims Data . . . 41
2.3.4 Patient/Disease Registries . . . 42
2.3.5 Health Surveys . . . 43
2.3.6 Clinical Trial Data . . . 44
2.4 Why Is Data Important in Healthcare? . . . 44
2.4.1 Market Competitiveness . . . 44
2.4.2 Well-Informed Decisions . . . 45
2.4.3 Reduction of Costs . . . 45
2.4.4 Implementation of Artificial Intelligence (AI) . . . 45
2.4.5 Outbreak Prevention . . . 46
2.4.6 Storage of All Data in One Place . . . 46
2.4.7 Why Are Databases Critical in Healthcare? . . . 46
2.5 Data Analytics . . . 47
2.5.1 Insurance Claims KPI Dashboards . . . 48
2.5.2 Medical Claims Fraud . . . 48
2.5.3 Inpatient Readmissions . . . 48
2.5.4 Bed Utilization Forecasting . . . 49
2.5.5 Comorbidity Predictions and Analysis . . . 49
2.6 Tips for Accessing Different Types of Healthcare Data . . . 49

2.6.1 Define the Healthcare Data Needed for Your Project 49
2.6.2 Develop Multi-Sector Partnerships 50
2.6.3 Keep in Mind the Regulations on Healthcare Data Sharing.................................... 50
2.7 Collecting Meaningful Data in a Clinical Research Study 50
2.7.1 Questionnaire Surveys and Patient-Reported Data 51
2.7.2 Proxy/Informant Data 52
2.7.3 Review of Ambulatory or Hospital Medical Records 53
2.7.4 Collection of Biological Material 54
2.8 Clinical Data Management (CDM) 56
2.8.1 Why Is CDM Essential?........................... 59
2.9 Data Management Requirements for AI in Medical Devices 59
2.9.1 Data Collection 60
2.9.2 Data Cleaning/Processing 61
2.9.3 Reference Standard 61
2.9.4 Data Annotation................................. 62
2.9.5 Data Storage.................................... 62
2.9.6 Management and Independence of Data 62
2.9.7 Representativeness............................... 63
2.10 Data Quality Framework for AI/ML per EU AI Act 63
2.10.1 Data Lifecycle Model 64
2.10.2 Data Quality Plan................................ 66
2.10.3 Data Quality Management Lifecycle.................. 67
2.11 Summary 68
2.12 Points to Ponder.. 68
2.13 Answers 68
References.. 71

3 Machine Learning Algorithms.............................. 73
3.1 Introduction 74
3.2 Supervised Machine Learning Algorithms 76
3.2.1 Regression 77
3.2.2 Classification 77
3.3 Machine Learning Algorithms That Use Supervised Learning 77
3.3.1 Unsupervised Machine Learning Algorithms 77
3.3.2 Clustering...................................... 78
3.3.3 Dimension Reduction............................. 78
3.4 Machine Learning Algorithms That Use Unsupervised Learning 79
3.5 Considerations When Choosing an Algorithm................. 79
3.6 Performance Metrics of ML Algorithms 82
3.7 What Are the Most Common and Popular Machine Learning Algorithms? 82
3.7.1 Linear Regression 82
3.7.2 K-Nearest Neighbors (KNN)........................ 86
3.7.3 Logistic Regression 89
3.7.4 Naïve Bayes Classifier Algorithm 92

3.7.5 Support Vector Machine Algorithm 96
3.7.6 Decision Trees 99
3.7.7 Ensemble Learning 104
3.7.8 Random Forests 105
3.7.9 AdaBoost 108
3.7.10 Apriori Algorithm 109
3.7.11 K Means Clustering Algorithm (Unsupervised Learning—Clustering) 111
3.7.12 Deep Learning 114
3.8 Summary 127
3.9 Points to Ponder 128
3.10 Answers 128
References 129

4 Performance Metrics and Training of AI/ML Clinical Model 131
4.1 Machine Learning Algorithms 131
4.1.1 Supervised Machine Learning 132
4.1.2 Unsupervised Machine Learning 132
4.1.3 Reinforcement Machine Learning 133
4.1.4 Semi-Supervised Machine Learning 134
4.1.5 Deep Learning 134
4.1.6 Ensemble Machine Learning 134
4.2 AI/ML Algorithm for Medical Device Intended Use 135
4.3 Performance Metrics for AI/ML-Enabled Medical Devices 136
4.4 Performance Metrics for Supervised Machine Learning Models 136
4.4.1 Classification Metrics 137
4.4.2 Regression Metrics 141
4.4.3 Selection of Regression Performance Metrics 144
4.4.4 Overfitting in Machine Learning 145
4.4.5 Outliers Affect the Performance Metrics in Regression Models 146
4.4.6 Strategies to Address Outlier-Induced Overfitting and Underfitting 147
4.4.7 Significance of the Confusion Matrix 147
4.4.8 Binary and Multiclass Confusion Matrices 148
4.4.9 One-Vs-All Approach in Multiclass Classification 149
4.5 Performance Metrics for Unsupervised Machine Learning Models 149
4.5.1 Clustering Metrics 149
4.5.2 Dimensionality Reduction Metrics 150
4.5.3 General Metrics 150
4.5.4 Challenges in Evaluating Unsupervised Learning Models 151
4.6 Performance Metrics for Gen AI Models 151
4.6.1 Perplexity 152
4.6.2 BLEU (Bilingual Evaluation Understudy) 152

4.6.3 ROUGE (Recall-Oriented Understudy for Gisting Evaluation) 152
4.6.4 Human Evaluation 153
4.6.5 Fluency 153
4.6.6 Coherence 153
4.6.7 Diversity 154
4.6.8 Factual Accuracy 154
4.6.9 Relevance 154
4.6.10 Toxicity and Bias 154
4.6.11 Inference Time 155
4.6.12 Hallucination Rate 155
4.6.13 Zero-Shot and Few-Shot Performance 156
4.6.14 Fine-Tuning Efficiency 156
4.6.15 RAG for Long Language Models 156
4.6.16 RAG and Fine-Tuning 158
4.6.17 Limitations of Using RAG for Real-Time Applications 158
4.6.18 Hallucination Index Report [68] 159
4.7 Performance Metrics for Natural Language Processing 160
4.7.1 Classification Metrics 160
4.7.2 Text Generation Metrics 161
4.7.3 Language Modeling Metrics 161
4.7.4 Sequence Labeling Metrics 161
4.7.5 Other Important Metrics 161
4.8 Performance Metrics for Reinforcement Learning 162
4.8.1 Pivotal Role of Reward Functions in Reinforcement Learning 163
4.8.2 Visualizing the Learning Process in Reinforcement Learning 164
4.9 Monotonicity in Machine Learning Models 165
4.10 Train Supervised AI/ML Model 167
4.10.1 Improving the Accuracy of AI/ML Model 168
4.10.2 Hyperparameter Tuning to Improve Model Accuracy 168
4.10.3 Best Practices for Splitting the Dataset into Training and Test 169
4.10.4 Training Ensemble Machine Learning Models 170
4.11 Training Large Language Model 171
4.11.1 Challenges of Training Large Language Models 172
4.11.2 Techniques to Mitigate Overfitting in Large Language Models 173
4.12 Training Unsupervised Machine Learning Models 174
4.13 Train a Deep Learning Neural Network 175
4.13.1 Learning Rate for Neural Networks 176
4.14 Cross-Validation in AI/ML Model Training 176
4.14.1 Balancing Bias and Variance During Cross-Validation 177
4.15 Early Stopping and Regularization Techniques 179
4.16 Optimize Model Training to Reduce Computational Costs 181

4.16.1 Model Pruning to Reduce Resource Usage During Training . . . 181
4.17 Differences in Training and Real-World Populations . . . 182
4.18 Summary . . . 183
4.19 Points to Ponder . . . 184
4.20 Answers . . . 184
References . . . 186

5 Transfer Learning and Multimodal Artificial Intelligence . . . 191
5.1 Transfer Learning in Medical Device AI/ML Models . . . 191
5.1.1 Data Efficiency . . . 192
5.1.2 Model Architecture . . . 192
5.1.3 Fine-Tuning Approaches . . . 192
5.1.4 Task-Specific Optimization . . . 193
5.1.5 Regulatory Considerations . . . 193
5.1.6 Transfer Learning Methods in AI/ML-Enabled Medical Devices . . . 194
5.1.7 Limitations of Transfer Learning for Medical Data . . . 195
5.2 Selection of Foundation AI/ML Models . . . 195
5.2.1 Task Requirements . . . 196
5.2.2 Model Characteristics . . . 197
5.2.3 Resources and Constraints . . . 197
5.2.4 Data Considerations . . . 198
5.2.5 Customization and Integration . . . 198
5.2.6 Legal and Ethical Considerations . . . 198
5.2.7 Challenges with Fine-Tuning a Foundation Model . . . 199
5.2.8 Catastrophic Forgetting in Fine-Tuning . . . 199
5.3 Multimodal AI/ML in Healthcare . . . 200
5.4 Challenges in Multimodal AI/ML Algorithm Development . . . 205
5.5 Multimodal Generative AI . . . 207
5.6 4M: Massively Multimodal Masked Modeling . . . 207
5.7 Summary . . . 208
5.8 Points to Ponder . . . 208
5.9 Answers . . . 209
References . . . 210

6 Clinical AI/ML Model Testing and Validation . . . 213
6.1 Testing of AI/ML Model for Medical Device Application . . . 214
6.1.1 Data Management . . . 214
6.1.2 Model Evaluation . . . 214
6.1.3 Clinical Validation . . . 215
6.1.4 Human–AI Interaction . . . 215
6.1.5 Bias and Fairness . . . 215
6.1.6 Continuous Monitoring and Improvement . . . 216
6.1.7 Regulatory Considerations . . . 216
6.2 Testing the Robustness of AI Models . . . 217

6.3 Transparency, Explainability, and Interpretability . . . 218
6.3.1 Transparency for ML-Enabled Medical Devices . . . 219
6.3.2 AI Model Interpretability . . . 220
6.3.3 Implications of Lack of Transparency in Healthcare AI . . . 224
6.3.4 AI Decision-Making in Healthcare . . . 226
6.4 Bias in AI Algorithms for Medical Devices . . . 227
6.4.1 Algorithmic Bias in Medical Devices . . . 229
6.4.2 Testing for Bias . . . 230
6.5 Design for Interpretability . . . 232
6.6 Human in Loop and Human in Command . . . 234
6.6.1 Benefits of Human-in-Command Design over Human-in-Loop . . . 236
6.7 Clinical Validation of AI-Enabled Medical Devices . . . 237
6.7.1 Retrospective and Prospective Clinical Validation Study . . . 238
6.7.2 Randomized Controlled Clinical Trials . . . 239
6.8 Generalizability of AI Algorithm Performance . . . 240
6.8.1 Best Practices for AI Model Generalization . . . 241
6.9 Clinical Study Documentation . . . 242
6.10 Real-World Evidence . . . 243
6.11 EU MDR Clinical Evaluation Guidance . . . 243
6.12 US FDA Guidance on SaMD Clinical Evaluation . . . 244
6.13 Summary . . . 246
6.14 Points to Ponder . . . 246
6.15 Answers . . . 246
References . . . 247

7 AI/ML Model Integration and Deployment . . . 251
7.1 Domain Knowledge for AI Integration in Medical Devices . . . 251
7.2 Integrating Artificial Intelligence Models in Medical Devices . . . 252
7.3 Integrating Multimodal AI Models in Medical Devices . . . 254
7.4 Integration of Large Language Models . . . 255
7.5 Integration of Vision AI Models . . . 255
7.6 Integrating Human Expertise into AI Workflows . . . 256
7.7 Human in the Loop and Human in Command . . . 257
7.8 Interoperability Issue in AI Integration . . . 258
7.9 Deployment of AI Models on Medical Devices . . . 259
7.10 AI Model Deployment Time . . . 260
7.11 AI Model Deployment Architecture . . . 262
7.12 AI Model Deployment Strategies . . . 263
7.12.1 Shadow Model Deployment Strategy . . . 263
7.12.2 A/B Testing Model Deployment Strategy . . . 264
7.12.3 Multi-Armed Bandit Model Deployment Strategy . . . 264
7.12.4 Blue-Green Model Deployment Strategy . . . 265
7.12.5 Canary Model Deployment Strategy . . . 265
7.12.6 Feature Flag Model Deployment Strategy . . . 265

7.12.7 Rolling Model Deployment Strategy . . . 266
7.12.8 Recreate Model Deployment Strategy . . . 267
7.13 AI Model Deployment on Cloud and On-Premise . . . 268
7.14 Multimodal AI Model Deployment on Cloud and On-Premise . . . 269
7.15 AI Model Deployment on Hybrid Environment . . . 270
7.16 AI Model Deployment on Mobile Devices . . . 270
7.17 AI Model Deployment on Edge Devices . . . 271
7.18 Containerizing Machine Learning Models Deployment . . . 272
7.19 Performance of AI Models in Cloud and On-Premises . . . 274
7.19.1 Model Latency and User Experience . . . 275
7.20 Medical Imaging AI System Architecture for Integration and Deployment . . . 276
7.20.1 Key Architectural Components . . . 276
7.21 Deep Neural Networks (DNNs) Integration and Deployment . . . 279
7.22 An Example of Real-Time ML Operational Deployment Pipeline . . . 279
7.23 Data and Model Versioning for Deployment . . . 281
7.24 Summary . . . 282
7.25 Points to Ponder . . . 282
7.26 Answers . . . 283
References . . . 286

8 Security and Privacy for AI/ML-Enabled Medical Devices . . . 289
8.1 Introduction . . . 289
8.2 Clinical Model Security and Privacy Considerations . . . 290
8.3 IT Cybersecurity Requirement . . . 291
8.4 Security Using Encryption Keys . . . 296
8.5 Medical Data Management . . . 297
8.6 Regulations for Medical Applications . . . 300
8.7 Cybersecurity Certifications for AI/ML-Enabled Medical Devices . . . 302
8.8 EU AI Act Implications for Medical Devices . . . 305
8.9 US FDA Requirements . . . 306
8.10 Evolutions and Risks of LLMs . . . 307
8.11 Manufacturer Disclosure Statement for Medical Devices . . . 309
8.12 Cybersecurity Risks of Legacy Medical Devices . . . 310
8.13 Summary . . . 311
8.14 Points to Ponder . . . 311
8.15 Answers . . . 312
References . . . 313

9 Medical Device Machine Learning Operation . . . 315
9.1 Introduction . . . 315
9.2 Why Does MLOps Matter? . . . 317
9.3 DevOps Versus MLOps . . . 318
9.4 MLOps Best Practices . . . 318

9.5 AI/Machine Learning Project Lifecycle 319
9.5.1 Data Collection 319
9.5.2 Data Hygiene 321
9.5.3 Model Design/Exploration 321
9.5.4 Model Training 322
9.5.5 Model Tuning and Testing 322
9.6 MLOps General Concepts 324
9.6.1 Version Control 324
9.6.2 Automation 325
9.6.3 Idempotency 325
9.6.4 Deployment Strategies 325
9.7 The Google MLOps Levels 326
9.7.1 MLOps Level 0 (All Manual) 327
9.7.2 MLOps Level 1 328
9.7.3 MLOps Level 2 330
9.8 How to Automate ML Workflow? 331
9.8.1 Ethical Considerations 333
9.8.2 Fairness and Bias 333
9.8.3 Customer Data and Privacy 333
9.8.4 Bias Amplification 333
9.9 A Use Case for an AI System in Production 334
9.9.1 Development, Validation, and Production Environments 335
9.9.2 Data Workflows 337
9.9.3 Data Workflows 339
9.10 Tools Used in MLOps 340
9.10.1 Docker 340
9.10.2 Kubernetes 341
9.11 Most Popular Machine Learning Software Tools 341
9.12 Machine Learning Platforms 342
9.12.1 Benefits of Machine Learning Platforms 343
9.12.2 Types of Machine Learning Platforms 343
9.12.3 Components of a Machine Learning Platform 344
9.12.4 Popular Machine Learning Platforms 344
9.13 Predetermined Change Control Plan 347
9.14 Summary 348
References 348

10 Regulatory Requirements and Risk Management for AI/ML-Enabled Medical Devices 349
10.1 US FDA Marketing Submission for AI-Device Software Function (AI-DSF) 350
10.1.1 AI Lifecycle Framework 351
10.1.2 AI-DSF Labeling 352
10.1.3 Risk Management 354
10.1.4 Data Management 354
10.1.5 Model Description and Development 355

10.1.6 Validation . . . 355
10.1.7 Cybersecurity Requirements for AI-Enabled Medical Devices . . . 357
10.1.8 Public Information Summary . . . 357
10.2 Health Canada Pre-market Guidance for Machine Learning-Enabled Medical Devices . . . 358
10.2.1 Guidance for Implementation . . . 359
10.2.2 Design . . . 359
10.2.3 Risk Management . . . 360
10.2.4 Data Selection and Management . . . 360
10.2.5 Development, Training, Testing, and Evaluation . . . 361
10.2.6 Clinical Validation . . . 361
10.2.7 Transparency . . . 362
10.2.8 Post-market Monitoring . . . 363
10.3 Predetermined Change Control Plan . . . 363
10.3.1 Modification Protocol . . . 364
10.4 European Union's Artificial Intelligence Act (AIA) . . . 367
10.4.1 AI System Definition as per AI Act . . . 368
10.4.2 AI Act and Medical Devices . . . 369
10.5 Regulatory Implications of Autonomy in Medical Devices . . . 370
10.6 Structure of the EU AI Act and Quality Management System . . . 371
10.6.1 Prohibited AI System in AI Act . . . 371
10.6.2 Quality Management System for AI Act . . . 372
10.6.3 ISO/IEC JTC 1 and CEN/CLC/JTC 21: Artificial Intelligence . . . 373
10.7 Risk Management . . . 374
10.8 Identify Characteristics Related to Safety for AI Software . . . 375
10.9 Risk Control Measures for AI-Enabled Medical Devices . . . 376
10.9.1 Risk Control Associated with Data Quality . . . 376
10.9.2 Issues with AI/ML Applications . . . 379
10.10 Usability and Human Factor Requirements . . . 380
10.10.1 Usability, Transparency, and Explainability . . . 380
10.10.2 Use Related Risks . . . 381
10.11 Summary . . . 382
10.12 Points to Ponder . . . 382
10.13 Answers . . . 383
References . . . 384

11 AI-Based Health Care Applications . . . 385
11.1 Introduction . . . 385
11.2 Heart Care Data in the Cloud . . . 386
11.3 Cloud-Based Collaborative Tools . . . 388
11.4 Multi-Cloud Solutions . . . 389
11.5 Multimodal Glaucoma Detection . . . 391
11.6 AI Application for Glaucoma . . . 396
11.6.1 Deep Learning for Glaucoma . . . 398

11.6.2 Available Glaucoma Datasets 399
11.6.3 Future Challenges for Glaucoma Detection 401
11.7 Role of GenAI in Healthcare 403
11.8 Case Study for Bias Avoidance and Transparency 403
11.9 Future of AI-Based Health Care 404
11.10 Summary 406
11.11 Points to Ponder 406
11.12 Answers 406
References 408

12 AI-Enabled Medical Devices Approved/Cleared by the US FDA 409
12.1 Analysis of the US FDA AI/ML-Enabled Medical Devices 409
12.2 Public Information Summary from US FDA Database 413
12.3 Sleep Apnea Notification Feature (SANF) by Apple Inc. (K240929) 413
12.3.1 Indication for Use 413
12.3.2 Device Description 413
12.3.3 Comparison with the Predicate Device 414
12.3.4 Summary of Non-clinical Testing 415
12.3.5 Summary of Clinical Testing 416
12.3.6 Predetermined Change Control Plan 417
12.3.7 Conclusion 418
12.4 Notal Vision Home Optical Coherence Tomography (OCT) System 418
12.4.1 Indications for Use 419
12.4.2 Device Description 419
12.4.3 Artificial Intelligence Algorithm 420
12.4.4 Summary of Nonclinical/Bench Studies 420
12.4.5 Summary of Clinical Information 421
12.4.6 Training 422
12.4.7 Labeling 422
12.4.8 Risk to Health 422
12.4.9 Special Controls 424
12.4.10 Benefit-Risk Determination 424
12.4.11 Conclusion 425
12.5 Summary 425
References 425

Glossary 427

Index 437

Acronyms

4M	Massively Multimodal Masked Modeling
AI	Artificial Intelligence
AIA	Artificial Intelligence Act
ANN	Artificial neural network
AUC	Area under curve
AWS	Amazon web services
BERT	Bidirectional Encoder Representations from Transformers
BLEU	Bilingual evaluation understudy
CNNs	Convolutional neural networks
CPU	Central processing unit
CV	Cross-validation
DDOS	Distributed denial of service
DevOps	Development and Operations
DICOM	Digital imaging and communications in medicine
DL	Deep learning
DNN	Deep neural network
DOS	Denial of service
EHR	Electronic health record
FDA	Food and Drug Administration
FPGA	Field programmable gate array
GAN	Generative adversarial networks
GCP	Google Cloud Platform
GDPR	General data protection regulation
GenAI	Generative artificial intelligence
GMLP	Good machine learning practices
GPT	Generative pre-trained transformer
GPU	Graphics processing unit
HE	Homomorphic encryption
HIC	Human-in-command
HIPAA	Health Insurance Portability and Accountability Act
HITL	Human-in-the-loop

HL7	Health level seven
HPC	High performance computing
IaaS	Infrastructure as a Service
InfoSec	Information Security
IoT	Internet of Things
IVDR	In vitro diagnostic medical devices regulation
LDA	Linear discriminant analysis
LLMs	Large language models
LSTM	Long short-term memory
MAE	Mean absolute error
MDCG	Medical device coordination group
MDR	Medical device regulation
MDS2	Manufacturer Disclosure Statement for Medical Device Security
MDSW	Medical device software
ML	Machine learning
MLOps	Machine learning operations
MSE	Mean squared error
NIST	National Institute of Standards and Technology
NLP	Natural language processing
NMF	Non-negative matrix factorization
NNs	Neural networks
OCT	Optical coherence tomography
PaaS	Platform as a Service
PACS	Picture archiving and communication systems
PCA	Principal component analysis
PCCP	Predetermined change control plan
PHI	Protected health information
PII	Personally identifiable information
PMA	Premarket approval
QMS	Quality management system
QSR	Quality system regulation
RAG	Retrieval-augmented generation
RL	Reinforcement learning
RMSE	Root mean squared error
RNN	Recurrent neural networks
ROC	Receiver operating characteristic curve
SaaS	Software as a Service
SaMD	Software as a medical device
SBOM	Software bill of materials
SiMD	Software in medical device
SMPC	Secure multi-party computing
SOC2	Systems and Organization Controls 2
TEE	Trusted execution environment
TPU	Tensor processing unit
VPU	Vision processing unit

Chapter 1
Introduction

Abstract This chapter provides a comprehensive overview of artificial intelligence (AI) and machine learning (ML) in medical device software development, as well as the AI/ML Lifecycle. It begins by introducing AI/ML-enabled medical device software, highlighting its potential to transform healthcare by deriving new insights from vast amounts of data generated during healthcare delivery. It explores how Agile Scrum methodologies can be adapted to address the unique challenges of AI/ML development in the medical device industry, enabling iterative progress and rapid response to changing requirements. The chapter then shifts focus to the regulatory landscape, providing an overview of the medical device regulatory framework. It examines the FDA's approach to AI/ML-enabled medical devices, including the concept of Predetermined Change Control Plans (PCCPs) and the agency's efforts to adapt its traditional paradigm to the dynamic nature of AI/ML technologies. The chapter then delves into AI/ML lifecycle management, emphasizing the importance of continuous monitoring and improvement. Security and privacy considerations are addressed, reflecting their critical importance in the healthcare sector. Finally, the chapter introduces the European Union's AI Act, a landmark legislation with significant implications for AI/ML medical device developers. By covering these crucial aspects, the chapter equips readers with a comprehensive understanding of the complexities of developing AI/ML-enabled medical device software, emphasizing the need for a multidisciplinary approach that balances innovation with regulatory compliance.

Keywords FDA · QSR · MDR · AI/ML · SiMD · SaMD · PMA · 510K · IEC62304 · TIR45 · AI/ML lifecycle · GMLP · MLOps · EU AI Act · Cybersecurity · Privacy · Clinical evaluation · Regulatory framework · Agile scrum · Waterfall · Risk management · Human factor · Deployment · Gen AI · Validation · Multimodal · Software safety class · Device classification · Change control · PCCP · HIPAA · GDPR · FDA guidance · MDCG

A. Pandey et al., *Artificial Intelligence in Medical Software*,
https://doi.org/10.1007/978-3-032-10808-1_1

1.1 Introduction

Over the past few decades, there has been a significant evolution in software development for medical devices [1], closely following advancements in computer technology and data management. From the 1980s onwards, the advent of personal computers sparked the creation of companion software programs that could be linked with medical equipment and installed on external computers [2]. During this time, increasingly complex software programs in conjunction with medical equipment began to emerge. In the 1990s and 2000s, the development of smartphones and laptops, Cloud computing, and novel software technologies, along with the expansion of the Internet and the adoption of 5G technology, further changed medical device software technology.

Medical device software development is heavily regulated to ensure safety and efficacy, given its potential impact on patient health. In particular, medical device software regulations in the United States have undergone significant evolution. A crucial turning point was the Medical Device Amendments of 1976 [3], which established a risk-based classification system for medical devices and laid the framework for controlling the software in medical devices. The Food and Drug Administration (FDA) introduced the Quality System Regulation (QSR) in 1997 [4] to set quality standards for developing software for medical devices. One of the many ways the agency has modified its regulatory frameworks over time is to address the unique challenges software presents. With the introduction of the European Union Medical Device Regulation (EU MDR) [5], the development of medical device software has become more challenging for device manufacturers [6].

The IEC 62304 standard (Medical Device Software Lifecycle Management) [7] is a critical framework that outlines the necessary activities in the software development process, including planning, requirements analysis, design, implementation, testing, and release. Because the dependability of medical software significantly affects people's lives, this standard ensures that the software development process adheres to the highest standards. Medical device software development has always taken a linear, waterfall method [8, 9]. Agile approaches, on the other hand, are gaining popularity as they support incremental and iterative development [9]. Standards like AAMI TIR 45 [10], which offer guidelines on applying Agile methods while still adhering to regulatory obligations, help to support this transition. Agile methodologies help manage changes more effectively and improve the responsiveness of development teams to new requirements.

The medical device industry has historically invested a significant portion of its sales in research and development, with a focus on both basic and applied research. To meet the varied demands of the medical device market, which includes a broad range of devices, this investment has proven essential in promoting innovation. Despite significant advancements, several issues persist in the development and regulation of medical device software. These relate to Artificial Intelligence and machine learning (AI/ML) enabled medical devices, cybersecurity, interoperability, and privacy. Global regulatory agencies are refining their frameworks to accommodate the distinct needs of AI/ML-enabled medical devices. The role of software in medical devices is expected to increase with the adoption of artificial intelligence and machine learning algorithms.

1.2 Types of Medical Device Software

IEC 62304 [7] is an international standard that outlines the lifecycle management requirements for developing medical device software. IEC82304-1 (Health software—General requirements for product safety) [11] is an international standard for any health software. Figure 1.1 below illustrates four types of medical software and how IEC82304-1 and IEC 62304 standards are applied to health software.

1.2.1 Software as Part of a Medical Device, also Known as Software in Medical Device (SiMD)

This software is integrated into the device and is crucial for its operation. Advanced programming skills are required to manage the embedded electronics of complex machines. This software is an integral component of a hardware medical device, either as a host, embedded software, or a combination of both. SiMD cannot

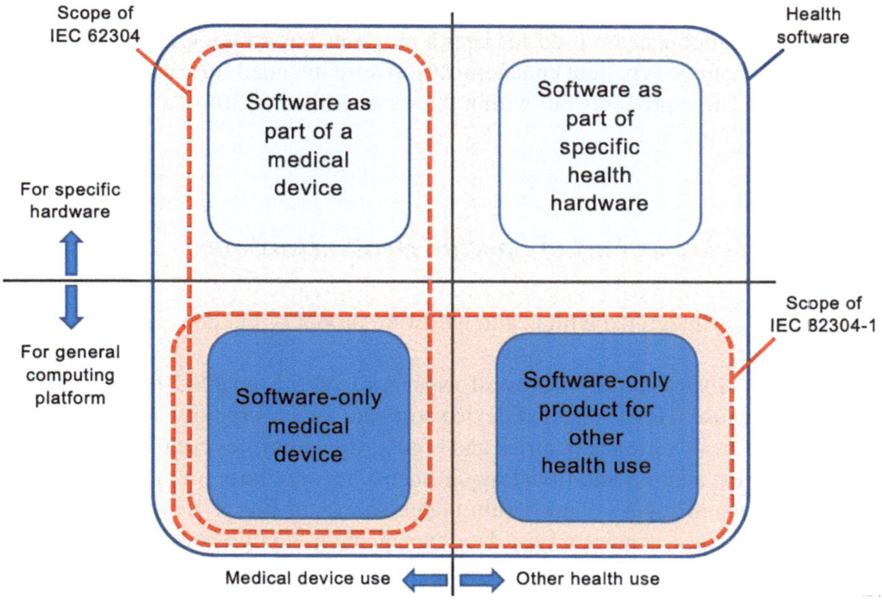

Fig. 1.1 Health software application domains and scope of related standards [11]

function independently and is necessary for the device's operation. Examples include firmware in medical instruments or embedded software in diagnostic equipment. SiMDs are required to provide evidence of compliance with IEC62304. In addition to IEC 62304, SiMD products must comply with product standards based on a product's intended use and indication for use. For example, the IEC 60601 series of standards is a product standard for medical electrical devices.

1.2.2 Software-Only Medical Device, also Known as Software as a Medical Device (SaMD)

A SaMD has one or more medical purposes that it performs without being part of a hardware medical device. SaMD applications include remote patient monitoring systems, diagnostic tools, and health-tracking apps. It performs functions such as diagnosing, treating, or managing health conditions without being part of a hardware device. SaMD products are required to provide evidence of compliance with IEC82304-1 and IEC62304. In addition to IEC 62304 and IEC 82304-1, SaMD products must comply with product standards based on their intended use and indication for use.

1.2.3 Software-Only Product for Other Health Use

Software-only products are used for health purposes but do not qualify as medical devices. An example is patient engagement software intended for patient scheduling and training. These products fall within the scope of IEC 82304-1 and are required to comply with it.

1.2.4 Software as Part of Specific Health Hardware

This software is integrated with health hardware products, but does not qualify as a medical device.

Figure 1.1 does not fully cover all aspects of AI/ML-enabled medical device software. Because AI/ML-enabled device software evolves rapidly and regulations and guidance from various countries and regulatory authorities continue to change, developers must address additional requirements beyond those outlined in the software development process and requirements shown in Fig. 1.1. Specifically, AI/ML-enabled devices require robust data management, comprehensive training and testing, careful integration and deployment, and thorough post-production monitoring and operations. This book will discuss these essential concepts for AI/ML-enabled device software in detail.

1.3 Medical Device Software Development Process

The development process for medical device software typically involves several key steps [1, 7], as discussed below:

- **Design Planning and Input Gathering:** This involves developing software development plans, defining user and software requirements, and ensuring compliance with regulatory standards.
- **Usability, Human Factors, and User Experience**: Developing intuitive and user-friendly software is crucial. A well-designed interface enhances productivity, reduces errors, and improves customer experience. Developers must create software that meets user needs, provides value to both healthcare providers and patients, and complies with quality and regulatory requirements.
- **Design and Development:** Developers finalize requirements and architecture, design the user interface and core modules, and ensure integration with other medical systems and data security. Development should utilize AAMI TIR 45 and agile methodologies in regulated medical device development [10], allowing for iterative improvements.
- **Design Verification and Validation:** Comprehensive testing and validation are crucial to ensure the software is reliable, safe, and meets quality standards. Testing medical device software is complex due to its interconnected nature and the need for rigorous validation of algorithms, particularly those used for critical functionalities such as patient monitoring.
- **Interoperability:** Medical device software must often integrate with other systems, which can be challenging due to data protocol incompatibility. Ensuring seamless data exchange requires adherence to HL7 [12], DICOM [13], and Medical Device Communication standards (example: ISO/IEEE 11073 [14]). Interoperability issues can lead to safety concerns, inefficiencies, and data loss. Developers should perform interoperability assessments and work with technology partners to integrate necessary standards. Medical device software often needs to integrate with other healthcare systems, which can be difficult due to data protocol incompatibility and the asynchronous evolution of interfaced components [15]. Ensuring seamless data exchange while maintaining data integrity and security is a complex task.
- **Product Security and Privacy:** Cybersecurity and privacy have become a critical concern as medical devices increasingly rely on software and networking. Developers must establish a cybersecurity risk management approach that includes identifying assets, threats, and vulnerabilities, assessing their impact, and determining suitable mitigation strategies. Premarket and post-market strategies are essential for managing cybersecurity and privacy risks effectively. Developers must implement robust security measures, including data encryption and access controls, to protect sensitive patient data from breaches and malicious attacks [16, 17].
- **Regulatory Approval:** Medical device software must comply with strict regulations to ensure patient safety. This involves obtaining necessary certifications

and meeting market approval standards. Medical device software must adhere to strict regulations that vary by region, such as FDA regulations in the United States, CE marking in Europe, and the GDPR for data protection. Noncompliance can lead to recalls, fines, and reputational damage.
- **Deployment and Lifecycle Management:** Once developed, the software is deployed and requires ongoing support and updates to address any issues and incorporate new features. Updating medical device software is a persistent challenge. Regular updates are necessary to address security issues and improve device performance, but they can also introduce new risks that require careful management.

1.4 Regulatory Framework for Medical Device Software

Global regulations for Software as a Medical Device (SaMD) and Software in a Medical Device (SiMD) vary in several key aspects, primarily due to differences in how these software types are classified, assessed, and regulated across regions [3, 5, 7]. SiMD is considered part of the medical device, meaning its regulatory assessment is integrated with the hardware. This integration ensures the software functions correctly within the device, contributing to its safety and effectiveness. The regulatory requirements for SiMD are less distinct from those for hardware, as it is considered an integral component of the device's operation. The risk assessment for SiMD involves evaluating how the software interacts with the hardware and the potential risks associated with its malfunction. This integrated assessment ensures that software and hardware meet the regulatory requirements [18].

1.4.1 United States (US)

The FDA categorizes medical devices into three risk classes: Class I (low risk), Class II (moderate risk), and Class III (high risk) [3]. The classification depends on the medical device's intended use and potential impact on patient safety. Depending on their risk class, medical devices may require different regulatory clearances and approvals in the US market, such as 510(k) premarket notifications [18], premarket approval (PMA) [19], or De Novo classification [20]. The 510(k) pathway enables manufacturers to demonstrate that their device is substantially equivalent to a legally marketed predicate device, thereby expediting the approval process for specific devices. The FDA is a centralized governmental authority regulating medical devices in the United States. It oversees the approval of medical devices, from premarket submissions to post-market surveillance. Compliance with international standards and the FDA's Quality System regulations (QSR) [4] is crucial for ensuring the quality and reliability of medical device software. The FDA's centralized structure allows for a more uniform application of regulatory standards across all medical devices.

1.4.2 European Union (EU)

In Europe, the development and regulation of medical device software are governed by Medical Device Regulation (MDR) 2017/745 [5] to ensure safety, effectiveness, and compliance. MDR replaced previous directives, setting new standards for medical devices, including software. The regulation categorizes medical devices into four classes based on risk: Class I, Class IIa, Class IIb, and Class III. Software intended to provide diagnostic or therapeutic information is generally classified as Class IIa unless it poses a higher risk, in which case it may be classified as Class IIb or III. Under MDR Rule 11, software used for diagnostic or therapeutic purposes is classified based on the potential impact on health [21]. This classification determines the level of scrutiny required for CE marking, ranging from self-certification for lower-risk devices to a full review by a notified body for higher-risk devices.

The European Commission and the Medical Devices Coordination Group (MDCG) have issued guidance documents to clarify requirements for medical device software [5]. These documents address the integration of software with hardware devices and outline the responsibilities of manufacturers in ensuring compliance with MDR [22, 23]. Manufacturers must conduct clinical evaluations to demonstrate the safety and performance of medical device software. The clinical evaluation involves validating clinical associations, technical performance, and clinical performance according to the intended purpose. The MDCG guides the generation of data to support these evaluations [24]. Compliance with international standards such as ISO 13485 [25] is crucial for ensuring the quality and reliability of medical device software. These standards emphasize the need for robust quality management systems and transparency in the development process.

1.4.3 Global Harmonization

The International Medical Device Regulators Forum (IMDRF) plays a significant role in harmonizing international regulations to facilitate faster product approvals and market access. The IMDRF comprises representatives from major regulatory authorities, including those from the United States, the European Union, Japan, China, and other countries. It provides guidance on regulatory pathways and clinical evidence requirements, promoting consistency across borders. The World Health Organization (WHO) also contributes to global harmonization through its Global Model Regulatory Framework for Medical Devices. This framework provides guiding principles and harmonized definitions to help countries with little or no regulation develop effective regulatory systems. In conclusion, the global regulatory

Table 1.1 IMDRF SaMD risk-based categorization [26]

SaMD categories			
	Significance of information provided by SaMD to healthcare decision		
State of Healthcare situation or condition	Treat or diagnose	Drive clinical management	Inform clinical management
Critical	IV	III	II
Serious	III	II	I
Non-serious	II	I	I

environment for medical devices is evolving toward greater harmonization, with international organizations and regulatory authorities collaborating to establish consistent standards. This trend facilitates international trade, ensures high safety and quality standards, and supports innovation in medical device development.

The International Medical Device Regulators Forum (IMDRF) plays a significant role in harmonizing regulations for SaMD globally, aiming to streamline processes and ensure high standards of safety and efficacy. IMDRF has introduced SaMD categorization [26], as shown in Table 1.1.

1.5 Medical Device Software Lifecycle Management

IEC 62304 (Medical Device Software Lifecycle Management) is an international standard that outlines the lifecycle requirements for both SiMD and SaMD medical device software development types.

1.5.1 Overview of IEC 62304

IEC 62304 [7] defines the processes required for designing and maintaining medical device software. This standard is structured to ensure that software is developed, verified, validated, and maintained to minimize risks to patients and users. IEC 62304 is a process standard that focuses explicitly on the software lifecycle processes for medical device software. In contrast, ISO 13485 [25] and ISO 14971 [27] cover broader aspects of medical device quality management and risk management, respectively.

IEC 62304 requires a risk-based approach to determining the software safety class of medical device software. Figure 1.2 contains the process for deciding this software safety class. This determination considers risk control measures external to the software system. The software safety class determines the extent of rigor and deliverables necessary for software development and lifecycle management.

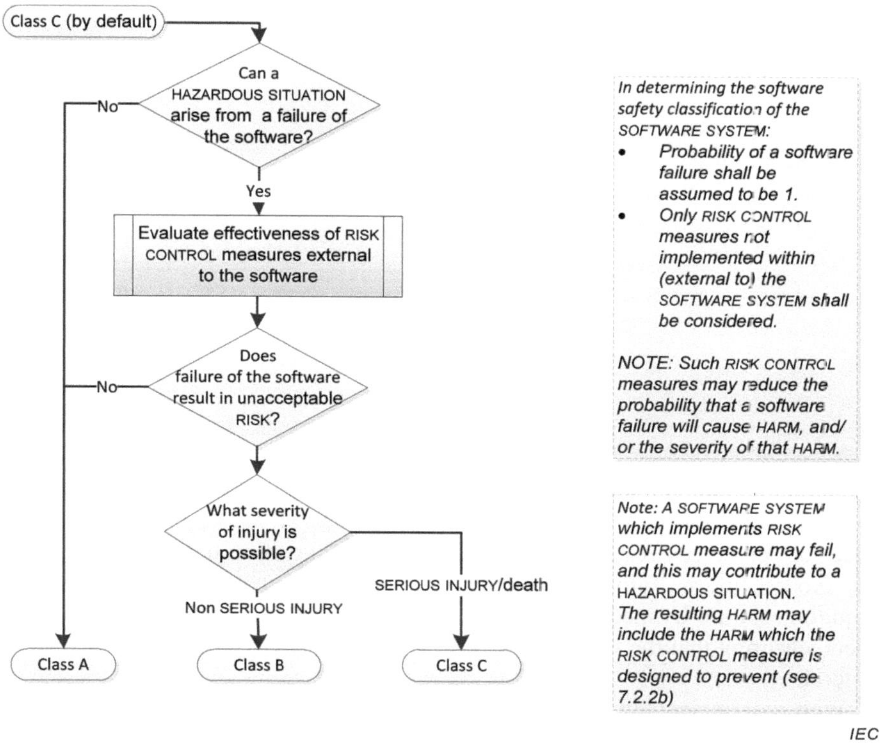

Fig. 1.2 Medical device software safety class determination process [7]

IEC 62304 defines three safety classes for software based on the potential risk to patients and users:

- Class A: No injury or damage to health is possible.
- Class B: Injury is possible but not serious.
- Class C: Death or serious injury is possible.

1.5.2 Critical Components of IEC 62304

The software development process outlined in IEC 62304 is a comprehensive framework designed to ensure the creation of safe, reliable, and effective software, particularly in the context of medical devices. As illustrated in Fig. 1.3, this process encompasses several critical stages and activities that work in concert to produce high-quality software products.

The journey begins with meticulous planning, where a detailed software development plan is crafted. This plan serves as a roadmap, outlining the project's

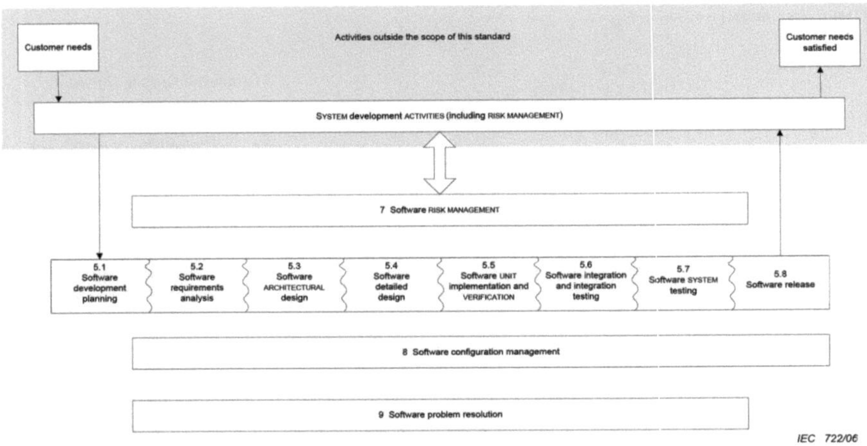

Fig. 1.3 Overview of software development processes and activities [7]

necessary processes, resources, and timeline. Following this crucial planning phase, the process moves into requirements development and analysis. The software requirements are defined based on its intended use and the specific needs of its users, ensuring that the final product will meet all necessary functional and safety criteria.

With requirements in hand, the design and implementation phase commences. This stage involves creating the software architecture and detailed design, followed by the actual coding and unit testing of individual components. As these components take shape, the process transitions into integration and testing. During this phase, software units are combined and subjected to rigorous integration testing to ensure they function cohesively. Subsequently, system testing is conducted to verify that the software meets all requirements at a comprehensive system level.

An essential aspect of the development process highlighted in IEC 62304 is the management of Software of Unknown Provenance (SOUP) [28]. This includes third-party or open-source software, which can introduce vulnerabilities if not properly evaluated. The standard emphasizes the importance of a thorough evaluation process for SOUP, particularly in terms of cybersecurity, to mitigate potential risks.

Once the software is deployed, the focus shifts to maintenance, as illustrated in Fig. 1.4. The software maintenance process ensures the software's continued functionality, security, and reliability throughout its lifecycle. This process includes several key activities:

- Maintenance Planning involves developing a comprehensive plan for ongoing maintenance activities, including regular updates and cybersecurity measures.
- Software Change Analysis: This step identifies, analyzes, and resolves software issues and necessary modifications.
- Modification Deployment: The changes are implemented, and their effectiveness is verified without introducing new risks.

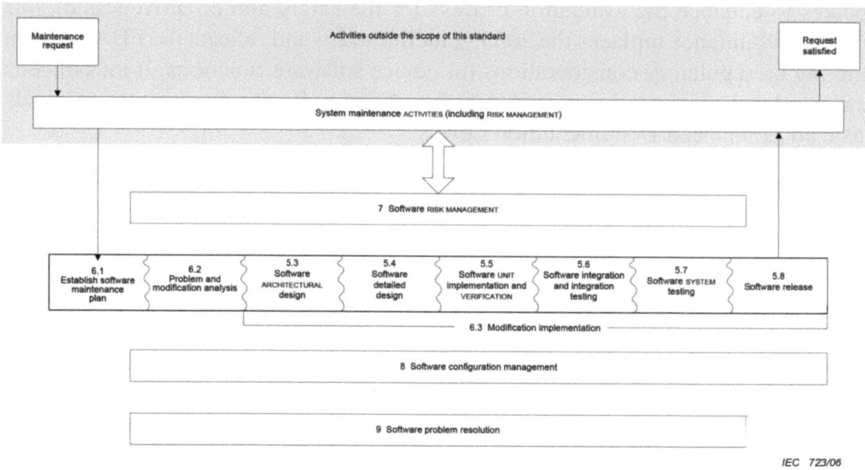

Fig. 1.4 Overview of software maintenance processes and activities [7]

Three other crucial processes described below complement the software development and lifecycle process:

- The Risk Management Process involves identifying potential hazards associated with software use, assessing their impact, implementing risk control measures, and continuously managing the risks associated with software updates and changes.
- The Configuration Management Process ensures consistency and traceability by defining and managing software configurations, establishing change control procedures, and maintaining accurate records of configurations and changes.
- Finally, the Problem Resolution Process involves documenting and tracking software issues, analyzing and resolving problems, and conducting trend analysis to identify and prevent recurring issues. Together, these interconnected processes form a robust framework for developing, maintaining, and improving software systems throughout their lifecycle, ensuring they remain safe, effective, and fit for purpose in the ever-evolving landscape of medical technology.

1.6 US FDA Premarket Submission Guidance for Medical Device Software Functions

The US FDA introduced premarket submission guidance for device software functions in 2023 [1]. It shifts from a three-tier risk-based approach [29] to a two-tier software documentation level classification [1] to determine the extent of rigor and deliverables required for medical device software. The 2023 FDA guidance on premarket submissions for medical device software functions introduces significant updates to enhance the evaluation process for the safety and effectiveness of software. This guidance replaces the 2005 guidelines [29] and reflects the FDA's current thinking on regulatory considerations for device software functions. It introduced a new two-level risk-based approach for classifying software documentation levels: Basic and Enhanced Documentation Levels.

Risk-Based Documentation Levels The 2023 guidance introduces a new risk-based system of Basic and Enhanced documentation requirements, replacing the previous Major, Moderate, or Minor Levels of Concern.

Enhanced Documentation It is required for software functions where a failure could lead to death or serious injury. This level emphasizes evaluating potential injuries before applying any risk control measures. However, this definition does not include language similar to IEC 62304, which considers risk controls implemented externally to software systems. Risk controls external to software systems must be considered, as this guidance from the FDA only considers medical device software function failure. If the risk is already mitigated as far as possible by risk controls external to software systems, it must be regarded as at the documentation level.

Basic Documentation It applies to software functions that do not meet the criteria for Enhanced Documentation.

As previously discussed, FDA guidance continues to align with the international standard IEC 62304, which outlines software lifecycle processes. IEC62304 remains a recognized framework for meeting the FDA's software design control requirements. Notably, the 2023 guidance now requires a Software Design Specification, Unit, and Integration testing for both Basic and Enhanced documentation levels, which shall be available to support premarket device submissions.

The FDA Premarket Submission Guidance also addresses emerging technologies, such as Artificial Intelligence and Machine Learning (AI/ML)-enabled medical devices. While the guidance does not directly reference the FDA's draft guidance on the Pre-determined Change Control Plan and premarket submission guidance for AI/ML-enabled device software functions [30], it acknowledges the evolving landscape of digital health technologies. The FDA encourages the use of the pre-submission process to include Predetermined Change Control Plans (PCCPs) in submissions, allowing for changes that might otherwise require new submissions. The guidance emphasizes that manufacturers establish software documentation prospectively rather than retrospectively. This means documentation should be defined and implemented throughout the software development lifecycle rather than after the completion of development activities.

The 2023 FDA Premarket Guidance for Device Software Functions reflects the FDA's commitment to modernizing its regulatory approach to medical device software. By adopting a risk-based documentation system and aligning with international standards, the FDA aims to streamline the pre-market review process while ensuring adherence to safety and effectiveness standards. The changes introduced by the 2023 guidance are expected to have a substantial impact on submission documentation for software in medical devices (SiMD) and software as a medical device (SaMD), as well as on Design History Files (DHFs) subject to FDA premarket submission. Overall, IEC 62304 is applicable in the EU and other global regulatory jurisdictions, whereas the 2023 FDA guidance is applicable in the US. Both approaches and processes address current technological advancements and the need for robust regulatory oversight.

1.7 AAMI TIR45: Agile Process in Medical Device Software Development

A typical medical device development process follows a waterfall method, shown in Fig. 1.5, [8, 9], which follows design planning, design input, design output, and design transfer processes.

Both IEC 62304 [7] and the US FDA Software Guidance [1] do not mandate a specific software development process. However, the medical device industry has adopted the Agile Scrum software development process, utilizing guidance from AAMI TIR 45 [10]. A typical Agile Scrum process [31] is shown in Fig. 1.6. The product backlog and initiatives are broken into sprints, which are small increments of manageable software work items. Design input progression, design output progression, and design verification are completed by the end of each sprint. The process continues with the completion of documentation, sprint review (design review progression), software increment release, and sprint retrospective.

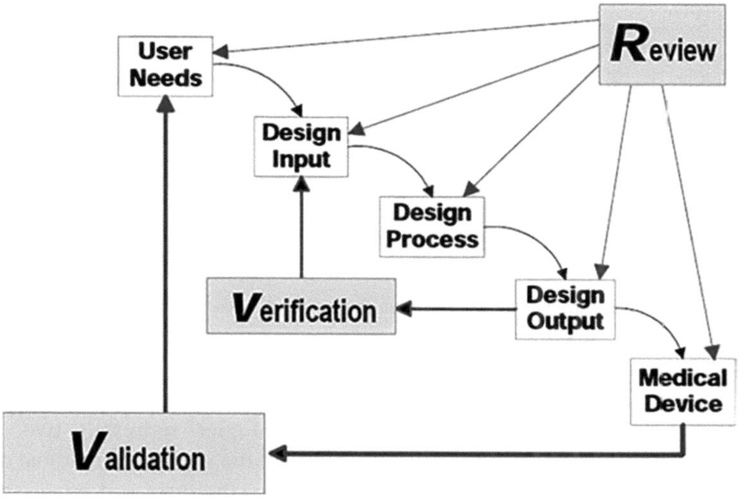

Fig. 1.5 Waterfall product development process [8]

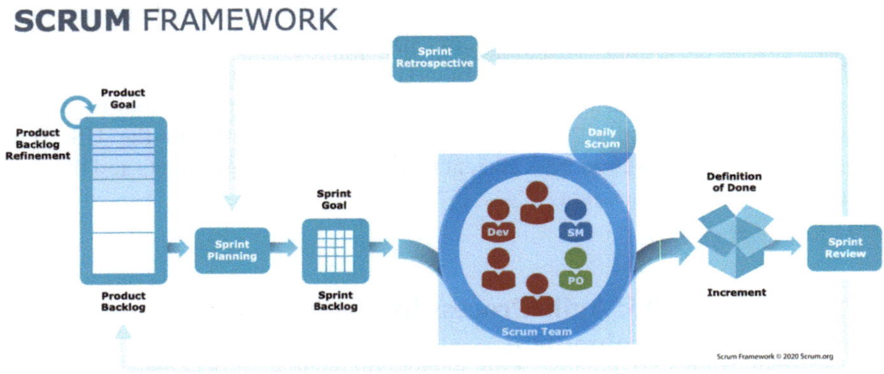

Fig. 1.6 Agile Scrum framework from scrum.org [31]

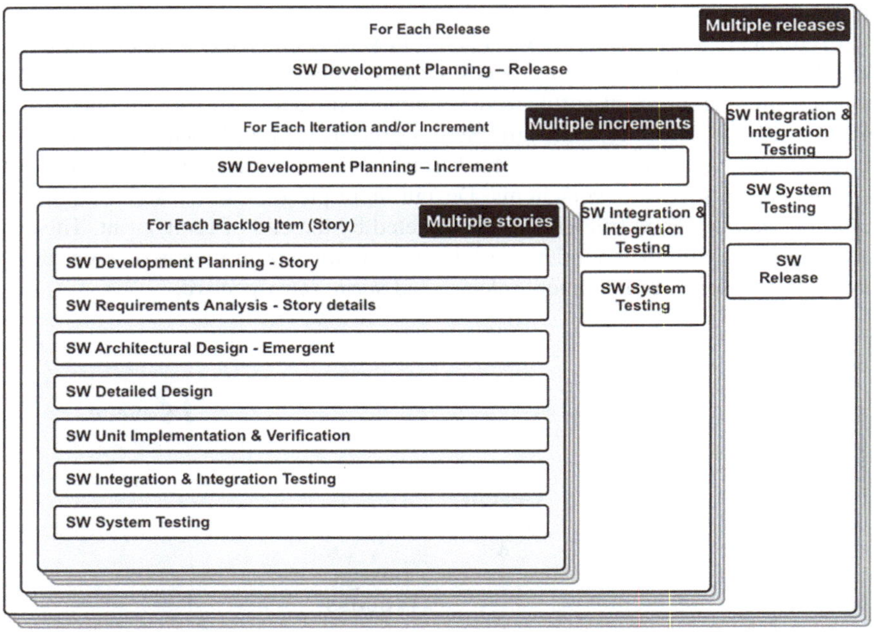

Fig. 1.7 Story, increment, and release layers in agile scrum methodology [10]

Figure 1.7 illustrates an agile process flow, demonstrating how an agile approach can be applied to medical device software development, utilizing AAMI TIR 45 guidance. Agile Scrum and Waterfall methodologies represent two distinct approaches to software development, particularly in the context of medical devices. While both aim to deliver a successful product, their structures and processes differ significantly.

In Agile Scrum for medical device software development, there are three primary layers: the Story, Increment, and Release layers. These layers form a hierarchical structure, each building upon the previous one. At the foundation, we have individual stories—small, manageable work units. Multiple stories combine to form an increment, representing more substantial functionality. Several increments are then merged into a release, and numerous releases ultimately culminate in a complete product release.

When comparing Agile Scrum to the traditional Waterfall approach, we can identify some parallels in their processes. Agile Scrum‘s product backlog and sprint planning stages serve a similar purpose to the design planning phase in the Waterfall method. Both aim to outline the project’s scope and requirements. Similarly, the Story review in Agile Scrum corresponds to the design input review in Waterfall, while the Sprint demonstration aligns with the design output review. These analogous stages suggest that both methodologies share certain fundamental principles of project management.

However, a key distinction between Agile Scrum and Waterfall lies in their approach to executing activities. Waterfall follows a linear, sequential approach where each phase must be completed before moving on to the next. In contrast, Agile Scrum employs a multi-threaded model, enabling simultaneous activities. This means that development, verification, and documentation occur concurrently, rather than in distinct and separate phases.

This simultaneous approach in Agile Scrum offers several advantages. It promotes flexibility, allowing teams to adapt to changes more readily. It also facilitates faster feedback loops as various aspects of the project are continuously reviewed and refined. Furthermore, this parallel processing can lead to more efficient use of resources and potentially shorter overall development time. In medical device software development, where regulatory compliance and safety are paramount, Agile Scrum‘s iterative nature can be particularly beneficial. It enables ongoing risk assessment and quality control throughout the development process, rather than relegating these crucial aspects to later stages, as often occurs in the waterfall model. While Agile Scrum and Waterfall share some common elements in their approach to software development for medical devices, Agile Scrum‘s layered structure and multi-threaded execution set it apart. This methodology offers a more dynamic and adaptable framework, potentially leading to more efficient and responsive development processes in the complex and highly regulated field of medical device software.

1.7.1 Project Success with Agile Software Development

Agile is significantly more flexible than the waterfall method, allowing teams to respond quickly to changes in requirements or priorities throughout the development process [9]. This adaptability ensures that the final product meets the evolving needs of customers and better aligns with market conditions. The CHAOS study [9] analyzed the success rate of over 10,000 software projects. The study results in Table 1.2 show that Agile projects have approximately four times the success rate compared to the Waterfall model. Additionally, Waterfall projects have a three times

Table 1.2 Analysis of software development project success rate with waterfall and agile development processes [9]

Chaos resolution by Agile versus Waterfall				
Size	Method	Successful (%)	Challenged (%)	Failed (%)
All size projects	Agile	39	52	9
	Waterfall	11	60	29
Large size projects	Agile	18	59	23
	Waterfall	3	55	42
Medium size projects	Agile	27	62	11
	Waterfall	7	68	25
Small size projects	Agile	58	38	4
	Waterfall	44	45	11

The resolution of all software projects from FY2011–2015 within the new CHAOS database, segmented by the Agile process and Waterfall method. The total number of software projects is over 10,000

higher failure rate than Agile projects. If project sizes are decomposed, these results show similar differences persist, except for small-sized projects.

While agile may not be suitable for every project, its emphasis on flexibility, collaboration, and continuous improvement makes it a powerful approach for many software development initiatives. Agile offers clear advantages over the more rigid waterfall model in dynamic environments where requirements are likely to change, and rapid delivery is crucial.

In the following sections of this chapter, we will explore embedded and application software, Cloud-based device software, and artificial intelligence/machine learning (AI/ML) software in medical devices.

1.8 Cloud Software Development for Medical Devices

Cloud software development [9] involves creating applications and services that run on Cloud platforms. Integrating Cloud computing into medical device software development has revolutionized the healthcare industry, offering enhanced diagnostics, monitoring, and treatment capabilities. Cloud-based solutions provide several significant advantages, including scalability, real-time data access, and improved collaboration.

1.8.1 Benefits of Cloud-Based Medical Device Software

- **Scalability and Flexibility:** Cloud computing enables medical device software to scale resources up or down according to demand, ensuring optimal performance without requiring significant upfront investments in hardware. This flexibility is particularly beneficial for handling large volumes of data generated by medical devices.

- **Real-Time Data Access and Collaboration:** Cloud-based solutions offer real-time access to patient data, facilitating timely decision-making and improving patient outcomes. Healthcare providers can collaborate more effectively, sharing data and insights across different locations and departments.
- **Cost Efficiency:** Healthcare organizations can reduce capital expenditures on hardware and maintenance by leveraging Cloud infrastructure. Cloud services operate on a pay-as-you-go model, converting capital expenses into operational expenses and providing cost savings.
- **Enhanced Data Storage and Analysis**: The Cloud offers vast storage capabilities, allowing for the aggregation and analysis of large datasets. Advanced analytics and AI/ML algorithms can be applied to this data to derive actionable insights, improving diagnostics and treatment plans.
- **Remote Monitoring and Telehealth:** Cloud-based medical devices support remote patient monitoring and telehealth services, extending healthcare access to patients in remote or underserved areas. This capability is essential for managing chronic conditions and delivering ongoing care.

1.8.2 Challenges in Cloud-Based Medical Device Software Development

- **Regulatory Compliance:** Ensuring compliance with regulatory requirements from the FDA, MDR, and GDPR is a significant challenge. These regulations mandate rigorous validation, data protection, and security measures, which must be integrated into the Cloud-based development process.
- **Cybersecurity Risks:** Cloud environments are susceptible to cybersecurity threats, including data breaches and unauthorized access. Robust security protocols, including encryption, multi-factor authentication, and regular security audits, are crucial for protecting sensitive patient data [16, 17].
- **Data Privacy and Protection:** Compliance with data protection regulations, such as GDPR (for EU-based data) and HIPAA (for US-based data), is critical. Medical device software must ensure that patient data is collected, processed, and stored securely, with mechanisms in place for data anonymization and patient consent management [16, 17].
- **Interoperability:** Seamless data exchange with other medical devices and information systems is essential for interoperability. Standardized data formats and communication protocols, such as HL7, FHIR, and DICOM, must be adopted to ensure compatibility and integration [15].
- **Validation and Verification (V&V):** The V&V process is more complex in a Cloud environment due to the dynamic nature of Cloud services. Continuous validation is required to ensure that software updates and changes by Cloud

service providers do not compromise the safety and functionality of the medical device.

Cloud-based medical device software development offers significant benefits, including scalability, real-time data access, cost efficiency, and enhanced patient care. However, it also presents challenges related to regulatory compliance, cybersecurity, data privacy, interoperability, and validation. By adopting best practices and adhering to regulatory standards, medical device manufacturers can leverage the power of Cloud computing to deliver innovative, safe, and effective healthcare solutions.

1.9 Embedded and Application Software Development

The embedded and application software development process focuses on creating software integrated into hardware devices. The embedded software is designed to perform specific functions on dedicated hardware systems. It may or may not include an operating system; "deep embedded" refers to software that operates without one. The application software runs on general-purpose computers and requires an operating system. It encompasses various types, including games, spreadsheets, and database systems. Embedded and application software development for medical devices is a specialized field that requires a deep understanding of software engineering principles and stringent regulatory requirements. Embedded software, often referred to as firmware, is integral to the operation of various medical devices, ranging from simple diagnostic tools to complex therapeutic systems. Embedded and application software are designed to control the functions of specific hardware devices. These are tailored to meet the unique requirements of medical devices, ensuring reliability, efficiency, real-time performance, cost, and compliance with industry standards.

Developing embedded and application software for medical devices follows a structured lifecycle, often guided by standards such as IEC 62304. Like Cloud software applications, host and embedded software development must follow IEC 62304, FDA submission guidance, risk management, human factors, and other relevant product standards.

1.9.1 Challenges in Embedded Software Development

- **Regulatory Compliance:** Ensuring compliance with stringent regulatory standards such as IEC 62304 and FDA guidelines can be challenging. These regulations mandate rigorous validation, documentation, and risk management processes.

- **Cybersecurity Risks:** Embedded medical devices are susceptible to cybersecurity threats. Robust security measures, such as encryption and secure boot mechanisms, are crucial for protecting patient data and ensuring device functionality [16, 17]. This topic will be discussed in detail in a later chapter.
- **Resource Constraints:** Embedded systems often need more computational resources and memory. Developing efficient, reliable software within these constraints requires careful optimization and design.
- **Interoperability:** For seamless data exchange, interoperability with other medical devices and information systems is crucial. Standardized data formats and communication protocols must be adopted to ensure compatibility [15].

Embedded software development for medical devices is a complex and highly regulated field requiring a deep understanding of software engineering principles and healthcare industry standards. By following a structured development process, adhering to regulatory requirements, and implementing best practices, manufacturers can develop safe, effective, and reliable embedded software that enhances patient care and meets the stringent demands of the healthcare industry.

Overall, while both embedded and Cloud software integration aim to create seamless workflows and enhance functionality, the challenges and approaches differ due to the inherent characteristics of the systems involved. Embedded integration focuses on hardware compatibility and real-time constraints, whereas Cloud integration emphasizes scalability, standardization, and remote management capabilities.

There is a growing trend toward hybrid solutions that combine the strengths of both Cloud and embedded technologies. These solutions aim to leverage the computational power and flexibility of the Cloud while maintaining the efficiency and real-time capabilities of embedded systems.

1.10 AI/ML Software in Medical Devices

Artificial Intelligence (AI) is increasingly being integrated into medical devices, enhancing their capabilities and improving patient care. Integrating AI/ML into medical device software development revolutionizes healthcare by enabling more accurate diagnostics, personalized treatments, and improved patient monitoring. However, developing AI/ML-enabled medical devices involves navigating complex technical, regulatory, and ethical landscapes [30]. One of the most significant benefits of AI/ML in medical devices is their ability to learn from real-world data and improve over time. This continuous learning capability enables devices to adapt to new information and enhance their performance, ultimately leading to improved healthcare delivery.

1.10.1 Cloud-Based AI/ML Medical Device Software

The development of Cloud-based AI/ML-enabled medical devices represents a significant advancement in healthcare technology, offering the potential for improved diagnostics, personalized treatment, and enhanced patient monitoring. These devices leverage the computational power of Cloud platforms to process vast amounts of data and deliver real-time insights. Still, they also present unique challenges that must be addressed to ensure their safe and effective use. As Cloud-based AI/ML-enabled medical devices evolve, collaboration among developers, healthcare providers, regulators, and patients will maximize device effectiveness. Ongoing advancements in AI/ML algorithms and improvements in Cloud computing infrastructure will drive further innovation in personalized medicine and remote care delivery.

1.10.2 On-Prem and Embedded AI/ML Medical Device Software

The development of on-premise and embedded AI/ML-enabled medical devices represents a significant shift in how healthcare technologies are designed and deployed. These devices leverage AI/ML to enhance diagnostics, treatment, and patient monitoring, offering real-time insights and personalized care. However, they also present unique challenges and opportunities that must be navigated carefully [30]. On-premise AI/ML-enabled medical devices process data locally rather than relying on Cloud-based systems. This approach offers several advantages, including reduced latency, enhanced data security, increased reliability, and greater control over data processing. By analyzing data close to its source, these devices can provide immediate insights and support critical decision-making processes in healthcare settings. Embedded AI/ML-enabled medical devices integrate AI/ML capabilities directly into the device hardware, allowing for continuous monitoring and analysis of patient data. These devices are often used in wearable and implantable devices, providing personalized health insights and enabling proactive healthcare management.

1.11 Artificial Intelligence Lifecycle Framework

The AI lifecycle management wheel in Fig. 1.8 [32] represents a comprehensive framework for developing and deploying artificial intelligence solutions. This model builds upon traditional software development processes, including the IEC 62304 Process, AAMI TIR-45, and the Waterfall development process [7, 8, 10], while incorporating specific considerations unique to AI projects. The AI lifecycle

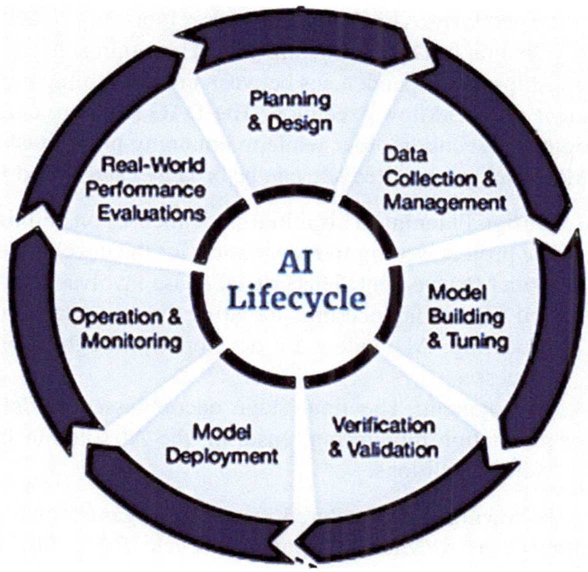

Fig. 1.8 AI lifecycle framework [32]

comprises seven key phases: planning and design, data collection and management, model building and tuning, verification and validation, model deployment, operations and monitoring, and real-world performance evaluation. These phases are then mapped to a medical device product development process lifecycle.

Each phase of the AI lifecycle [32] corresponds to specific technical and procedural considerations. For instance, the planning and design phase involves defining problems, planning data collection, and addressing ethical considerations. Data collection and management focus on ensuring data quality, privacy, and governance. Model building and tuning encompass algorithm selection, feature engineering, and hyperparameter optimization. Verification and validation emphasize rigorous testing and error analysis. Model deployment addresses scalability and integration challenges. Operations and monitoring involve real-time performance tracking and incident response. Ultimately, real-world performance evaluation entails defining key performance indicators and implementing continuous improvement strategies.

As shown in Fig. 1.8, this AI lifecycle aligns with traditional waterfall development processes, mapping to stages such as design planning, design input, design output, design verification and validation, design transfer, and post-production monitoring. However, incorporating AI-specific elements throughout the process goes beyond these conventional approaches.

The AI/ML Software Development Process builds upon established frameworks, such as the Software Lifecycle Process (IEC 62304) and the US FDA Design Control Process. Integrating these principles creates a robust development strategy that meets regulatory requirements and industry best practices for AI/ML-based medical software.

Importantly, AI/ML algorithm development projects typically follow a directed acyclic graph (DAG) structure [33]. This approach acknowledges the intricate relationships and dependencies between tasks, enabling more effective project management and workflow execution. The DAG structure enables teams to determine the optimal execution order while maintaining proper task interdependencies. The AI/ML development lifecycle can be broadly categorized into four main stages:

1. Design Planning: This initial stage focuses on defining the problem and planning the project, setting the foundation for the development process.
2. Data Management: This crucial phase involves data acquisition, pre-processing, and feature engineering, preparing the raw data for use in AI models.
3. Modeling: AI models are developed through training, testing, and validation processes.
4. Deployment: The final stage encompasses model implementation and post-production monitoring, ensuring the AI solution performs effectively in real-world conditions.

Following this structured approach, organizations can systematically progress through the AI/ML development lifecycle, from initial concept to final deployment and ongoing maintenance. This framework ensures that all critical aspects of AI development are addressed, leading to more robust, reliable, and effective AI solutions. The US FDA introduced a draft guidance [34] in January 2025 for "Artificial Intelligence-Enabled Device Software Functions: Lifecycle Management and Marketing Submission Recommendations." Although this guidance is not yet effective as of June 2025, it outlines specific requirements for AI-DSF marketing submissions and reflects the FDA's current thinking on the topic. Chapter 10 contains a detailed discussion on this draft guidance.

The AI/ML lifecycle [32] encompasses a comprehensive series of phases, each with its own set of critical considerations. The planning and design phase lays the foundation for the entire process, addressing crucial aspects such as problem definition, data collection strategies, ethical considerations, algorithm selection, and model design. This phase also emphasizes the importance of feature engineering, evaluation metrics, scalability, interpretability, and deployment planning.

Moving into the data collection and management phase, the focus shifts to ensuring the suitability, quality, and integrity of the data. This phase prioritizes data privacy and security, establishes robust governance practices, and implements strategies to mitigate bias. It also addresses the technical aspects of data versioning, storage infrastructure, and accessibility while considering the nuances of data labeling and annotation.

The model-building and tuning phase is where the core of the AI/ML system takes shape. This phase involves careful model selection, hyperparameter optimization, and advanced feature engineering [35]. It emphasizes the importance of rigorous validation techniques, ensemble methods, and the delicate balance between model complexity and performance. Explainability and interpretability remain critical considerations throughout this phase.

Verification and validation form a critical checkpoint in the lifecycle, focusing on comprehensive evaluation metrics, data verification, and deployment testing. This phase employs various validation strategies, conducts thorough error analyses, and ensures model robustness and interpretability. Documentation and reporting play a vital role in maintaining transparency and accountability.

The deployment phase transitions the model from development to production, addressing concerns related to scalability, performance, and reliability. It emphasizes seamless integration, version control, and the establishment of robust monitoring and logging systems. Compliance with regulatory requirements and comprehensive documentation are paramount at this stage.

Once deployed, the operations and monitoring phase focuses on real-time performance tracking, alert systems, and continuous optimization. This phase emphasizes scalability, resource management, and the implementation of feedback mechanisms. It also addresses ongoing security and compliance concerns, focusing on detecting model drift and implementing effective incident response protocols.

The final phase, real-world performance evaluation, provides a holistic view of the model's effectiveness in production. It involves defining and tracking key performance indicators, collecting and analyzing production data, and comparing results against established baselines. This phase emphasizes continuous improvement through the collection of feedback, error analysis, and comprehensive documentation.

Risk management [27, 36] and cybersecurity considerations [16, 17, 37] permeate every phase of the lifecycle, underscoring their critical importance in the development and deployment of AI/ML. Regulatory bodies, recognizing the unique challenges of AI/ML technologies, have developed specialized guidelines. The FDA's draft guidance on Predetermined Change Control Plans (PCCPs) [28] aims to facilitate safe and effective iterative improvements while maintaining regulatory compliance. Developing Good Machine Learning Practices (GMLP) [38] in collaboration with international partners emphasizes the importance of transparency, diversity, fairness, and health equity in AI/ML-enabled devices. These regulatory frameworks provide a structured approach to the ethical and responsible development of AI/ML technologies, ensuring their safe and effective implementation in medical devices.

1.11.1 US FDA's Good Machine Learning Practices (GMLP)

- **Multi-Disciplinary Expertise Is Leveraged Throughout the Total Product Life Cycle:** An in-depth understanding of a model's intended integration into clinical workflow, along with the desired benefits and associated patient risks, can help ensure that ML-enabled medical devices are safe, effective, and address clinically meaningful needs throughout the device's lifecycle. The figure below outlines the multi-disciplinary expertise required for developing AI/ML-enabled medical device software.

- **Good Software Engineering and Security Practices Are Implemented:** Model design is implemented with attention to the "fundamentals": good software engineering practices, data quality assurance, data management, and robust cybersecurity practices. These practices include systematic risk management and design processes that can appropriately capture and communicate design, implementation, and risk management decisions and rationale and ensure data authenticity and integrity.
- As depicted in Fig. 1.9, the core algorithms for learning and prediction, shown in black box, constitute only a tiny portion of the overall codebase in many machine learning (ML) systems. The bulk of the code is dedicated to what industry insiders often refer to as "scaffolding"—the essential supporting infrastructure that enables ML models to function effectively in real-world scenarios [39].
- This surrounding infrastructure is both extensive and intricate, encompassing various critical components:
  - Data preprocessing and management
  - Model deployment and serving mechanisms
  - Monitoring and logging systems
  - Scalability solutions
  - Integration with existing software ecosystems

Developing robust ML systems demands Strong software engineering skills, rigorous security practices, efficient system architecture design, comprehensive testing, and quality assurance. These elements ensure that ML models can operate reliably, securely, and at scale in production environments. While the learning algorithms are at the heart of ML systems, the surrounding infrastructure plays a crucial role in their practical implementation and success.

- **Clinical Study Participants and Data Sets aRepresentative of the Intended Patient Population:** Data collection protocols should ensure that the relevant characteristics of the intended patient population (for example, in terms of age, gender, sex, race, and ethnicity), use, and measurement inputs are sufficiently represented in a sample of adequate size in the clinical study and training and test datasets so that results can be reasonably generalized to the population of inter-

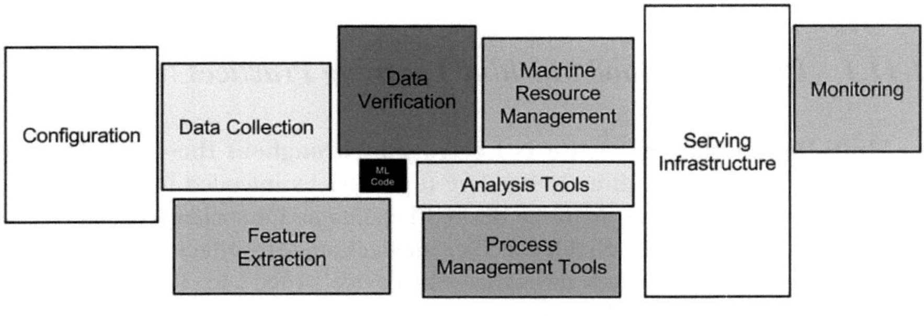

Fig. 1.9 Vast and complex surrounding infrastructure of a real-world ML system [39]

est. This is crucial for managing bias, promoting appropriate and generalizable performance across the intended patient population, assessing usability, and identifying circumstances where the model may underperform.

- **Training Data Sets Are Independent of Test Sets:** Training and test datasets are selected and maintained to be appropriately independent. All potential sources of dependence, including patient, data acquisition, and site factors, are considered and addressed to assure independence.
- **Selected Reference Datasets Are Based Upon Best Available Methods:** Accepted, best-available methods for developing a reference dataset (a reference standard) ensure that clinically relevant and well-characterized data are collected and that the reference's limitations are understood. If available, accepted reference datasets are used in model development and testing to promote and demonstrate model robustness and generalizability across the intended patient population.
- **Model Design Is Tailored to the Available Data and Reflects the Device's Intended Use:** The model design is tailored to the available data and supports the active mitigation of known risks, including overfitting, performance degradation, and security risks. The clinical benefits and risks of the product are well understood, used to derive clinically meaningful performance goals for testing, and support the idea that the product can be used safely and effectively for its intended purpose. Considerations include the impact of both global and local performance, as well as uncertainty and variability in device inputs, outputs, intended patient populations, and clinical use conditions.
- **Focus Is Placed on the Performance of the Human-AI Team:** Where the model has a "human in the loop," human factors considerations, and the human interpretability of the model outputs are addressed with emphasis on the performance of the Human-AI team rather than just the performance of the model in isolation.
- **Testing Demonstrates Device Performance During Clinically Relevant Conditions:** Statistically sound test plans are developed and executed to generate clinically relevant device performance information independently of the training dataset. Considerations include the intended patient population, important subgroups, clinical environment, and use by the Human-AI team, measurement inputs, and potential confounding factors.
- **Users Are Provided Clear, Essential Information:** Users are provided ready access to transparent, contextually relevant information that is appropriate for the intended audience (such as health care providers or patients), including the product's intended use and indications for use, performance of the model for appropriate subgroups, characteristics of the data used to train and test the model, acceptable inputs, known limitations, user interface interpretation, and clinical workflow integration of the model. Users are also aware of device modifications and updates through real-world performance monitoring, which serves as the basis for decision-making when available and provides a means to communicate product concerns to the developer.

- **Deployed Models Are Monitored for Performance, and Re-training Risks Are Managed:** Deployed models can be monitored in "real world" use, focusing on maintained or improved safety and performance. Additionally, when models are periodically or continually trained after deployment, appropriate controls are in place to manage risks of overfitting, unintended bias, or degradation of the model (for example, dataset drift) that may impact the safety and performance of the model as the human-AI team uses it.

1.11.2 Machine Learning Operation (MLOps)

MLOps, or Machine Learning Operations, is an emerging engineering practice that aims to streamline and automate the end-to-end lifecycle of machine learning solutions. It addresses the challenges faced by data scientists who often manage ML workflows manually, resulting in operational issues. As shown in Fig. 1.10, by integrating principles from machine learning, software engineering (particularly DevOps), and data engineering, MLOps bridges the gap between development and operations in ML projects [40].

The core objective of MLOps is to enable more efficient productionization of machine learning models. It encompasses best practices, concepts, and a development culture that facilitate the design, implementation, monitoring, deployment,

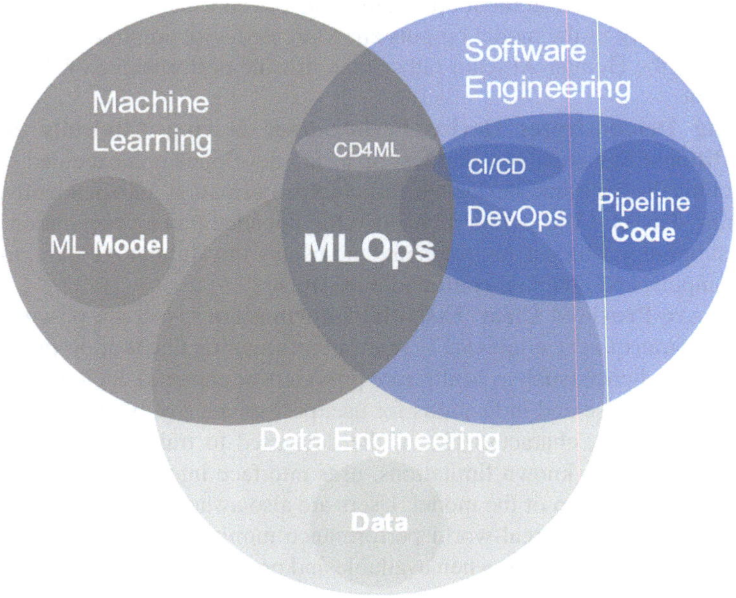

Fig. 1.10 Intersection of ML, Software Engineering disciplines of the MLOps paradigm [40]

and scalability of ML products. Key principles of MLOps include CI/CD automation, workflow orchestration, reproducibility, data versioning, model and code management, collaboration, continuous training and evaluation, metadata tracking, monitoring, and feedback loops. By adopting MLOps practices, organizations can overcome the hurdles of manual ML processes, allowing more proofs of concept to transition successfully into production. This approach not only enhances the efficiency of ML workflows but also ensures the reliability, scalability, and maintainability of machine learning systems in real-world applications [41–43].

A critical component of MLOps extends beyond initial deployment, focusing on the ongoing management and optimization of machine learning models in production environments [32, 37]. This aspect of MLOps is integral to AI Lifecycle Management, encompassing two key activities:

Monitoring Continuous monitoring of deployed models is essential to ensure their performance remains consistent and reliable over time [32, 37, 39]. This involves tracking various metrics, including:

- Model accuracy and prediction quality.
- Data drift and concept drift.
- System performance and resource utilization.
- Operational health and uptime.

Real-World Performance Evaluation Regularly assessing models' performance in actual production scenarios is crucial [28, 32]. This involves:

- Comparing predicted outcomes against real-world results.
- Analyzing user feedback and interaction data.
- Identifying discrepancies between test and production environments.
- Assessing the business impact and value delivered by the model.

By emphasizing these post-production activities, MLOps ensures that AI systems remain adequate, relevant, and aligned with business objectives throughout their lifecycle. This approach enables organizations to maintain high-quality AI solutions and make data-driven decisions for continuous improvement. This topic will be discussed in detail in a later chapter.

1.12 AI/ML-Enabled Medical Device Software Clinical Evaluations

The FDA emphasizes the importance of clinical evidence to support the safety and effectiveness of medical devices. Depending on the device's classification and risk profile, the process may involve clinical trials or the collection of real-world evidence [44]. The FDA and EU have released specific guidance documents for SaMD and medical devices [24, 45, 46], focusing on premarket submissions and clinical evaluation requirements tailored to the software's risk classification. The FDA requires clinical data that supports the safety and effectiveness of the device, with the amount and type of data varying based on the device's risk classification.

AI/ML devices must comply with data protection regulations, such as GDPR and HIPAA, which mandate robust security measures to prevent data breaches and unauthorized access. AI/ML algorithms can inadvertently perpetuate biases in the training data, leading to disparities in healthcare outcomes. Developers must ensure their algorithms are trained on diverse and representative datasets to minimize bias and promote fairness. The "black box" nature of some AI/ML models can make it difficult to understand how decisions are made. Ensuring transparency and explainability is crucial for gaining the trust of healthcare providers and patients, as well as for meeting regulatory requirements. The development of AI/ML-enabled medical device software offers significant benefits, including enhanced diagnostics, personalized medicine, continuous learning, and remote monitoring. However, it also presents challenges related to regulatory compliance, data privacy, bias, and transparency. By adopting best practices and adhering to regulatory standards, medical device manufacturers can leverage the power of AI/ML to deliver innovative, safe, and effective healthcare solutions.

The clinical evaluation process in the EU involves several stages, including proving a valid association between the medical device output and its target clinical condition, ensuring analytical and technical validation, and demonstrating that the clinical data produced by medical device is clinically meaningful and beneficial. The European Commission provides detailed guidance documents, such as MDCG 2020-1 [24, 47], which outline the clinical and performance evaluation requirements for medical device software. The MDR requires comprehensive clinical evaluations, including clinical investigations for higher-risk devices. The regulation emphasizes the importance of continuous postmarket surveillance and the submission of periodic safety update reports. This topic will be discussed in detail in a later chapter.

1.13 Cybersecurity in Medical Device Software

Cybersecurity is crucial to the development of medical device software, especially as devices become increasingly interconnected and reliant on digital technologies. The FDA has issued guidance to help manufacturers integrate cybersecurity measures into the design and development of medical devices. The latest FDA guidelines on cybersecurity for medical devices, issued on September 27, 2023, provide comprehensive recommendations for ensuring the cybersecurity of medical devices throughout their lifecycle [16]. This includes premarket submissions that require information on potential cybersecurity risks and the controls implemented to mitigate them. Under the latest guidelines, the FDA defines a "cyber device" as a medical device that includes software, can connect to the Internet or other networks, and contains technology that could be vulnerable to cybersecurity threats. This definition is part of the Food and Drug Omnibus Reform Act (FDORA) [48], which mandates that cybersecurity considerations be integrated into the development and regulatory review of such devices.

The US FDA's criteria for a cyber device include:

- The presence of software that is validated, verified, or intended for use in a medical device.
- The ability to connect to the Internet or another network.
- Technology that could be susceptible to cybersecurity vulnerabilities.

These guidelines emphasize the need for manufacturers to incorporate cybersecurity measures throughout the device's lifecycle, including premarket submissions that detail cybersecurity risk assessments and plans for postmarket cybersecurity vigilance. The EU's Medical Device Regulation (MDR) also emphasizes the importance of cybersecurity, requiring manufacturers to address cybersecurity risks throughout the device lifecycle [17]. This topic will be discussed in detail in a later chapter.

1.13.1 Privacy in AI/ML-Enabled Medical Device Software

The General Data Protection Regulation (GDPR) [49] and the Health Insurance Portability and Accountability Act (HIPAA) [50] are two prominent regulatory frameworks that govern data privacy. Still, they differ significantly in scope, applicability, and specific requirements, particularly regarding medical device data privacy. Here are the key differences between GDPR and HIPAA in terms of medical device data privacy:

General Data Protection Regulation (GDPR) The GDPR is a comprehensive data protection regulation that applies to any organization processing the personal data of individuals within the European Union, regardless of the organization's location. It covers a wide range of personal data, including any information that can identify an individual, directly or indirectly. This includes names, addresses, identification numbers, online identifiers, and sensitive data, such as genetic or biometric information.

Health Insurance Portability and Accountability Act (HIPAA) HIPAA is a US federal law that specifically targets the protection of Protected Health Information (PHI) within the healthcare sector. It applies to covered entities, such as healthcare providers, health plans, healthcare clearinghouses, and business associates. PHI includes medical records, diagnoses, treatment information, and insurance details.

The European Union (EU) has established comprehensive privacy requirements for medical device software, primarily governed by the General Data Protection Regulation (GDPR). These requirements are designed to protect the personal data of EU citizens, particularly sensitive health information, and apply to any organization, regardless of location, that processes such data. Here is a detailed exploration of the EU privacy requirements for medical device software:

1.13.2 Consent and Data Subject Rights

GDPR strongly emphasizes obtaining explicit and informed consent from individuals to process their data. Consent must be freely given, specific, informed, and unambiguous. Individuals can withdraw their consent at any time, and organizations must make this process easy. GDPR also grants individuals the "right to be forgotten," allowing them to request the deletion of their data under certain conditions. HIPAA does not require explicit consent to use and disclose PHI for treatment, payment, and healthcare operations. Instead, it allows for sharing PHI within the boundaries of permitted uses and disclosures outlined by the regulation. Patients are notified of privacy practices, but explicit consent is not typically required for these standard operations.

1.13.3 Data Protection Measures

GDPR mandates comprehensive data protection measures, including data minimization, pseudonymization, and the implementation of technical and organizational measures to ensure data security. It requires Data Protection Impact Assessments (DPIAs) for high-risk processing activities and mandates the appointment of a Data Protection Officer (DPO) for specific organizations. HIPAA requires covered entities to implement administrative, physical, and technical safeguards to protect protected health information (PHI). This includes conducting risk assessments, implementing access controls, and ensuring the integrity and confidentiality of data. HIPAA also requires notification of breaches to affected individuals and the Department of Health and Human Services (HHS) in the event of a data breach.

In summary, while GDPR and HIPAA aim to protect personal health information, they differ in scope, consent requirements, data protection measures, and penalties. GDPR has a broader reach and more stringent consent and data subject rights requirements, while HIPAA focuses on health-related information within the US healthcare sector. Understanding these differences is crucial for organizations handling medical device data to ensure compliance with the relevant regulations. This topic will be discussed in detail in a later chapter.

1.14 The EU AI Act and AI/ML-Enabled Medical Device Software

The European Union's Artificial Intelligence Act (EU AI Act) [51] represents a significant milestone in regulating AI technologies, including their application in medical devices. This comprehensive legal framework aims to ensure the safe and ethical deployment of AI, fostering innovation while protecting fundamental rights and ensuring user safety. As AI and machine learning (ML) become increasingly integral to medical devices, understanding the implications of the AI Act is crucial for developers and manufacturers.

The European Union's Artificial Intelligence Act (AI Act) represents a pioneering effort to regulate AI systems based on their potential risks. This comprehensive legislation categorizes AI applications into four risk levels, shown in Fig. 1.11, each with corresponding regulatory requirements. At the highest concern are AI systems deemed to pose unacceptable risks. These are outright prohibited and include applications such as social scoring systems and manipulative AI technologies that could threaten fundamental rights and societal values. This ban reflects the EU's commitment to safeguarding its citizens from potentially harmful AI applications. The bulk of the AI Act focuses on high-risk AI systems, which are subject to stringent regulations. While not banned, these systems have significant potential for harm and thus require careful oversight. High-risk AI applications span various sectors, including critical infrastructure, education, employment, and law enforcement. For high-risk AI Systems requiring conformity assessment, medical devices under the MDR regulations should undergo conformity assessment.

A third category addresses AI systems with limited risk, primarily focusing on transparency requirements. For instance, developers and deployers of chatbots and deepfake technologies must ensure that end-users are aware they are interacting with AI. This provision aims to maintain user trust and informed decision-making in AI interactions. The fourth category encompasses minimal-risk AI applications,

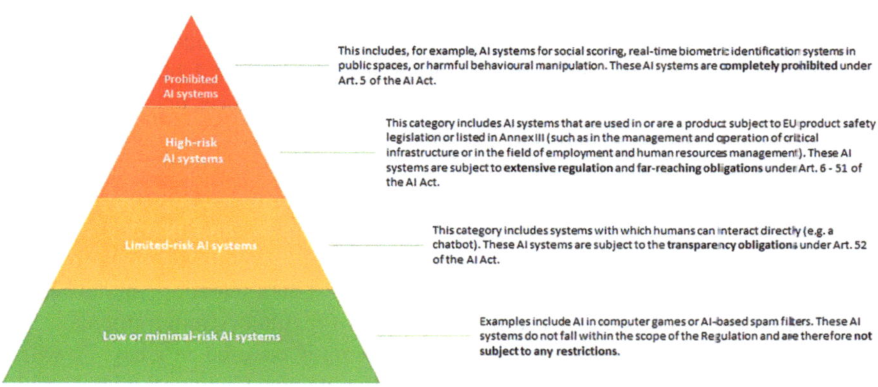

Fig. 1.11 Risk classification as per EU AI ACT [52]

which remain largely unregulated. This includes many AI-enabled consumer health products currently available in the EU market, such as video games and spam filters. However, it's worth noting that the landscape is evolving, particularly with the advent of generative AI technologies.

The AI Act places most compliance obligations on the providers or developers of high-risk AI systems. This applies to all entities intending to market or deploy such systems in the European Union, regardless of their geographical location. Users or deployers of high-risk AI systems also bear some responsibilities, albeit to a lesser extent than providers. Notably, the Act also addresses general-purpose AI (GPAI) models. All GPAI providers must furnish technical documentation and usage instructions and comply with copyright directives. Those offering free and open-license GPAI models have fewer obligations unless their systems present systemic risks. However, all providers of GPAI models that could pose systemic risks must conduct thorough evaluations, perform adversarial testing, and implement robust cybersecurity measures.

The EU AI Act establishes a nuanced, risk-based framework for regulating AI, striking a balance between innovation and the protection of fundamental rights and societal values. This approach aims to foster responsible AI development while maintaining Europe's competitiveness in the global AI landscape. A later chapter will discuss this topic in detail.

1.15 This Book's Organization

This comprehensive book on AI/ML-enabled medical device software development is structured into 12 chapters, thoroughly exploring the field from concept to deployment. Chapter 1 provides an overview of medical device software development, with a specific focus on AI/ML-enabled systems. Chapter 2 examines the fundamental aspects of clinical data collection, pre-processing, and feature engineering, laying the groundwork for the effective implementation of AI/ML in medical devices.

The technical core of the book begins with Chap. 3, which discusses the selection of appropriate machine learning algorithms, considering factors such as bias, transparency, integration, and deployment strategies. Chapter 4 builds on this by exploring the intricacies of AI/ML model training and evaluation, covering standard techniques, performance metrics, hyperparameter tuning, and the concepts of human-in-loop and human-in-command. Chapter 5 discusses multimodal artificial intelligence. Ethical and practical considerations are addressed in Chap. 6, which examines the transparency, interpretability, bias, fairness, and privacy concerns associated with AI/ML algorithms. Chapter 6 also focuses on the critical processes of model testing and clinical validation, discussing test data selection, model robustness, and clinical study design within the context of regulatory requirements.

The practical implementation of AI/ML systems is covered in Chap. 7, which explores integration and production deployment strategies. Chapter 8 addresses the vital topics of cybersecurity and privacy, outlining considerations that must be integrated throughout the development process. The book then shifts to operational aspects, with Chap. 9 discussing machine learning operations frameworks and change control plans. Chapter 10 covers premarket regulatory submissions, EU AI Act, risk management, human factors, and AI/ML-related medical device standards, emphasizing the importance of developing state-of-the-art evidence. Chapter 11 provides real-world context by presenting examples of AI/ML-enabled medical devices and exploring the role of generative AI in healthcare, including its potential applications in various stages of medical device software development. Chapter 12 summarizes the current state of US FDA clearance and approval of AI/ML-enabled medical device development.

Readers interested in regulatory submissions and related requirements can focus on Chaps. 10 and 12. For those interested in AI testing and validation, they can concentrate on Chaps. 4 and 6. For those interested in deployment and operations, they can focus on Chaps. 7 and 9. For research and development for artificial intelligence, they can concentrate on Chaps. 2–5, and 11. Product security professionals may find Chaps. 8 and 9 of interest.

1.16 Summary

This chapter provided a comprehensive overview of artificial intelligence and machine learning in medical device software development, guiding readers through the development and AI/ML Lifecycle. It began by introducing AI/ML-enabled medical device software, highlighting its potential to transform healthcare by deriving new insights from vast amounts of data generated during healthcare delivery. The chapter then explored how Agile Scrum methodologies could be adapted to the unique challenges of AI/ML development in the medical device industry, allowing for iterative progress and rapid response to changing requirements.

The focus then shifted to the regulatory landscape, providing an overview of the medical device regulatory framework. It examined the FDA's approach to AI/ML-enabled medical devices, including the concept of Predetermined Change Control Plans (PCCPs) and the agency's efforts to adapt its traditional paradigm to the dynamic nature of AI/ML technologies. The chapter delved into AI/ML lifecycle management, emphasizing the importance of continuous monitoring and improvement. Security and privacy considerations were addressed, reflecting their critical importance in the healthcare sector. Finally, the chapter introduced the European Union's AI Act, a landmark legislation with significant implications for AI/ML medical device developers.

By covering these crucial aspects, the chapter gave readers a comprehensive understanding of the complexities of developing AI/ML-enabled medical device software. It emphasizes the need for a multidisciplinary approach that strikes a balance between innovation and regulatory compliance. This summary encapsulates the key points discussed in the chapter, providing a concise overview of the content and its significance in the development of AI/ML-enabled medical device software.

1.17 Points to Ponder

1. Identify Software Safety Class as per IEC62304 for the following indications for use

   SmartChest is a radiological computer-assisted triage and notification software that analyzes frontal chest X-ray images (Postero-Anterior (PA) or Antero-Posterior (AP)) of transitional adolescents (18 21 yo but treated like adults) and adults (222 yo) for the presence of suspected pleural effusion and/or pneumothorax. SmartChest uses an artificial intelligence algorithm to analyze the images for features suggestive of critical findings and provides case-level output available to a PACS (or other DICOM storage platforms) for worklist prioritization. As a passive notification for prioritization-only software tool within the standard of care workflow, SmartChest does not send a proactive alert directly to a trained medical specialist. SmartChest is not intended to direct attention to a specific portion of an image. Its results are not intended to be used on a stand-alone basis for clinical decision-making [53].
2. Do the Software Safety Classes of IEC 62304 and FDA documentation levels correlate?

1.18 Answers

1. This medical device's indication states, "SmartChest is not intended to direct attention to a specific portion of an image. Its results are not intended to be used on a stand-alone basis for clinical decision-making." This clearly shows that the clinician is responsible for interpreting the results and that the device does not provide direct clinical benefits. This medical device software should be classified as Class A Software Safety Class.
2. These two software classifications, for development rigor and documentation level methodologies, differ significantly. FDA's definition of Enhanced Documentation Level does not explicitly mention that risk control measures external to software shall be considered to determine documentation level requirements. However, the definition of documentation level only focuses on software system failures. If risk control measures external to software reduce the risk as far as possible, software failure will not lead to unacceptable risk. So, Risk Control external to software shall be considered in the determination. Additionally, IEC 62304 classifies medical device software into three safety classes: A, B, and C. At the same time, the US FDA documentation level for software is at two levels: Basic and Enhanced levels of documentation.

References

1. Guidance for Industry and Food and Drug Administration Staff (2023) Content of premarket submissions for device software functions
2. US FDA (2022) Global approach to software as a medical device. https://www.fda.gov/medical-devices/software-medical-device-samd/global-approach-software-medical-device
3. US FDA. A history of medical device regulation & oversight in the United States. https://www.govinfo.gov/content/pkg/STATUTE-90/pdf/STATUTE-90-Pg539.pdf
4. US FDA. Quality system (QS) regulation/medical device current good manufacturing practices (CGMP). https://www.fda.gov/medical-devices/postmarket-requirements-devices/quality-system-qs-regulationmedical-device-current-good-manufacturing-practices-cgmp
5. Regulation (EU) 2017/745 of the European Parliament and of the Council of 5 April 2017 on medical devices, amending Directive 2001/83/EC, Regulation (EC) No 178/2002 and Regulation (EC) No 1223/2009 and repealing Council Directives 90/385/EEC and 93/42/EEC. https://eur-lex.europa.eu/legal-content/EN/TXT/?uri=celex%3A32017R0745
6. Guidance on qualification and classification of software in regulation (EU) 2017/745—MDR and Regulation (EU) 2017/746—IVDR
7. IEC 62304:2006+AMD1:2015 Medical device software—Software life cycle processes
8. US FDA (1997) Design control guidance for medical device manufacturers
9. Bhatt PC, Sehgal NK. Cloud software project management
10. AAMI TIR45:2023; Guidance on the use of agile practices in the development of medical device software
11. IEC 82304-1: Health software—General requirements for product safety
12. Introduction to HL7 Standards. https://www.hl7.org/implement/standards/index.cfm?ref=nav
13. The DICOM Standards. https://www.dicomstandard.org/current
14. ISO/IEEE 11073-10101:2020(en) Health informatics—device interoperability—Part 10101: Point-of-care medical device communication—Nomenclature. https://www.iso.org/obp/ui#iso:std:iso-ieee:11073:-10101:ed-2:v1:en
15. Guidance for Industry and Food and Drug Administration Staff (2017) Design considerations and pre-market submission recommendations for interoperable medical devices
16. Guidance for Industry and Food and Drug Administration Staff (2023) Cybersecurity in medical devices: quality system considerations and content of premarket submissions
17. MDCG 2019-16. Guidance on cybersecurity for medical devices. https://health.ec.europa.eu/document/download/b23b362f-8a56-434c-922a-5b3ca4d0a7a1_en?filename=md_cybersecurity_en.pdf
18. US FDA. Premarket notification orocess, 510K. https://www.fda.gov/medical-devices/device-approvals-and-clearances/510k-clearances
19. US FDA. Premarket approval, PMA. https://www.fda.gov/medical-devices/premarket-submissions-selecting-and-preparing-correct-submission/premarket-approval-pma
20. US FDA. De Novo classification request. https://www.fda.gov/medical-devices/premarket-submissions-selecting-and-preparing-correct-submission/de-novo-classification-request
21. MDCG 2019-11. Qualification and classification of software—Regulation (EU) 2017/745 and Regulation (EU) 2017/746
22. MDCG-2023-4 (2023) Medical device software (MDSW)—hardware combinations: guidance on MDSW intended to work in combination with hardware or hardware components
23. Is your software a Medical Device? (2021) https://health.ec.europa.eu/document/download/b865d8e9-081a-4601-a91a-f120321c0491_en?filename=md_mdcg_2021_mdsw_en.pdf
24. MDCG 2020-1. Guidance on clinical evaluation (MDR)/Performance evaluation (IVDR) of medical device software
25. ISO 13485—Medical devices—quality management systems—Requirements for regulatory purposes

26. IMDRF Software as a Medical Device (SaMD) Working Group (2014) "Software as a medical device": possible framework for risk categorization and corresponding considerations
27. ISO 14971: Medical devices—Application of risk management to medical devices
28. US FDA Guidance (2023) Off-the-shelf software use in medical devices
29. US FDA Guidance (2005) Guidance for the content of premarket submissions for software contained in medical devices
30. Guidance for Industry and Food and Drug Administration Staff (2024) Marketing submission recommendations for predetermined change control plan for artificial intelligence/machine learning (AI/ML)-enabled device software functions
31. Scrum Framework (2020). www.scrum.org
32. US FDA Blog. A lifecycle management approach toward delivering safe, effective AI-enabled health care. https://www.fda.gov/medical-devices/digital-health-center-excellence/blog-lifecycle-management-approach-toward-delivering-safe-effective-ai-enabled-health-care
33. Directed acyclic graph. https://en.wikipedia.org/wiki/Directed_acyclic_graph
34. US FDA (Draft Guidance) (2025) Artificial intelligence-enabled device software functions: lifecycle management and marketing submission recommendations
35. Gupta P, Sehgal N (2025) Introduction to machine learning with security. Springer
36. BS/AAMI 34971:2023. Application of ISO 14971 to machine learning in artificial intelligence. Guide 2023
37. ISO/IEC TR 27563:2023. Security and privacy in artificial intelligence use cases—Best practices
38. US FDA, Good machine learning practice for medical device development: guiding principles. https://www.fda.gov/medical-devices/software-medical-device-samd/good-machine-learning-practice-medical-device-development-guiding-principles
39. Sculley D et al (2014) Hidden technical debt in machine learning systems. Google Inc.
40. Kreuzberger D et al. Machine learning operations (MLOps): overview, definition, and architecture
41. ASW MLOps. https://aws.amazon.com/what-is/mlops/
42. Databricks MLOps. https://www.databricks.com/glossary/mlops
43. Azure microsoft machine learning operation. https://azure.microsoft.com/en-us/solutions/machine-learning-ops/
44. US FDA for Industry and Food and Drug Administration Staff (2017) Use of real-world evidence to support regulatory decision-making for medical devices
45. Guidance for Industry and Food and Drug Administration Staff (2017) Software as a medical device (SAMD): clinical evaluation
46. Guidance for Industry and Food and Drug Administration Staff (2018) Acceptance of clinical data to support medical device applications and submissions frequently asked questions
47. MEDDEV 2.7/1 Rev. 4: Clinical evaluation: guide for manufacturers and notified bodies (applicable to all MD only)
48. Food and Drug Omnibus Reform Act (FDORA) of 2022. https://www.fda.gov/regulatory-information/selected-amendments-fdc-act/food-and-drug-omnibus-reform-act-fdora-2022
49. General Data Protection Regulation. https://gdpr-info.eu/
50. The HIPAA privacy rule. https://www.hhs.gov/hipaa/for-professionals/privacy/index.html
51. The EU Artificial Intelligence Act. https://artificialintelligenceact.eu/
52. A milestone for AI regulation: European Parliament approves new legal framework for artificial intelligence. https://www2.deloitte.com/dl/en/pages/legal/articles/ki-verordnung-eu.html
53. https://www.accessdata.fda.gov/cdrh_docs/pdf23/K232410.pdf

Chapter 2
Data Management for AI/ML-Enabled Medical Devices

Abstract Data management is foundational in developing artificial intelligence-based algorithms and solutions for the healthcare industry. This chapter provides a comprehensive overview of clinical data management, beginning with an examination of various healthcare data types and their corresponding collection protocols. It emphasizes the critical importance of data security, privacy, regulatory compliance, and the necessary agreements to protect sensitive information. A strong focus is placed on assessing data quality and completeness, which are essential for the success of AI applications in healthcare. The chapter also details key data preprocessing and manipulation techniques required to prepare raw data for analysis. It further explores the Data Quality Lifecycle and Data Quality Management Lifecycle frameworks, as introduced by EU AI Act-aligned standards, highlighting their role in maintaining data quality and integrity throughout the training and validation of AI algorithms. The chapter concludes by emphasizing the need for AI solutions in healthcare to consistently meet the rigorous standards set by regulatory bodies.

Keywords Data collection approaches`Clinical research · Observational studies · Electronic health records (EHR) · HL7 · Artificial intelligence (AI) · Machine learning (ML) · Comorbidity · Regulations · Compliance · Data quality · Data preprocessing · Data lifecycle · Reference standards · Data annotation · EU AI Act

2.1 Introduction

Clinical data has immense potential to drive progress in health care by providing the means to measure and track processes and outcomes, to develop and refine best clinical practices, and to enable rapid discovery and innovation. Over the past decade, the amount of clinical data captured in electronic form has increased exponentially. Given the federal government's encouragement of the adoption of electronic health records (EHRs) and private industry's adoption of digital technology, these data will increasingly be augmented by clinically rich, course-of-care data. With the potential to be stored securely, aggregated, and shared, this data can enable

A. Pandey et al., *Artificial Intelligence in Medical Software*,
https://doi.org/10.1007/978-3-032-10808-1_2

rapid learning and continuous improvement in the efficiency and effectiveness of care practices. Harnessing the power of these data, however, will require efforts to address barriers to their access and use [1]. This chapter examines clinical data management (CDM) to enhance healthcare research through the application of artificial intelligence (AI) and machine learning (ML). However, CDM also introduces additional challenges in terms of privacy, security, and proprietary concerns. Privacy and security of data are discussed in the book later.

Clinical data is information about patients that is used for various purposes, including research, clinical decision-making, and tracking patient trends. Clinical data is a collection of structured and unstructured information about patients, including their electronic health records, symptoms, disease codes, demographics, and laboratory test results. Clinical data can be collected from various sources, including electronic health records, administrative datasets, government payer datasets, large-scale registries, and clinical trials. Clinical data can be used to develop new knowledge, guide best practices, and track patient trends. For example, doctors can use clinical data to help select treatments that have been effective for patients with similar conditions.

2.2 Clinical Data Needs

One approach to clarifying key data needs for a learning health system is to agree on a core data set—sufficient, if not perfect, for several information needs. Known as the 80/20 rule, this approach advocates starting with something simple that can be achieved now and developing clever approaches over time [2]. The American Recovery and Reinvestment Act (ARRA) presented a significant opportunity for health information technology (HIT) in this respect, as it gave patients the right to their records in electronic format. Even if only a small fraction of the patients in any given system choose to exercise this right, every healthcare provider and electronic health record (EHR) vendor must produce a patient summary document in a standard format. This could be a key opportunity to utilize the clinical care summary, which includes the medication list, problem list, allergies, lab values, and patient encounters, as the foundational 80 percent—the standard data core on which to build a unified HIT infrastructure. These data can be represented in multiple modalities, such as images, text, video, and speech, which requires considerable effort to secure and access the data. This also imposes liability on data owners due to stringent regulations.

Data users have a wide range of needs and require a tailored strategy for collecting data that matches their intended usage. For quality reporting, for example, detailed information is needed to calculate exclusions. For public health case reporting, information on whether a specific infection was central line-associated may be required; For drug safety, the first date of prescribing the medication may be needed. We will discuss a few of the issues.

2.2.1 *Creating Infrastructure for Targeted Data Collection*

For AI and ML, we require a substantial amount of relevant data that is aligned with the intended use of the product. For example, suppose a company is developing a product to identify patients who are more susceptible to a heart attack. In that case, it will have different data requirements than a product that diagnoses cancer. This intended use of the product will have a direct implication on the infrastructure required for targeted data collection. There are two types of requirements: data needed to make the product, and the second one is to test the product.

One cannot expect a provider to use an EHR in the routine delivery of care while also collecting all the information needed to determine patient eligibility for a multitude of clinical trials. As one example, an individual's occupation may be a data element necessary to determine eligibility for a clinical trial. Although those with an interest in this information may think that requiring it as a data element should be elementary, a provider's front-office staff is not going to determine and select each patient's occupation from among the hundreds of potential Census Bureau categories as a routine part of delivering care.

A more feasible approach is to collect a limited amount of information during routine care delivery, which also serves as an opportunity to identify the need for additional data collection. Instead of asking each provider to record each patient's occupation and expecting these data to be collected in a structured form, a system could be developed to trigger deeper data collection when appropriate (i.e., manual rather than routine data collection on a small subset of patients). Suppose a patient is diagnosed with hepatitis A, for example. In that case, it is entirely appropriate to prompt the clinician to inquire about the patient's occupation, specifically, whether the patient is a healthcare worker, daycare worker, or food worker. This type of approach makes sense to providers, as the information requested is relevant to the person's care and is limited in scope. For clinical trials, minimal information can also serve as an initial screening tool, with follow-up questions about trial participation asked only as necessary.

2.2.2 *Adding Value for Providers and Patients*

To ensure data quality, it is also essential to consider the value proposition for capturing and utilizing data. The data provider's work may involve caring for the patient, creating, and documenting structured data. This data is needed for analysis, diagnosis, and report preparation. A business case needs to be made to ensure that the documentation of the data collection is completed appropriately. As part of the data collection agreement with the data provider, a fair market value of the collected data should be estimated, and a commitment to pay the provider after the data collection is made. As the owner of the data, this notion also extends to the patient.

Perhaps patients should also benefit financially from the use of their health information to incentivize more data sharing.

2.3 Types of Healthcare Data

Data is key to the development of healthcare technology, enabling the creation of more effective treatments and improving patient care. High-quality patient data is essential for informing the clinical care team, and everyone should have access to information about their healthcare to help manage their conditions effectively. The patient care team, whether in a clinic or hospital, needs to have access to the most current data and healthcare history to provide the best possible care.

This section will introduce various types of data [3, 4], ranging from medical records to personal health information, and explain how they can contribute to making healthcare more effective and patient-centered. Clinical data can be categorized into different types, as discussed below.

2.3.1 Electronic Health Records and Electronic Medical Records

Electronic health records (EHRs) and electronic medical records (EMRs) are two key types of healthcare data that have fundamentally transformed how healthcare providers manage patient information. EMR data refers to the digital version of a patient's paper chart in a clinician's office, encompassing medical and treatment histories within one practice. EHR data, on the other hand, takes this concept a step further by offering a more comprehensive and holistic view of the patient's health. EHR comprises any medical data that is recorded during a patient's visit.

EHRs are designed to share information with other healthcare providers and organizations, such as laboratories, specialists, medical imaging facilities, pharmacies, emergency facilities, and clinics, ensuring that a patient's health information is accessible and securely shared across different healthcare settings. This seamless exchange of information not only improves the accuracy of diagnoses and treatment plans but also enhances the overall quality of care by making the patient's health history easily accessible to all authorized providers involved in their care. Examples of data that can be accessed in an EHR include a patient's name, date of birth, address, any vital signs, test results, healthcare provider notes, and billing information. However, EHR data was created to save time and resources, as it supports the easy and quick management of all available patient information. This type of data is rarely cross-checked or improved, due to time constraints in hospitals. Thus, the data quality is unacceptable for making any meaningful conclusions.

2.3.2 *Administrative Data*

Healthcare administrative data is generated during each interaction with the healthcare system, whether it involves a visit to a doctor's office, a diagnostic procedure, a hospital admission, interactions with care facilities, home care, or the filling of pharmaceutical prescriptions.

This data is gathered for administrative or billing purposes, but can also be used to analyze healthcare delivery, including its advantages, disadvantages, and cost-effectiveness. In certain areas, such data includes details on hospital discharge information that will later be reported to the government or other entities. These types of healthcare data primarily cover financial and activity-tracking data, which serve as key components for various monitoring strategies. This can include, for example, admission information and discharge data, and is intended to facilitate the sharing of information across facilities.

2.3.3 *Claims Data*

Claims data include the billable interactions between insured patients and the healthcare delivery system, primarily for purposes of coverage or compensation. Claims contain information on patient diagnoses, procedures and tests conducted, dates of services, costs, and where services were rendered. Claims data fall into four general categories: inpatient, outpatient, pharmacy, and enrollment. The sources of claims can be obtained from the government (e.g., Medicare) and/or commercial health firms (e.g., UnitedHealthcare).

2.3.3.1 Basic Stand Alone (BSA) Medicare Claims Public Use Files (PUFs)

This is the Basic Stand-Alone (BSA) Public Use File (PUF) for Medicare claims. PUF is a claim-level file in which each record is a claim incurred by a 5% sample of Medicare beneficiaries. These claims include inpatient/outpatient care, prescription drugs, DME, SNF, hospice, etc. Demographic and claim-related variables are provided in every PUF.

2.3.3.2 Medicare Provider Utilization and Payment Data

This refers to the data that summarize utilization and payments for procedures, services, and prescription drugs provided to Medicare beneficiaries by specific inpatient and outpatient hospitals, physicians, and other suppliers.

2.3.3.3 Medicaid Data Sources

The Medicaid Analytics extracted data contains state-submitted data on Medicaid eligibility, service utilization, and payments. The CMS-64 [5] provides data on Medicaid and SCHIP Budget and Expenditure Systems.

2.3.3.4 Medicaid Statistical Information System (MSIS)

MSIS is the basic source of state-submitted eligibility and claims data on the Medicaid population, their characteristics, utilization, and payments. This data typically includes:

- **Diagnosis Codes ("ICD-10"):** These are standardized codes that describe patient conditions, facilitating the analysis of trends and the allocation of resources [6].
- **Procedure Codes ("CPT"):** These are standardized codes for medical services, assisting in efficient billing, resource allocation, and assessment of treatments [7].
- **National Provider Identifier (NPI):** These are distinct identifiers that guarantee accurate reimbursement and clarity for healthcare providers during transactions [8].
- **Service Dates:** These document services are rendered to help monitor patient care, streamline workflows, and improve health outcomes [9].

Leveraging claims data alongside other real-world data opens new avenues for healthcare research. This enables a range of innovative applications, including evaluating the effectiveness of treatments, tracking disease outbreaks, and enhancing medication safety. These insights provide a deeper understanding of patients' real-life experiences and healthcare protocols.

2.3.4 Patient/Disease Registries

Patient/disease registries are collections of secondary data associated with patients who are diagnosed with a specific condition, disease, or procedure. This data is essential for post-marketing surveillance of pharmaceuticals.

Clinical information systems monitor essential data for chronic conditions, including Alzheimer's, cancer, diabetes, heart disease, and asthma. These systems offer crucial insights for handling and monitoring patient conditions. Examples of disease registries are the Global Alzheimer's Association Interactive Network [10] or the National Program of Cancer Registries [11]. Typically, these registries vary in complexity, ranging from basic spreadsheets limited to a small group of physicians to intricate online databases accessible across multiple organizations. These registries can also serve as reminders for healthcare providers (and sometimes patients) to conduct specific tests in response to specific healthcare needs.

2.3.5 *Health Surveys*

National health surveys assess the health of the population and estimate the prevalence of diseases. These data outlets are curated for research objectives and are broadly available. Examples of this data type include the National Medical Expenditure Survey [12], the Medicare Current Beneficiary Survey [13], and the National Health & Nutrition Examination Survey [14].

Health surveys play a pivotal role in the decision-making process of formulating health plans, providing accurate data on epidemiological status, health patterns, lifestyle behaviors, and patients' experiences with healthcare services. Public health researchers utilize health survey data to analyze health-related behaviors and psychological wellness. Health surveys are also valuable for identifying specific health conditions or risk factors within the community near healthcare facilities, including tobacco and alcohol use, unhealthy diet plans, and physical inactivity. Patients complete health surveys and evaluate their satisfaction with the health services they received, or track vaccination rates along with other demographic information. Thus, this type of data—despite being affordable and easy to obtain—is not considered a reliable source of information due to the potential of false self-reporting. National surveys are one of the few types of data collected specifically for research purposes, thus making it more widely accessible.

2.3.5.1 Medicare Current Beneficiary Survey

The Medicare Current Beneficiary Survey (MCBS) is a continuous, multipurpose survey of a nationally representative sample of the Medicare population. The central goals of MCBS are to determine expenditures and sources of payment for all services used by Medicare beneficiaries.

2.3.5.2 National Health and Nutrition Examination Survey (NHANES)

The National Health and Nutrition Examination Survey (NHANES) is a program of studies designed to assess the health and nutritional status of adults and children in the United States. The survey is unique in that it combines interviews and physical examinations.

2.3.5.3 National Medical Expenditure Survey

The Medical Expenditure Panel Survey (MEPS) [12] is a series of large-scale surveys conducted among families and individuals, their medical providers, and employers across the United States. MEPS is the most comprehensive source of data on the cost and use of healthcare and health insurance coverage.

2.3.6 *Clinical Trial Data*

Clinical data is collected in trials that help experts to understand the effects, efficacy, and safety of new treatments. Manufacturers are required to conduct such trials to demonstrate the quality of their products, such as drugs or medical instruments, under real-world conditions. These trials are necessary to obtain formal approval for using a new product in healthcare. Clinical trial data stems from research studies that evaluate medical, surgical, or behavioral treatments in patients. These data points serve as the primary method for researchers to:

- Explore methods for early disease detection, often before symptoms
- Discover strategies to take preventive actions for specific health issues, even for seemingly healthy people
- Enhance the quality of life for patients coping with severe illness or chronic health conditions
- Improve the performance of health professionals and caregivers

Throughout a clinical trial, various types of data are collected and converted into analyzable datasets to tackle different research inquiries. Clinical trial data may then be utilized in multiple publications and reports for varying audiences. Various registries provide access to databases of clinical trials, such as ClinicalTrials.gov, OpenTrials, and the WHO International Clinical Trials Registry Platform [15]. This data is collected to demonstrate how genetic data influences the growth and operation of organisms, thereby aiding research in life sciences, diagnosing genetic diseases, and developing medicines.

2.4 Why Is Data Important in Healthcare?

Data collection is crucial because it allows providers to make more informed decisions about a patient's care. With applications designed for sharing patient data across channels, every care provider has access to all the necessary information to treat their patients. Continue reading to learn more about why data is essential in healthcare.

2.4.1 *Market Competitiveness*

Community health is a complex system that involves governments, hospitals, and third-party agencies, such as registries and health plans. Proper patient data collection can significantly enhance the quality of services offered to patients, leading to improved market positions and increased revenues. To achieve these results, care providers must utilize the latest data collection and analysis tools, as well as specialized software offerings.

2.4.2 Well-Informed Decisions

The importance of accurate data in healthcare can't be overstated, and there are many reasons why doctors, patients, insurance plans, and other institutions need to maintain accurate data collection in healthcare. One of the main reasons is decision-making. What is patient information suitable for if providers and organizations can't get the information they need to understand their patients' needs? With the aid of medical data collection tools, it is possible to efficiently collect and evaluate patient data and results. Doctors, caregivers, insurance plans, and other stakeholders must exchange information securely and promptly, which can save lives and lead to improved care.

2.4.3 Reduction of Costs

According to a McKinsey study published in the National Library of Medicine [16], efficient data collection and analysis can save the US medical industry $300 billion per year. The importance of accurate data in healthcare notably increases when individual healthcare is involved. For hospitals and other medical institutions, detailed and comprehensive diagnosis and treatment may lead not only to better results but also to lower costs.

2.4.4 Implementation of Artificial Intelligence (AI)

AI in healthcare has been introduced for some time, and with its help, patient data collection can be more efficient, and systems can be improved automatically. The AI technologies that are becoming increasingly present in everyday life are also being applied to healthcare. The use of AI has the potential to assist providers in many aspects of patient care and administrative processes. The importance of data collection in healthcare means saving lives and improving the quality of life; therefore, companies and governments are working diligently to offer innovative solutions. Artificial intelligence has enough capacity to store, process, and analyze vast volumes of patient data. AI in healthcare can educate itself as it goes. The more data there is to learn from and analyze, the faster and more accurately AI tools can make suggestions. Even offering evidence from the data that supports a particular decision or suggestion. Some software systems today can process a patient's data across the healthcare system using proprietary medical algorithms and provide clinical decision support to physicians at the point of care. Using artificial intelligence in healthcare, such as machine learning and natural language processing, enables care providers and health plans to process the vast volumes of unstructured patient data that is input into EHR systems daily.

2.4.5 Outbreak Prevention

One of the benefits of having accurate data in healthcare is the opportunity to prevent epidemics before they occur. For example, a woman in California has unrelated symptoms that are treated separately. However, with better tools that can analyze data collection in healthcare and offer suggestions, doctors may be able to detect other individuals with the same unrelated symptoms in the area and prevent problems before they occur. With a shared database, doctors and epidemiologists can identify risk factors and trends across different groups and locations. This can significantly help society and save lives, especially among at-risk populations.

2.4.6 Storage of All Data in One Place

For individuals, the importance of data collection in healthcare is also evident. The collection and storage of complex patient data in one place can reduce the frequency of repeated doctor visits and help prevent unnecessary and costly readmissions. There are still many companies and institutions that store medical data in various locations. When patient data is collected in a specific system, all healthcare participants can collaborate and significantly enhance the quality of the services provided.

For example, HL7's newest standard is Fast Healthcare Interoperability Resources (HL7 FHIR, pronounced HL7 "fire") [17]. The goal of the FHIR data model, according to HL7, is to help ease the challenges of data sharing and interoperability by simplifying implementation without sacrificing information integrity. By having regulations define and require interoperability using the FHIR standard, we can expect a more unified and easier integration process across EHR applications [18] and data collection systems over the next few years, facilitated by HL7 "fire".

Before the robust database systems of today, records were kept inside a single facility that a person needed to visit and request their medical record. In a way, it was beneficial for privacy, but it could also be viewed as significantly slowing the development of the healthcare industry. When healthcare providers have access to a complete picture of a patient's health history, they can offer a thorough treatment plan. If patient data is not shared among doctors, researchers, and hospitals, it hinders development and can cost lives.

2.4.7 Why Are Databases Critical in Healthcare?

In today's healthcare industry, staying current is extremely important. Using the latest products and information systems available, doctors can collect the necessary patient data and chart concurrently [19] at the point of care or shortly thereafter. Fast and accurate database systems are critical to healthcare because they provide access to massive amounts of patient data information at the click of a mouse or the tap of a screen. The importance of accurate data in healthcare can mean the difference between life and death to that specific patient and the larger collective of patients as a whole. Today, doctors at every moment of the care process need instant access to the most up-to-date and accurate patient information possible.

Some question the problems that arise concerning the security of data collection in healthcare, and these issues shouldn't be dismissed. When deciding whether to implement a new data collection system, consider the following points.

2.4.7.1 Security of Data

Protecting and keeping sensitive data secure has always been a crucial issue, not only in the medical field. Companies must comply with numerous laws and regulations regarding data storage. For example, a stolen computer from an insurance company may result in a fine of up to $one million. A good alternative for keeping data secure is to utilize cloud-based storage, as most, if not all, security issues are typically managed by the vendor.

2.4.7.2 Low Data Quality

One of the biggest challenges is the quality of patient data. Some database systems are far superior to others, and costs vary greatly. Additionally, consider the process of collecting medical data itself. For example, handwritten notes that are then scanned can be relatively slow and prone to mistakes, resulting in low-quality patient data.

2.4.7.3 Difficulties in Collaboration

Doctors use different methods for entering patient data into their charts. Some use EHR systems [18], while others prefer traditional spreadsheets or even paper documents. Some physicians don't allow others access to certain documents, while others grant access only to specific specialists. The way medical data is written also matters because not all doctors speak the same way. Every doctor has a unique style and vision. This makes it quite challenging to cooperate, especially virtually.

2.5 Data Analytics

Analytics has become an integral part of the entire healthcare industry, accelerating decision-making, better aligning valuable resources, and enhancing the overall patient experience. We highlight five real-world examples of analytics playing a fundamental role in the healthcare industry:

- Insurance claims KPI dashboards
- Medical claims fraud
- Inpatient readmissions
- Bed utilization forecasting
- Comorbidity predictions and analysis

2.5.1 Insurance Claims KPI Dashboards

The simplest yet most widely adopted use of analytics in healthcare is medical claim KPI dashboards. Health care leadership teams depend on recurring monthly, weekly, and daily dashboards to help drive their decision-making. These dashboards highlight a range of information, from seasonal flu trends to daily inpatient bed inventory. Healthcare organizations and insurers worldwide depend on data scientists to analyze their vast trove of medical records and identify past, present, and future trends, needs, and obstacles.

2.5.2 Medical Claims Fraud

Medical claims fraud, waste, and abuse pose significant challenges for hospitals and insurers to manage. CMS [20] estimates that it spent more than $86 billion on fraudulent payments in 2020. Examples of medical fraud range from surgeons manipulating records to reflect work that was not done, thereby increasing reimbursements, to patients "shopping" for providers to obtain prescriptions they do not need. Through outlier detection and predictive models [21], data scientists can utilize claims data to identify areas of potential fraud. By identifying small clusters of outlier claims at the provider and member levels, auditors can pinpoint which claims to focus on and identify the parties involved in the fraudulent activity.

2.5.3 Inpatient Readmissions

Inpatient visits, especially inpatient readmissions, are often preventable and high-cost events for healthcare providers and patients. When a patient is admitted to an inpatient facility, analysts use the patient's medical history and other discharge information to create predictive models [22] to identify patients at risk of subsequent readmission. Hospitals can then use the output from these models to determine which flags are associated with a patient being at high risk for readmission. These flags can then be used to create preventative care strategies for different patient types and help the hospital decrease its readmission rates.

2.5.4 Bed Utilization Forecasting

Hospitals struggle to manage the number of available beds across their units. To efficiently manage bed utilization and forecast demand [23], hospitals enlist the help of data scientists to predict future demand for hospital beds and anticipate when occupied beds will become available.

2.5.5 Comorbidity Predictions and Analysis

Comorbidities are a leading indicator of a patient's general health. Acquiring multiple comorbidities can have a devastating impact on a patient's health and the cost of care. Hospitals and insurers utilize data mining and machine learning [24] to identify which combinations of comorbidities have the most significant impact on a patient's health and to pinpoint which patients are at the highest risk of developing those combinations. Using this information, providers and insurers can create preventative care strategies tailored to their at-risk population to manage the quality of care that their patients receive and mitigate their risk of further complications.

2.6 Tips for Accessing Different Types of Healthcare Data

2.6.1 Define the Healthcare Data Needed for Your Project

Defining the correct healthcare data needed and what is available can help keep the project objectives in sight, rather than getting caught up in the initial data access or sharing phase. Healthcare data infrastructure can vary significantly among organizations, underscoring the importance of evaluating the systems within your specific setting.

Reviewing current data sources also involves evaluating the data for its quality, comprehensiveness, and suitability for your specific needs. Consider the data owner's ability to provide information, data dictionaries, tracking mechanisms, data collecting approaches, as well as data validity and reliability. Ensure your organization is equipped with the necessary infrastructure to securely handle shared data. If it's not ready to manage sensitive information like EHR data, which is strictly regulated by HIPAA [25], then consider exploring other types of data that align better with your organizational capabilities.

2.6.2 Develop Multi-Sector Partnerships

Finding healthcare data entities and systems in your area can pose challenges; however, collaborating with insurers, academic institutions, health departments, local governing bodies, and technology vendors can enhance your understanding of local health and public healthcare data. Within public health and healthcare, establish relationships with staff at various levels, including data analysts, population health managers, epidemiologists, community benefits directors, administrators, care managers, and other key stakeholders. For public health, start by learning about the community health improvement process and opportunities to get involved. Building collaborations can be time-consuming, but it is an essential step in setting a shared vision before specific projects are realized.

2.6.3 Keep in Mind the Regulations on Healthcare Data Sharing

Dealing with healthcare data-sharing rules can be challenging and may require guidance from experts. Instead of thinking that HIPAA rules will slow you down, take a close look to make sure your plans follow the rules. This helps you keep your data secure while advancing your project. The legal complexities related to sharing healthcare data may necessitate consultation from healthcare experts. A key tip here is that rather than assuming HIPAA will hinder your project's progress, thoroughly explore whether your strategies are compliant, so you can also build a secure environment for your data.

2.7 Collecting Meaningful Data in a Clinical Research Study

In this section, we provide an overview of the various data collection approaches commonly used in conducting clinical, public health, and translational research [26]. We discuss several factors that researchers need to consider when using data collected through questionnaire surveys, as well as from proxy informants and reviews of medical records, and in the collection of biological samples. With increasing attention being paid to patient-reported outcomes in observational, comparative effectiveness, and clinical trials research, data collection approaches that combine medical record abstraction, patient interviews, and administrative data will be more commonly utilized in the future. We discuss several issues related to the collection of high-quality data in the conduct of clinical, translational, and epidemiological research projects, as well as ways to enhance the collection of reliable and meaningful data. We also discuss issues related to the accuracy of these data, and factors to consider in the possible independent confirmation of information collected from different data sources. The data collection instruments reviewed include questionnaire surveys, patient self-reported data, use of proxy/informant information, hospital and ambulatory medical records, and analysis of biological materials.

2.7.1 *Questionnaire Surveys and Patient-Reported Data*

Much of the information collected in observational epidemiologic studies is obtained through patient/participant self-reports on standardized questionnaires, which are administered either self-administered or interviewer-administered in person, by phone, or via mail or the internet. The factors on which information is routinely collected in these studies include socio-demographic characteristics, lifestyle practices, medical history, and use of prescribed and/or over-the-counter medications. Questions are also often asked about participants' knowledge and attitudes toward various lifestyle and disease-predisposing factors. With increasing attention being paid to patient reported outcomes by funding agencies such as the National Institutes of Health (NIH) [27], Agency for Healthcare Research and Quality (AHRQ) [28], and the newly formed Patient-Centered Outcomes Research Institute (PCORI) [29], measures of patient-centered factors such as Quality of Life (QoL), depression, anxiety, cognitive, and functional status are increasingly included in these surveys.

Surveyed individuals are typically asked to respond to these questions in one of three ways: yes/no, on a Likert-type scale (e.g., very often, not at all often), or with open-ended responses. The investigator dictates the choice of responses, and, of course, the standardized instrument (if one is used). The selection of the type of response desired is often made based on the difficulty of the question asked and the depth of knowledge and level of precision the investigator would like to have about a particular factor.

Standardized instruments often have different forms that vary in length so that an investigator can decide whether a "long" (e.g., SF-36) [30, 31] or "short" (SF-12) version is best suited for their study. Tests with multiple length versions typically have published psychometric properties (e.g., sensitivity and specificity of screening tests), which guide investigators in choosing a test version. For example, a consenting study participant might be asked a series of questions about their level of physical activity, either in the present or during a recent period of pertinent exposure. The number and depth of these questions would be determined, in part, by how this variable would be used in subsequent analyses and presented in peer-reviewed publications. Suppose the factor of physical activity were to be used as a controlling variable in either stratified or multivariable-adjusted regression analyses. In that case, a briefer assessment of physical activity might be more acceptable, with the added benefit of reduced respondent burden. On the other hand, if an investigator is particularly interested in the role of type of aerobic activity, level of exercise intensity, or duration of physical activity, then a more extensive battery of questions

Table 2.1 Advantages and disadvantages of questionnaire survey data [32]

Advantages	Disadvantages
Can collect personal and/or risk factor data not typically contained in hospital/ambulatory care records	Validating individual survey responses can be difficult, burdensome, costly, and of questionable utility
Can elicit information in an analytically desirable and standardized manner	If response rates are less than desirable, one may question the representativeness of the study sample and its generalizability
Can maintain high survey response rates through various financial or other incentives	Responses might differ if questions are asked in-person, by phone, by mail or internet

might be asked about this factor, with objective validation of self-reported activity carried out or a standardized instrument used.

The advantages and disadvantages of questionnaire data are summarized in Table 2.1.

2.7.2 *Proxy/Informant Data*

Collecting information about study participants through the use of proxy respondents can be one of the more challenging tasks for an investigator. Moreover, the accuracy and validity of the proxy's responses, as well as their level of knowledge about various health-related aspects of the study participant, need to be carefully considered when determining the type and quantity of information to be elicited from the proxy respondent. On the other hand, especially in observational studies where cases or controls in a retrospective study may have died or are unable to provide their responses, information from proxies may be the only available source of data. In certain situations, informant perspectives are essential data elements, even if they differ from those of the patient. For instance, family member reports of the type and amount of assistance a patient requires with activities of daily living may be qualitatively different, but equally important, as those reported by the patient.

Informal caregivers are increasingly being recognized as "stakeholders" in many research studies, particularly those that focus on patient-reported outcomes such as quality of life. In cases of questionable mental status or a non-communicative state of a patient, informants can be very helpful and essential in providing information to help establish a "baseline" for a patient. In these situations, informants can report on the patient's level of cognitive and physical function as well as level of independence, which are important outcomes in many contemporary clinical research studies. For some domains, validated informant questionnaires exist. For instance, the Informant Questionnaire on Cognitive Decline in the Elderly is an informant measure of cognitive function, and informant responses are also included on the SF-36 and in assessments of activities of daily living [32]. These scales have been used as assessments of health-related quality of life and functional status, with varying results.

2.7.3 Review of Ambulatory or Hospital Medical Records

Due to its ubiquity and the abundance of high-quality data embedded within it, the medical record is a commonly used source of information in clinical research studies. Information contained in hospital or ambulatory care records may be used either as the sole source of data or as a complement to other instruments used to elicit information. Decisions about the adequacy of using the medical record as the sole or primary source of data for a given study hinge on the investigator's hypotheses, study sample size, budget, and timeline, as well as the extent and type of data available in each record system. Medical records can be essential sources of information that can reliably document participants' medical history, clinical, laboratory, or physiologic profile at varying time points in a cost-efficient manner. On the other hand, the data contained in medical records can be frustrating to use and, in some cases, conflicting or of questionable accuracy, due to the non-standardized way this information is collected, recorded, and/or abstracted by various health care professionals and members of research teams. The increasing use of electronic medical records and their integration with administrative data have facilitated data abstraction efforts, and with the growing adoption of standardized data entry sets, reduced data heterogeneity.

One major limitation of using the medical record as a primary data source is that potentially crucial patient-reported information is often lacking, typically limited to the reporting of a "chief complaint" or symptoms directly related to the current complaint. If clinical information is stigmatized (e.g., sexual history, alcohol or drug use), or difficult to systematically assess in primary care settings (e.g., cognitive status, depression), it is often under-reported in the medical record. It is also important to note that factors (e.g., medication use) are defined by clinicians, not by trained study staff or study participants, and certain variables may not be accurately coded. Moreover, the extent of documentation about key medical history or clinical variables can vary significantly between providers (including conflicting data) and healthcare systems. Heterogeneity can create considerable difficulties in either constructing key study variables or using them.

For example, in studying a purported association between macular degeneration and various dietary components, it would be essential to document the presence of different medical conditions that may affect an individual's nutritional practices, as well as the development of macular degeneration. In this example, we are particularly interested in determining the presence of a history of type 2 diabetes mellitus based on the information contained in medical records. Since one needs to consider how this condition and related chronic medical conditions would be classified based on information contained in medical records. For example, is diabetes considered present if there is a simple notation of this condition in the patient's medical history by a sole provider? On the other hand, might there be a need for the documentation of various key elements of each condition to be noted in the medical records (e.g., multiple elevated serum glucose levels obtained under fasting conditions) before a diagnosis of diabetes can be accepted? For several relatively common conditions,

Table 2.2 Advantages and disadvantages of hospital/ambulatory care records [32]

Advantages	Disadvantages
Readily available and contain much useful demographic and clinical information	Often data contained in medical records are non-standardized and inconsistently collected and recorded
Can be linked to other follow-up information sources	Information is often incomplete and/or missing
Can be used to characterize the medical history and clinical course of hospitalized and outpatient individuals	Independent checks on validity and/or reliability are atypically performed
Can provide data on medication intensity and duration	Information on etiologic or prognostic factors of importance is often either not obtained or asked about or recorded in a standardized manner

such as heart failure and stroke, independently and extensively validated algorithms have been developed to ascertain the presence of these essential chronic diseases.

Depending on the central research questions under study, the available resources, and the level of variability or precision that can be tolerated in documenting the presence (or, equally importantly, the absence) of each of these comorbid conditions, rules of acceptance and rejection can be applied in considering these factors. Similarly, the investigator might also decide to ask the survey participant whether diabetes had ever been diagnosed in their past. This should be a straightforward task, but the investigator needs to consider beforehand how they will analyze the data if personal responses are not consistent with their medical record findings. Table 2.2 summarizes the advantages and disadvantages of using medical records.

2.7.4 Collection of Biological Material

An increasing number and array of contemporary clinical and translational research investigations involve the collection of biological samples from study participants. These include personal factors such as hair, saliva, urine, and serum. Biological samples are increasingly being used to profile participants' metabolic, proteomic, or genomic status, thereby enhancing our understanding of their underlying pathophysiology and their response to treatment or disease. Although it is beyond the scope of the present manuscript, the ethical implications of genetic research warrant careful consideration. Furthermore, various imaging modalities (e.g., computed tomography, magnetic resonance imaging, nuclear scans, ultrasonography) are being used to gain deeper insights into the underlying anatomic, pathologic, and biologic mechanisms involved in the development of disease, its prognosis, or response to treatment, and to suggest areas for future research endeavors.

Despite the critical information these biological samples provide about disease, its various causes, and natural history, several factors must be considered when collecting biological materials (Table 2.3). One crucial factor to consider when

Table 2.3 Advantages and disadvantages of biological data [32]

Advantages	Disadvantages
May provide novel insights into underlying disease pathophysiologic processes	Need to be collected under standardized conditions with considerable attention to detail, ongoing quality control procedures needed
Can serve as an important endpoint of relevance	Need to consider impact of possible biologic circadian variation for purposes of timing and frequency of data collection efforts
Can be linked to other sociodemographic, medical history, and clinical data to obtain insights into disease occurrence and prognosis	May need collection of multiple measures at baseline to adequately profile subsequent changes

obtaining biologic samples is the frequency of collection (often a balance between participant burden and pathophysiologic insights gained from the ability to assess change in a factor over time), timing of specimen collection (especially when this biologic variable has been shown to exhibit circadian variation), cost (both to the participant and investigator in terms of invasiveness and complexity, respectively), variability in test measurement (often presented as a coefficient of variation), and careful need for standardization of test methods and their interpretation (e.g., referencing vs. a "gold-standard").

For example, an investigator may consider conducting a prospective study to examine racial differences in serum biomarkers and echocardiographic determinants of atrial fibrillation. In addition to collecting clinical and demographic historical information, a baseline echocardiogram and serum levels of various biomarkers, such as B-type natriuretic peptide, will be assessed. Investigators need to balance the need for further information regarding changes in each of these parameters that lead to atrial fibrillation with the participant's burden.

Based on the current literature and existing clinical knowledge, the investigators in this study need to determine how echocardiographic atrial size and B-type natriuretic peptide levels change over key periods in patients with, or at risk for, atrial fibrillation. These concerns need to be built into data collection efforts, and there is a need for systematic assessment of serial changes in these factors. Depending on the degree of change in these parameters, this may entail collecting serial echocardiograms every 2 years, every 4 years, or more frequently, such as every 3 months, depending on the extent of change in left atrial size that may predispose an individual to the development of atrial fibrillation. On the other hand, since there may be more volatility and/or change in the serum biomarkers being examined, more frequent blood assays may be required, which must be balanced against participants' willingness to return to the clinic and the associated discomfort and burden. Inasmuch, compromises in the intensity of data collection efforts need to be balanced with patient-related concerns and the importance of keeping high rates of retention in a long-term longitudinal study.

Another primary consideration regarding the use of biologic data is the timing of sample collection regarding the definition of key study variables and outcomes (e.g., whether they are concurrent or separated by a considerable time). The following example illustrates the importance of timing in collecting various descriptive or risk factors. An investigator aims to perform metabolomic profiling to investigate differences between hepatic and circulating levels of a specific factor. To obtain in vivo hepatic tissue samples, the investigator conducts the study using patients undergoing a hepatic biopsy. However, he/she obtains blood samples in the pre-surgical holding area at the time of IV placement to minimize participant inconvenience. This study could be undermined, however, should the metabolomic profile of the liver be influenced by medications administered for procedural sedation, thereby confounding any comparisons between hepatic and circulating levels of factors of primary interest.

Storage of biologic samples, as well as technical factors relating to their measurement, also warrant special consideration when interpreting or performing studies involving biologic specimens.

Several factors must be considered when deciding which data and how much data to collect in any clinical research investigation. Investigators often believe that "more is better" and that it is essential to collect information on as many scientifically "interesting" factors as possible. This premise may be misguided and place an unnecessary burden on study participants, as well as lead to the collection of considerable data that would never be utilized, analyzed, or presented in a scientific publication.

It is often beneficial, and time well spent to identify those data elements that are essential from those that are academically "interesting" but may not be considered central to the key study hypothesis; this will significantly assist in narrowing down one's study questions and collecting data in as timely and rigorous a manner as possible. Moreover, it helps to create a list of the 5–10 major papers that might result from one's proposed research study and create an analysis plan for each manuscript. By doing so, we can separate the "data wheat" from the "data chaff". This enables focus on questions of key relevance and data elements that comprise key variables.

One also needs to carefully think about the independent validation of any self-reported responses and how intrusive, costly, and potentially burdensome this process may be. Validation of one's data, while important, can be a challenging and cumbersome process, with its attendant logistical and staffing complexities.

2.8 Clinical Data Management (CDM)

Clinical Data Management (CDM) is a crucial phase in clinical research, yielding high-quality, reliable, and statistically sound data from clinical trials. This helps to produce a drastic reduction in time from drug development to marketing. When introducing new products to the market through clinical research, adhering to regulations such as Good Laboratory Practice (GLP), Good Clinical Practice (GCP), and CFR Part 11 is mandatory. Clinical data, encompassing information for developing and maintaining software systems, databases, processes, and protocols, is crucial for these studies [33].

CDM is the process of collecting, cleaning, and managing subject data in compliance with regulatory standards. The primary objective of CDM processes is to provide high-quality data by minimizing errors and missing data while gathering as much data as possible for analysis [34]. To meet this objective, best practices are adopted to ensure that data is complete, reliable, and processed correctly [35]. This has been facilitated using software applications that maintain an audit trail and provide easy identification and resolution of data discrepancies. Sophisticated innovations have enabled CDM to handle large trials and ensure data quality even in complex trials.

How do we define "high-quality" data? High-quality data should be accurate and suitable for statistical analysis. These should meet the protocol-specified parameters and comply with the protocol requirements. This implies that in the event of a deviation from the protocol specifications, we may consider excluding the patient from the final database. It is worth noting that in certain situations, regulatory authorities may be interested in reviewing such data. Similarly, missing data is also a matter of concern for clinical researchers. High-quality data should have minimal or no errors or omissions. However, most importantly, high-quality data should exhibit only an arbitrarily acceptable level of variation that would not affect the study's conclusions in statistical analysis. The data should also meet the applicable regulatory requirements specified for data quality [36].

Clinical data contains information for developing and sustaining software systems, databases, processes, procedures, training, and protocols. Clinical data management enables organizations to maintain data integrity throughout a clinical research study. Correct data management ensures that a dataset is accurate, secure, reliable, and ready for analysis. By using a systematic process, a CDM team is responsible for collecting, entering, cleaning, and processing information gathered during a clinical research project, and ensuring the accuracy, completeness, and consistency of the clinical trial data, adhering to regulatory standards and guidelines.

The primary goal of CDM is to ensure the validity and reliability of the data, making it ready for regulatory submission and subsequent analysis. It serves as the backbone for producing high-quality, statistically sound results that inform medical decisions and regulatory approvals.

Regulatory compliance is a fundamental aspect of CDM, ensuring that all collected data adheres to stringent regulatory standards and guidelines set by authorities such as the FDA and EMA. By maintaining records and implementing robust validation processes, CDM ensures that the data adhere to regulatory requirements, not only facilitating the approval process for new treatments and interventions but also upholding the integrity of the research. Akin to other areas in clinical research, CDM has guidelines and standards that must be followed. Since the pharmaceutical industry relies on electronically captured data for the evaluation of medicines, there is a need to follow good practices in CDM and maintain standards in electronic data capture. These electronic records must comply with the Code of Federal Regulations (CFR) [36], 21 CFR Part 11. This regulation applies to records in electronic format that are created, modified, maintained, archived, retrieved, or transmitted. This requires the use of validated systems to ensure the accuracy, reliability, and consistency of data, utilizing secure, computer-generated, time-stamped audit trails to independently record the date and time of operator entries and actions that create, modify, or delete electronic records [37]. Adequate procedures and controls should be implemented to ensure the integrity, authenticity, and confidentiality of data. If data must be submitted to regulatory authorities, it should be entered and processed in 21 CFR part 11-compliant systems. Most CDM systems available are similar, and pharmaceutical companies, as well as contract research organizations, ensure this compliance.

Society for Clinical Data Management (SCDM) [38] publishes the Good Clinical Data Management Practices (GCDMP) guidelines, a document providing the standards of good practice within CDM. GCDMP was initially published in September 2000 and has undergone several revisions thereafter. The July 2009 version is the currently followed GCDMP document. GCDMP guides the accepted practices in CDM that are consistent with regulatory practices. Addressed in 20 chapters, it covers the CDM process by highlighting the minimum standards and best practices.

The Clinical Data Interchange Standards Consortium (CDISC) [39], a multidisciplinary nonprofit organization, has developed standards to support the acquisition, exchange, submission, and archiving of clinical research data and metadata. Metadata is the data of the data entered. This includes data about the individual who made the entry or change in the clinical data, the date and time of the entry/change, and details of the changes made. Among the standards, two important ones are the Study Data Tabulation Model Implementation Guide for Human Clinical Trials (SDTMIG) and the Clinical Data Acquisition Standards Harmonization (CDASH) standards, which are available for free download from the CDISC website [39]. The SDTMIG standard [31] outlines the details of the model and standard terminologies for the data, serving as a guide for the organization. CDASH v 1.1 [32] defines the basic standards for collecting data in a clinical trial and outlines the essential data information required from a clinical, regulatory, and scientific perspective. The emphasis on data quality is paramount, as it directly impacts the credibility of research outcomes.

High-quality data ensures that the results of clinical trials are statistically sound and dependable, forming a solid foundation for medical decisions and ultimately advancing patient care. In summary, the purpose of Clinical Data Management is to:

- Ensuring data quality and integrity
- Compliance with regulatory standards
- Data collection and management
- Facilitating accurate analysis and reporting

The main activities in clinical data management are:

- Data Management Plan (DMP) development
- Database design and setup
- Data capture

- Data validation
- Discrepancy management
- Medical coding
- Database lock
- Data anonymization, privacy, and security

2.8.1 *Why Is CDM Essential?*

Clinical research studies provide crucial information. For example, results should demonstrate that a product is safe and meets requirements [40]. Clinical data management offers:

- Assurance of data quality
- Accelerated development
- Protection from data loss
- Reduced expenses
- Security
- Complete and accurate collection of data
- A clean dataset to support statistical analysis and reporting
- A formatted dataset for optimal and timely usability
- Assurance of data integrity and quality during database transfer
- True representation of the trial in the study database

2.9 Data Management Requirements for AI in Medical Devices

Since the AI model is integral to an AI-enabled device's mechanism of action, thorough documentation and management of data practices—covering collection, processing, annotation, storage, control, and usage—are essential for evaluating the device's safety and effectiveness. The quality, diversity, and quantity of data used in training, tuning, and testing directly influence both model performance and the reliability of validation results; data management must align with intended use and device requirements. Characterizing and mitigating sources of bias is crucial, as AI models can inadvertently learn and amplify biases from limited or unrepresentative data, leading to systematic errors that undermine generalizability—for instance, overfitting to specific scanners or underrepresenting certain populations. Representative validation data are necessary to detect performance issues, especially in underrepresented groups. The validation data adequately reflect the target population, and the development and test datasets remain strictly separate and independent (e.g., sourced from different clinical sites and sequestered from model developers) to prevent data leakage and ensure accurate performance assessment.

In the following sections, information on data collection, development, and test data independence, as well as reference standards and representativeness [41], is discussed.

2.9.1 Data Collection

Data from the foundation of any AI system, so detailed documentation of the data collection process is essential and should address both the methods and constraints involved. Documentation must specify how data are gathered—such as through clinical study protocols with defined inclusion and exclusion criteria directly linked to the intended use population—and identify the clinical sites or institutions involved, each consistently referenced throughout the submission. The intended distribution of the dataset across patient demographics (such as sex, age, and race) and data acquisition conditions (including sites and protocols) should be described to ensure alignment with the target population and relevant intersectional groups. The protocol should clarify whether data are collected prospectively or retrospectively, confirm the inclusion of consecutive cases within a defined timeframe, or detail the random sampling method and its approach to mitigating bias. Additionally, it should assess the appropriateness of any pre-existing databases used. Plans for enrichment or stratified sampling targeting specific subgroups (such as disease subtypes or care settings) should be outlined to ensure diversity and relevance. The number and geographic distribution of data collection sites must also be provided.

The data collection should address the limitations of the dataset, outline quality assurance processes and controls to prevent human error in data acquisition (where applicable), and state the size of each dataset. Mechanisms to enhance enrollment diversity and ensure generalizability across patient populations and clinical sites should be described, along with measures to mitigate and address bias, including strategies to identify and manage biases arising from historical healthcare inequalities. The data collection should also explain how data relevance will be maintained over time as technology, clinical practices, or patient populations evolve. Compliance with regulatory standards for the protection of human subjects and clinical investigations must be confirmed. If real-world data (RWD) are utilized, the documentation should specify their source and collection methods, and include an evaluation of their relevance and reliability for the intended purpose. Finally, if synthetic data is used in development, its purpose and justification for being fit-for-purpose should be clearly explained. It is also essential to verify that the input data aligns with the acquisition systems and specifications of the intended device, including its interoperability requirements.

2.9.2 Data Cleaning/Processing

To ensure the accuracy and efficiency of data analytics, robust processes must be established to maintain data consistency, completeness, authenticity, transparency, and integrity across training, tuning, and test datasets. Given that real-world data are often imperfect, protocols should thoroughly describe how development data are cleaned—including addressing incorrect, duplicate, or incomplete records—and specify data quality factors, inclusion/exclusion criteria, and approaches to handling missing data, noting whether these steps are performed internally or externally to the AI/ML software. The protocol should also outline methods to minimize impacts on generalizability and testing accuracy if data are excluded during quality assurance, detail traceability methods for identifying and investigating data issues, and define criteria and strategies for removing obsolete data to support consistent data management. Controls to prevent unauthorized access or manipulation of datasets should be identified, and any automated data quality processes should be described. For testing data, processing should mirror real-world data conditions as encountered in practice, with all processing steps—including quality factors, inclusion/exclusion criteria, and strategies for handling missing data—justified as consistent with the pre-processing used in the final AI/ML software.

2.9.3 Reference Standard

The reference standard is defined as the best available truth used to establish the true condition of each patient, case, or record, and it may be applied at any stage, including device training, tuning, validation, or all stages. The protocol should include a clear rationale for selecting this standard, which may consider device risk. The reference standard must be validated by current medical and regulatory evidence, aligning with the clinical task (such as disease classification, image segmentation, detection, or localization). Documentation should describe how the reference standard was established, its inherent uncertainty, and strategies for managing equivocal or missing results. If clinician evaluations are used, the protocol must detail the grading protocol, the number and qualifications of participating clinicians, the data provided to them, and the process for collecting and adjudicating their evaluations (including blinding). Assessments of intra- and inter-clinician variability should be included, with an evaluation of whether observed variability meets accepted standards. The protocol should also confirm whether automated methods are used to determine the reference standard, and clarify any distinctions between reference standards applied to training, tuning, and testing phases.

2.9.4 Data Annotation

When data annotation is employed, comprehensive documentation should outline the data annotation approach, including a description of the annotators' expertise and the specific training, instructions, or guidelines they receive to inform their annotation decisions, noting whether annotators are blinded to each other's work. The documentation should also detail the methods used to assess the quality and consistency of annotations and to resolve disagreements, such as through consensus evaluation or sampling. Additionally, it is essential to develop a comprehensive plan for identifying and correcting any instances of incorrect data annotation.

2.9.5 Data Storage

Comprehensive documentation should detail storage practices for both training and test data, emphasizing dataset version control and robust security measures, and clarify how data—such as medical images accompanied by reference standards established through annotation—are used for training, tuning, and testing, with each dataset maintained in strict isolation to prevent overfitting and ensure accurate performance assessment. To further reduce bias and enhance algorithm robustness for diverse clinical conditions, methodologies should incorporate strategies such as cross-validation, bootstrapping, bagging, ensembling, and the use of synthetic or augmented data. When modifications to the AI/ML software are required, new data not previously used in the initial development should be collected. A Modification Protocol must specify how these data are acquired, annotated, curated, stored, retained, controlled, and utilized for each update. The protocol should also clarify the relationship among all datasets used across initial and modified versions of the AI/ML software and describe the controls in place to ensure test data remain separate and independent throughout model development and re-training. To maintain sequestration, manufacturers must not have access to test data for development purposes. Procedures should be defined to isolate test datasets during both initial collection and retraining. Measures should be applied to prevent bias if test data are reused for performance evaluation. Descriptive statistics—such as covariate range, mean, and median—should be provided for each dataset, with validation of their alignment to the intended use population.

2.9.6 Management and Independence of Data

To ensure robust model development and validation, the development data should be partitioned into distinct subsets for training, tuning, tuning evaluation, and any other necessary purposes. It is essential to implement controls that completely separate the data used for testing from the development process. Additionally, the rationale for why the validation data provides a strong external validation should be justified; for example, by describing the sources of the test data, since, in general, test data should be obtained from sites that are different from those used for developing the AI/ML software.

2.9.7 Representativeness

To ensure that data are representative of the intended use population and its indications, it is crucial to focus on relevant population characteristics such as disease conditions (including case status, severity, subtype, comorbidities, and disease spectrum distribution), patient demographics (sex, gender, age, race, ethnicity, height, weight), and data acquisition details (location, equipment, protocols, and environmental factors affecting signal analysis). Using multiple data collection sites is recommended to better reflect the diversity of clinical practice, ideally including at least three geographically diverse US sites for robust clinical validation of AI-enabled devices. The distribution of data across key covariates should be characterized, with subgroup or stratified analyses conducted as needed. When non-US (OUS) data are included in validation, justification should be provided regarding how these data compare to the U.S. population and medical practice in terms of disease presentation, prevalence, progression, and demographics. Given the data-driven nature of AI models and their sensitivity to training data patterns, a higher proportion of US data is generally required in clinical validation to ensure reliable performance for the target population.

2.10 Data Quality Framework for AI/ML per EU AI Act

The EU AI Act introduces comprehensive frameworks for data lifecycle management and data quality management, particularly for high-risk AI systems, to ensure safety, transparency, and compliance. ISO/IEC 5259-1 [42] lays the foundation for the series by establishing core concepts, terminology, and a governance framework for data quality in analytics and machine learning (ML), addressing common challenges such as missing data and bias, while emphasizing continuous improvement and integration with related standards. ISO/IEC 5259-2:2025 [43] builds on this by providing a flexible model for managing and measuring data quality throughout the data lifecycle, detailing both inherent and system-dependent characteristics and offering practical guidance for measurement and reporting. ISO/IEC 5259-3:2025 [44] further supports organizations of all sizes in establishing robust, adaptable data quality management systems, requiring structured processes, risk management, and ongoing improvement to ensure reliable analytics and ML outcomes, stakeholder confidence, and regulatory compliance.

ISO/IEC 5259-4:2025 [45] provides a structured process framework for managing and assessing data quality throughout the lifecycle of analytics and machine learning projects, ensuring data is fit for purpose, reliable, and trustworthy to support robust model development and deployment across various learning paradigms. The framework outlines systematic steps: planning data quality objectives, collecting representative and unbiased data, preparing and enriching data, validating against quality criteria, and continuously monitoring and maintaining data quality even after deployment. For supervised learning, it emphasizes label accuracy, feature relevance, rigorous labeling processes, and representative validation; for unsupervised learning, it focuses on accurate data distribution, relevant feature representation, and robust preprocessing; and for semi-supervised learning, it highlights the quality and representativeness of labeled and unlabeled data, with consistency checks and careful labeling strategies. The standard also details clear labeling guidelines, quality assurance measures (such as inter-rater reliability), and comprehensive documentation practices. Key benefits include improved model performance, support for regulatory compliance, and reduced risk of bias and errors due to low-quality data.

ISO/IEC 5259-5:2025 [46] is an international standard providing a comprehensive governance framework to ensure high-quality data throughout the lifecycle of AI and ML projects. It establishes structured approaches for governing data quality, supports organizations in defining and maintaining robust data management processes, and addresses challenges such as bias, incompleteness, inconsistency, and relevance. The framework is built on key principles—integrity, consistency, relevance, timeliness, and transparency—and defines clear roles and responsibilities, policies, accountability mechanisms, and continuous improvement processes. It includes regular data quality assessment, improvement, and monitoring, along with quantitative and qualitative metrics for measurement. The standard also emphasizes risk management, regulatory compliance, and incident management, while encouraging organizations to tailor the framework to their unique context and integrate it with existing systems and processes. Benefits include enhanced trust in AI/ML systems, reduced risk of errors and bias, improved decision-making, and scalable, high-quality data management.

2.10.1 Data Lifecycle Model

Analytics and machine learning (ML) both leverage data to generate predictions that inform organizational decision-making, making them highly dependent on the characteristics and quality of data within a specific context of use. While analytics and ML share a standard, overarching data life cycle, data quality requirements are shaped by the purpose of the analysis or ML application and should therefore be managed in alignment with that purpose and the data's life cycle. The data life cycle for analytics and ML offers an end-to-end view of how data is utilized and how additional data may be generated within these systems. This document outlines the

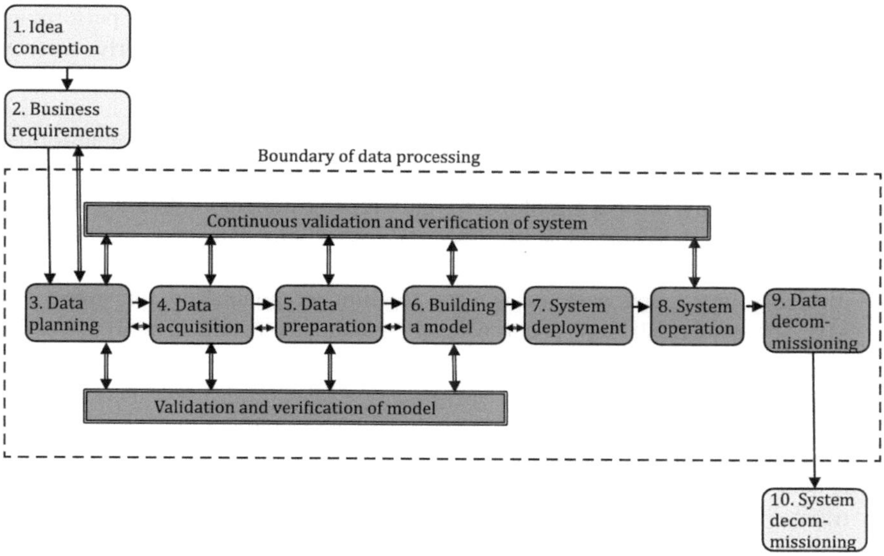

Fig. 2.1 Data lifecycle model as per ISO/IEC 8183 [47]

data life cycle (DLC) for analytics and machine learning (ML), utilizing a six-stage DLC model, shown in Fig. 2.1, that aligns with ISO/IEC 8183 [47]. This model serves as a framework to understand the processes involved in data quality management. In this model, single-headed arrows indicate the main progression through the stages, while double-headed arrows represent feedback to other stages of the DLC.

- **Stage 1: Data Requirements**
- The data requirements stage is where the necessary data for an analytics or machine learning project is identified. This process involves determining the specific types of data needed, checking their availability within the organization or external sources, and establishing a data quality model that outlines relevant data quality characteristics tailored to the project's objectives.
- **Stage 2: Data Planning**
- During the data planning stage, the focus is on ensuring that the data aligns with the requirements set in the previous stage and supports the goals of the analytics or ML initiative. This stage involves designing the data architecture, which entails defining the full scope and structure of the required data, as well as determining how it will be utilized and estimating the effort needed to acquire and prepare the data. This may include restructuring datasets, planning data transfers or collection, and building a comprehensive data quality model to guide the project.
- **Stage 3: Data Acquisition**
- The data acquisition stage involves gathering both live and historical data from the sources identified during planning, which may be delivered as streams or in batches depending on project needs. Key activities include protecting data pri-

vacy and security, testing and refining data collection methods, and performing data quality measurements to identify and address any issues early, thereby reducing downstream inconsistencies and workload for data users.

- **Stage 4: Data Preparation**
- In the data preparation stage, collected data are processed into a usable format for analytics or ML, playing a crucial role in meeting data quality requirements. This stage may involve transforming data representations, validating data for correctness and privacy, cleaning inaccurate or missing data, aggregating datasets, sampling data, creating or selecting features, enriching data by linking diverse sources, and performing data labelling or annotation for supervised ML. The stage can be repeated as needed, based on the results of analytics or the performance of the ML model.
- **Stage 5: Data Provisioning**
- During the data provisioning stage, prepared data is applied to analytics or ML projects, and their performance is evaluated against the expected results. If results fall short, root causes may be investigated in the training data or algorithms. Data quality issues identified at this stage are communicated to data originators and holders, enabling them to make improvements for future projects. If necessary, stages 2 through 4 may be repeated to enhance data quality before proceeding.
- **Stage 6: Data Decommissioning**
- The data decommissioning stage addresses the end-of-life for data used in analytics or ML projects. Depending on organizational or regulatory requirements, data may be stored or archived along with relevant metadata for future reference, or they may be destroyed or returned to the data holder. If archived, the original data requirements and usage context should be preserved to support future system-dependent data needs.

2.10.2 Data Quality Plan

The organization shall establish a comprehensive data quality plan that outlines processes, activities, and resources necessary to meet data specifications throughout the entire data life cycle. This plan should define data quality goals in alignment with data specifications, specify committed resource allocations, include provisions to ensure legal compliance, and detail processes for managing and updating the plan as needed. Additionally, it must incorporate provisions to ensure that all processes are traceable and reproducible. The data quality plan should also include process specifications for each stage of the data life cycle, such as a data acquisition plan, data preprocessing plan, data augmentation plan, data provisioning plan, and data decommissioning plan. Each process specification should describe the process, outline related activities and intended outcomes, and assign responsibility to a designated person.

2.10.3 Data Quality Management Lifecycle

Figure 2.2 contains the Data Quality Management Lifecycle and its relationship with the Data Lifecycle framework. Data quality management begins at the motivation and conceptualization stage, where potential data quality issues are identified and addressed as soon as the need for data arises in analytics and machine learning contexts. At this stage, the motivation and intended use of the data are validated and verified, ensuring that quality characteristics such as conformity and relevance are met. In the data specification stage, requirements are imposed on the creation of the data specification, including defining data formats, statistical properties, and divisibility. Quality management guidelines help identify erroneous, incomplete, or conflicting requirements and plans, ensuring that the data align with the needs of the analytics or ML context. Moving to the data planning stage, a detailed plan is formulated to meet these specifications, encompassing the allocation of specific tasks and resources for data acquisition and processing throughout the data life cycle, as well as defining evaluation methods and acceptance criteria.

During data acquisition, raw input data are gathered according to established plans. In the data preprocessing stage, quality management requirements and guidelines are applied to process this raw data, typically involving filtering and cleaning to remove problematic data items. Data augmentation then takes the preprocessed data and further organizes it according to quality management requirements, such as associating metadata, adding data labels, or performing other transformations to enrich the dataset. The data provisioning stage focuses on delivering the dataset to where it is needed, with quality management requirements including maintaining dataset integrity during delivery, access control, thorough documentation, and verification that the context of use matches the dataset's intended purpose. Finally, data quality management concludes with the data decommissioning stage, which

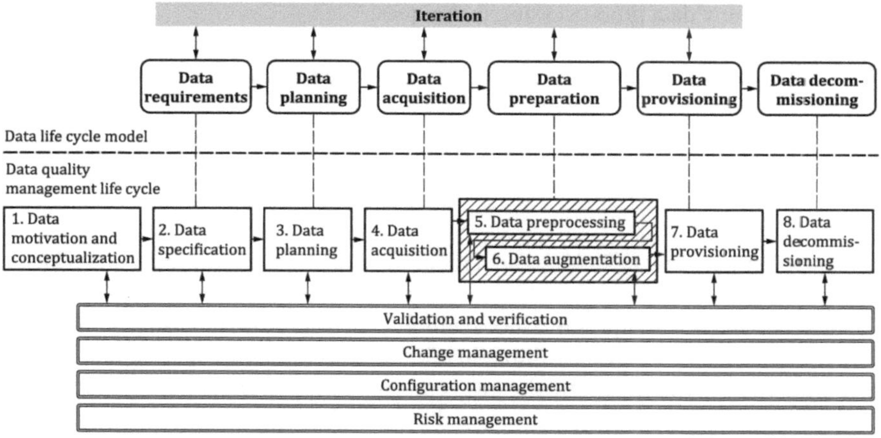

Fig. 2.2 Data quality management lifecycle [44]

involves either data deletion or transfer; in either case, responsibility for the data is released. Proper handling during decommissioning is crucial for maintaining organizational trust and preventing data misuse.

2.11 Summary

Data management stood as a cornerstone in the development of artificial intelligence-based algorithms and solutions for healthcare. This chapter provided a comprehensive overview of clinical data management, beginning with an exploration of various healthcare data types and the protocols for their collection. It delved into the critical aspects of data security and privacy, highlighting the importance of regulatory compliance and necessary agreements. Significant emphasis was placed on assessing data quality and completeness, as these were recognized as fundamental to the success of AI applications in healthcare. The chapter also examined essential data preprocessing techniques and manipulation methods, which were crucial for preparing raw data for analysis. The discussion extended to the Data Quality Lifecycle and Data Quality Management Lifecycle frameworks introduced by the EU AI Act-harmonized published standards, as well as the ISO/IEC 5259 series for data quality frameworks. These frameworks were essential for maintaining data quality and integrity during the training and validation of AI algorithms. The chapter concluded by outlining the importance of ensuring that AI solutions in healthcare met the stringent standards set by regulatory bodies.

2.12 Points to Ponder

1. Discuss a few data preprocessing techniques.
2. Discuss techniques to improve the data quality.
3. Discuss the key preprocessing steps for sensor and medical imaging data.

2.13 Answers

1. Data preprocessing techniques are crucial for transforming raw data into a clean, structured format that is suitable for machine learning and analytics. One of the most common techniques is data cleaning, which addresses missing values, outliers, and inconsistencies by filling in missing data through imputation or removing erroneous entries. Normalization and standardization are used to scale data features to a uniform range, ensuring that variables measured on different scales do not disproportionately influence model training. Feature encoding converts categorical variables into numerical formats, like one-hot or label encoding,

making them compatible with algorithms that require numerical input. Feature extraction and dimensionality reduction techniques, such as Principal Component Analysis, reduce the number of features while retaining essential information, which helps prevent overfitting and improves computational efficiency. Additionally, data transformation may involve detrending, smoothing, or applying mathematical functions to stabilize data distributions. These preprocessing steps collectively enhance data quality, reliability, and model performance in machine learning workflows.

Several Python libraries are widely used for data preprocessing tasks. One of the most popular is a library that provides powerful tools for data manipulation, cleaning, and transformation, such as filtering, grouping, and handling missing values. Another essential library focuses on numerical computations and array operations, making it suitable for basic data normalization. A comprehensive machine learning library offers utilities for standardizing data by adjusting features to have a mean of zero and a standard deviation of one, as well as for normalizing features to fit within a specified range. This library also includes tools for encoding categorical data and imputing missing values. Additional libraries extend data cleaning capabilities, making preprocessing workflows more efficient. Visualization libraries are often used in conjunction with these to understand data distributions and identify preprocessing requirements. Together, these libraries provide robust support for standardization, normalization, and many other preprocessing tasks.

- **Pandas**: Used for data manipulation, cleaning, and transformation (filtering, grouping, handling missing values).
- **NumPy**: Used for numerical computations, array operations, and basic normalization.
- **Scikit-learn**: Provides tools for standardization (StandardScaler), normalization (MinMaxScaler), robust scaling, categorical encoding (LabelEncoder, OneHotEncoder), and missing value imputation.
- **Feature-Engine**: Offers advanced feature engineering, missing value imputation, encoding, and scaling, and integrates with scikit-learn pipelines.
- **Pyjanitor**: Extends Pandas with convenient cleaning functions for streamlined preprocessing.
- **Matplotlib**: Used for data visualization to help understand data distributions during preprocessing.

2. Techniques to Improve the Data Quality.

   - **Data Cleaning**

   - Start by cleaning the data to identify and correct errors, remove duplicates, and manage missing values or outliers.
   - **Standardization and Normalization**
   - Standardization and normalization help ensure all features are on a similar scale, which is essential for many analytical methods.

- **Data Integration**
- Combining data from different sources can increase completeness and resolve inconsistencies.
- **Deduplication**
- Removing duplicate records helps maintain data integrity.
- **Formatting and Encoding**
- Consistent formatting, units, and encoding are essential for clear analysis, especially for categorical variables.
- **Handling Missing Values**
- Handling missing values through methods like imputation or removal ensures gaps in the data do not skew results.
- **Outlier Management**
- Detecting and addressing outliers prevents distortions in analysis.
- **Data Profiling and Monitoring**
- Regularly profiling and monitoring the data helps catch new issues as they arise.
- **Documentation and Validation**

- Documenting all preprocessing steps and validating their impact ensures the dataset remains accurate and reliable for analysis.

3. (A) Data Preprocessing Steps for Sensor Data

   - Noise Reduction: Apply filtering or smoothing techniques to minimize random fluctuations and sensor noise, improving signal clarity for analysis.
   - Outlier Detection and Removal: Identify and address anomalous readings that could distort results, using statistical or algorithmic methods.
   - Missing Value Handling: Fill in or interpolate missing sensor readings to maintain data continuity.
   - Synchronization: Align data from multiple sensors based on timestamps to ensure accurate temporal relationships.
   - Normalization/Standardization: Scale sensor values to a standard range or distribution, making them comparable across devices or sessions.
   - Data Transformation: Apply resampling, detrending, or aggregation to convert raw sensor signals into features suitable for modeling.

3. (B) Data Preprocessing Steps for Imaging Data [48]

   - Background Removal (ROI Segmentation): Isolate the region of interest by removing irrelevant background, improving focus on clinically significant areas.
   - Denoising: Reduce image noise using filters or advanced algorithms to enhance image quality and diagnostic accuracy.
   - Resampling: Adjust image resolution or voxel size to standardize images from different sources or modalities.
   - Registration: Align images to a standard coordinate system for consistent analysis, especially important in longitudinal or multi-modal studies.

- Intensity Normalization: Standardize pixel or voxel intensity values across images to reduce variability and improve comparability.
- Image Augmentation: Apply transformations such as rotation, flipping, or scaling to increase dataset diversity for training robust models.
- Artifact Removal: Eliminate unwanted elements or artifacts that could interfere with analysis.

References

1. https://www.foreseemed.com/importance-of-data-collection-in-healthcare#:~:text=When%20healthcare%20providers%20have%20access,development%20and%20can%20cost%20lives
2. https://www.ncbi.nlm.nih.gov/books/NBK92065/
3. https://kms-healthcare.com/blog/types-of-healthcare-data/
4. https://guides.lib.uw.edu/hsl/data/findclin
5. https://www.medicaid.gov/sites/default/files/2020-01/chip-cms64-expenditure-forms.pdf
6. https://www.icd10data.com/ICD10CM/Codes
7. https://www.cms.gov/medicare/regulations-guidance/physician-self-referral/list-cpt-hcpcs-codes
8. https://npiregistry.cms.hhs.gov/search
9. https://www.impacthealthsharing.com/knowledge/what-is-service-date
10. https://gaain.org/
11. https://www.cdc.gov/national-program-cancer-registries/index.html
12. https://meps.ahrq.gov/mepsweb/
13. https://www.cms.gov/data-research/research/medicare-current-beneficiary-survey
14. https://www.cdc.gov/nchs/nhanes/index.html
15. https://www.who.int/tools/clinical-trials-registry-platform
16. https://www.nlm.nih.gov/
17. https://www.foreseemed.com/blog/what-is-fhir
18. https://www.foreseemed.com/blog/ehr-software-and-interoperability
19. https://www.foreseemed.com/blog/concurrent-coding
20. https://www.cms.gov/
21. https://ieeexplore.ieee.org/abstract/document/7838167
22. https://www.sciencedirect.com/science/article/abs/pii/S0169260717313998
23. https://repository.rit.edu/cgi/viewcontent.cgi?referer=&httpsredir=1&article=8078&context=theses
24. https://www.zencos.com/blog/healthcare-analytics-use-cases
25. https://kms-healthcare.com/blog/guide-to-hipaa-compliant-software-development/
26. https://pmc.ncbi.nlm.nih.gov/articles/PMC3827694/#:~:text=The%20data%20collection%20instruments%20reviewed,and%20analysis%20of%20biologic%20materials
27. https://www.nih.gov/
28. https://www.ahrq.gov/
29. https://www.pcori.org/
30. https://pmc.ncbi.nlm.nih.gov/articles/PMC3827694/#R3
31. https://pmc.ncbi.nlm.nih.gov/articles/PMC3827694/#R4
32. https://pmc.ncbi.nlm.nih.gov/articles/PMC3827694/#R5
33. https://safetyculture.com/topics/good-laboratory-practice-glp/
34. https://www.quanticate.com/blog/what-is-clinical-data-management
35. Gupta P, Sehgal NK, Acken JM (2024) Introduction to machine leaning with security. Springer

36. Gerritsen MG, Sartorius OE, Veen FM, Meester GT (1993) Data management in multi-center clinical trials and the role of a nation-wide computer network. A 5 year evaluation. Proc Annu Symp Comput Appl Med Care:659–662
37. Lu Z, Su J (2010) Clinical data management: current status, challenges, and future directions from industry perspectives. Open Access J Clin Trials 2:93–105
38. https://scdm.org/
39. https://www.cdisc.org/standards
40. [https://www.scilife.io/glossary/clinical-data-management
41. US FDA Draft Guidance (2025) Artificial intelligence-enabled device software functions: life-cycle management and marketing submission recommendations
42. ISO/IEC 5259-1: Artificial intelligence—Data quality for analytics and machine learning—Part 1: Overview, terminology and examples
43. ISO/IEC 5259-2: Artificial intelligence—Data quality for analytics and machine learning—Part 2: Data quality model and data quality measures
44. ISO/IEC 5259-3: Artificial intelligence—Data quality for analytics and machine learning—Part 3: Data quality management requirements and guidelines
45. ISO/IEC 5259-4: Artificial intelligence—Data quality for analytics and machine learning—Part 4: Data quality process framework
46. ISO/IEC 5259-5: Artificial intelligence—Data quality for analytics and machine learning—Part 5: Data quality governance framework
47. ISO/IEC 8183: Information technology—Artificial intelligence—Data life cycle framework
48. Trojani V, Bassi MC, Verzellesi L, Bertolini M (2024) Impact of preprocessing parameters in medical imaging-based Radiomic studies: a systematic review. Cancers (Basel) 16(15):2668. https://doi.org/10.3390/cancers16152668. PMID: 39123396; PMCID: PMC11311340

Chapter 3
Machine Learning Algorithms

Abstract Machine learning (ML) is arguably responsible for the most prominent and visible use cases of data science and artificial intelligence (AI). From Netflix recommendations to DeepMind's AlphaFold algorithm, machine-learning-based solutions have produced awe-inspiring results and generated considerable hype. But what is machine learning? How does it work? And most importantly, is it worth the hype? Machine learning (ML) is a branch of artificial intelligence (AI) that enables computers to self-learn from data and improve over time, without being explicitly programmed. These algorithms form the core of intelligent systems, empowering organizations to analyze large amounts of data for patterns, predict outcomes, and automate decision-making processes. Instead of following fixed commands, these algorithms detect patterns in data, allowing them to improve as they receive more information. They empower computers to become more intelligent by learning from experiences, much like humans do through experience and examples. Machine learning algorithms can detect patterns in data and learn from them, enabling them to make their predictions. In short, machine learning algorithms and models learn through experience. Machine learning is utilized in various real-world applications, including image and speech recognition, natural language processing, and recommender systems.

This chapter provides a comprehensive exploration of machine learning algorithm selection, focusing on regression and classification algorithms and their applicability to specific problems. It offers a balanced analysis of various machine learning algorithms, discussing their advantages and limitations. The chapter then delves into deep learning algorithms, covering basic architectures as well as more complex models like Convolutional Neural Networks (CNN), Recurrent Neural Networks (RNN), and Long Short-Term Memory (LSTM) networks. It addresses the intricacies and challenges associated with these advanced algorithms. Furthermore, the chapter outlines strategies for developing algorithm architectures that align with the entire machine learning lifecycle, including data collection, algorithm development, integration, deployment, and MLOps activities. This comprehensive approach ensures that readers gain a thorough understanding of not only the technical aspects of machine learning algorithms but also their practical implementation and optimization in real-world scenarios.

A. Pandey et al., *Artificial Intelligence in Medical Software*,
https://doi.org/10.1007/978-3-032-10808-1_3

Keywords Supervised learning · Unsupervised learning · Linear regression · Decision tree · Clustering · Anomaly detection · K-means · Support vector machines · Random forest · Naïve Bayes · Performance measure · Confusion matrix

3.1 Introduction

Advances in Science and Technology help improve the quality of our lives. Today, the use of Machine Learning (ML) systems, an integral part of Artificial Intelligence, has surged and is playing a remarkable role in our lives. For instance, widely popular virtual personal assistant being used for playing music or setting an alarm, face detection, or voice recognition applications are beneficial examples of machine learning systems. Machine learning models are algorithms that can identify patterns or make predictions on unseen datasets. Unlike rule-based programs, these models do not require explicit coding and can evolve as new data is introduced into the system.

In machine learning, there's a concept known as the "No Free Lunch" theorem, which essentially states that not every problem can be solved by the same machine learning algorithm—a set of instructions that helps machines complete tasks, particularly identifying patterns and making predictions. As a result, you should try various algorithms for your problem, using a hold-out "test set" of data to evaluate performance and select the best one. Of course, the algorithms you try must be appropriate for your problem, which is where picking the right machine learning task comes in. As an analogy, you might use a vacuum, broom, or mop to clean your house, but you wouldn't bust out a shovel and start digging.

Machine learning is an area that falls within the realm of Artificial Intelligence. Just as in AI, ML algorithms are fundamental instruments used to solve ML problems. In other words, ML algorithms are at the core when we try to derive inferences from data. ML systems can learn, predict users' needs, and perform an expected task without human intervention. The inputs for the desired predictions are taken from the user's previously performed tasks or relevant examples. Figure 3.1 depicts how a prediction task is undertaken in ML [1–3]. The process shown in the figure indicates that a desired prediction is achievable through the analysis of previous inputs and corresponding observations.

The machine learning process defines the workflow that a data science team executes to create and deliver a machine learning model. The machine learning process guides the development and deployment of machine learning models in a structured way. It consists of various steps. Each step plays a crucial role in ensuring the success and effectiveness of the machine learning model. By following the machine learning lifecycle, we can solve complex problems, get data-driven insights, and create scalable and sustainable models. The machine learning process

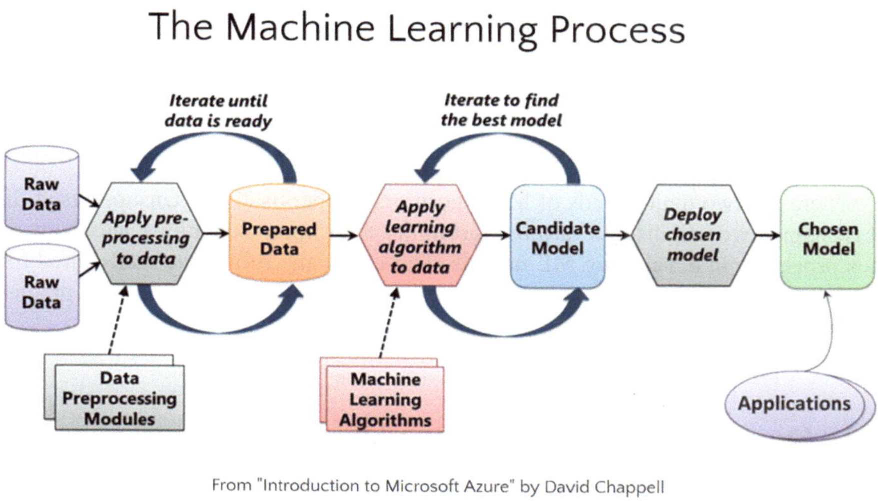

Fig. 3.1 Machine learning process [1]

involves collecting data, preparing the data, selecting a model, training the model, evaluating the model, and then deploying the model, as shown in Fig. 3.1. A Short description of each step is given below, and details are provided in Chap. 2.

Data collection

- Gather relevant data in a systematic way
- Ensure the data is of high quality

Data preparation

- Clean, transform, and structure the data
- Remove missing elements, redundancy, and unorganization

Model selection

- Choose an algorithm and model architecture that's best for the task
- Compare models of the same type or different types

Model training

- Train the algorithm on the prepared data

Model evaluation

- Use metrics to assess the model's performance
- Test the model's accuracy on new data
- Understand the model's strengths and weaknesses

Model deployment

- Implement the trained model for real-world use

Monitoring and maintenance

- Continuously evaluate and update the model as needed

There are two major kinds of ML algorithms: Supervised and Unsupervised. In this chapter, we will discuss various algorithms, their implementations, and the various processes in developing ML models. These ML algorithms are well-suited for creating predictive models as well as for carrying out classification and prediction tasks.

3.2 Supervised Machine Learning Algorithms

In the Supervised Learning method, to obtain the output for a new set of users' inputs, a model is trained to predict the results using an existing set of inputs and their corresponding known outputs. This is also called labeled data. In other words, the system learns from experience. A data scientist trains the system on identifying the features and variables it should analyze. After training, these models compare the new results to the old ones and update their data accordingly to improve the prediction accuracy. An example: If a set of data requires classifying gender into male and female, based on specifications such as color, BMI, foot size, height, and weight provided to the system, the model should be able to classify the gender accurately.

There is a common principle that underlies all supervised machine learning algorithms for predictive modeling:

> *Machine learning algorithms are described as learning a target function (f) that best maps input variables (X) to an output variable (Y):* $\boldsymbol{Y = f(X)}$

This is a general learning task where we aim to make predictions for future outcomes (Y) based on new examples of input variables (X). We don't know what the function (f) looks like or its form. If we did, we would use it directly and not need to learn it from data using machine learning algorithms.

The most common type of machine learning involves learning the mapping $Y = f(X)$ to make predictions of Y for new X values. This is called predictive modeling or predictive analytics, and our goal is to make the most accurate predictions possible.

There are two techniques in supervised machine learning. A technique to develop a model is chosen based on the type of data.

1. Regression
2. Classification

3.2.1 Regression

In regression, a numerical value or a continuous variable is predicted based on the relation between predictors (input variables) and the output variable. An example would be predicting a house price based on current price, school district, total area, number of rooms, locality, and crime rate, among other factors.

3.2.2 Classification

In classification, a categorical variable is predicted, i.e., input data can be categorized based on labels. For example, an email classification task, such as recognizing whether an email is spam or not, is a classification problem.

In summary, the regression technique is used when predictable data is quantified, and the classification technique is used when predictable data involves predicting a label.

3.3 Machine Learning Algorithms That Use Supervised Learning

Below are some of the popular machine learning algorithms:

- Linear regression
- Logistic regression
- Support vector machines
- Naive Bayes
- Decision trees.
- Random forest.
- Artificial neural networks
- K-nearest neighbors (KNN)

We shall discuss some of these algorithms in detail later in the chapter.

3.3.1 Unsupervised Machine Learning Algorithms

Unsupervised learning is a class of ML techniques used to find patterns in data. The data provided to unsupervised algorithms is unlabeled, meaning only the input variables (x) are given, with no corresponding output variables. In unsupervised learning, the algorithms are left to discover interesting structures in the data on their own.

The system is not trained by providing a set of inputs and corresponding outputs. Instead, the model itself will learn and predict the output based on its own observations.

For example, consider a basket that is not labeled or given any specifications this time. The model will only learn and organize them by comparing color, size, and shape. This is achieved by observing specific features and identifying similarities between them. The most prominent methods of unsupervised learning are cluster analysis and principal component analysis.

3.3.2 Clustering

Clustering is a method of dividing or grouping the data into similar groups based on similarities. Data is explored to make groups or subsets based on meaningful separations. Clustering is the process of separating different parts of data based on common characteristics. Disparate industries, including retail, finance, and healthcare, use clustering techniques for various analytical tasks. In retail, clustering can help identify distinct consumer populations, allowing a company to create targeted advertising based on consumer demographics that may be too complex to inspect manually. In finance, clustering can detect different forms of illegal market activity, such as orderbook spoofing, where traders deceitfully place large orders to pressure other traders into buying or selling an asset. In healthcare, clustering methods have been used to identify patient cost patterns, early-onset neurological disorders, and cancer gene expression.

Selecting an algorithm suited to the problem requires a method to evaluate how well these Python clustering algorithms perform. Typically, the average within-cluster-distance from the center is used to evaluate model performance. Specifically, the average distance of each observation from the cluster center, called the centroid, is used to measure the compactness of a cluster. This makes sense because a good Python clustering algorithm should generate groups of data that are tightly packed together. The closer the data points are to one another within a Python cluster, the better the algorithm's results. The sum within cluster distance plotted against the number of clusters used is a standard method for evaluating performance.

3.3.3 Dimension Reduction

If a dataset has too many features, it makes the process of data segregation more complex. To solve complex scenarios, the dimensional reduction technique is used. The basic idea is to reduce the number of variables or features in the given dataset without loss of important data. Image Classification can be considered the best example where this technique is frequently used.

Dimensionality reduction is the process of reducing the number of features within a dataset while preserving the critical properties of the data. This is done to reduce processing time, storage space, complexity, and overfitting in a machine learning model.

The two main methods for applying dimensionality reduction include feature selection and feature extraction. Feature selection involves selecting a subset of relevant features from the original feature set to use as input into a model, which helps simplify the model and improve the accuracy of outputs. Feature extraction involves extracting new, significant features from the original raw data for input, focusing on eliminating redundant data and selecting the features that will most improve the output. Popular dimensionality reduction algorithms include principal component analysis (PCA), non-negative matrix factorization (NMF), linear discriminant analysis (LDA), and generalized discriminant analysis (GDA).

3.4 Machine Learning Algorithms That Use Unsupervised Learning

Some of the common algorithms in unsupervised learning are:

- K-means clustering
- Hierarchical clustering
- DBSCAN
- Autoencoders
- Hebbian learning
- Deep belief nets
- Self-organizing map

We shall discuss some of these algorithms in detail later in this chapter.

3.5 Considerations When Choosing an Algorithm

Machine learning is both an art and a science. When examining machine learning algorithms, there is no single solution or approach that fits all. Depending upon the functionality expected from a machine, algorithms range from very basic to highly complex. One should be wise in making a selection of an algorithm that suits one's ML needs. Careful consideration and testing are needed before finalizing an algorithm for a purpose [4–6].

Some problems are very specific and require a unique approach, e.g., if we look at a recommender system, it's a very common type of machine learning algorithm, and it solves a very specific kind of problem. While some other problems are open and need a trial-and-error approach. They can be used in anomaly detection or to build more general types of predictive models [3].

With the advances in ML, there are several competing algorithms to choose from. Some algorithms are more practical than others in getting successful business results. Several factors can influence the decision to select an ML algorithm, which we will discuss in this section.

1. **Type of problem**: It is obvious that algorithms have been designed to solve specific problems. Therefore, it is essential to understand the type of problem we are dealing with and the type of algorithm that works best for each particular situation. At a high level, ML algorithms can be classified into supervised and unsupervised. Supervised learning can be further categorized into regression, classification, and anomaly detection.
2. **Understanding the nature of data:** For success in big data analytics, choosing the right data is paramount. The type and kind of data play a key role in deciding which algorithm to use. Some algorithms can work with smaller sample sets while others require a large number of samples. Some algorithms, such as Naïve Bayes, which work well with categorical input, are not sensitive to missing data. Therefore, it is important to understand the nature of the data. Different algorithms may have different feature engineering requirements. Some have built-in feature engineering. Time spent on data extraction and feature engineering generally accounts for a significant portion of the time budget for project development. If it is done properly, it is time well spent. There are several steps to take when preparing the data before feeding it to the ML algorithm. This process is known as pre-processing, which we will discuss later.
3. **Size of training set:** This is a major factor in our choice of algorithm. For a small training set, high bias/low variance classifiers (e.g., Naive Bayes) have an advantage over low bias/high variance classifiers (e.g., KNN), since the latter will overfit. But low bias/high variance classifiers give better results as the training set grows, i.e., they have lower asymptotic error, as high bias classifiers aren't powerful enough to provide accurate models [4].
4. **Accuracy:** The required accuracy varies depending on the application. Even an approximation is adequate sometimes. This may lead to a huge reduction in processing time. Additionally, approximate methods are highly robust against overfitting.
5. **Training time:** Various algorithms have different execution times. Training time is normally a function of dataset size and the desired accuracy.
6. **Linearity:** Lots of machine learning algorithms, such as linear regression, logistic regression, and support vector machines, make use of linearity. These assumptions are quite good for some problems. However, some other problems may reduce the accuracy. Despite these shortcomings, linear algorithms remain very popular as a first line of attack. They tend to be algorithmically simple and train quickly.
7. **Number of parameters:** Parameters affect the algorithm's behavior, such as error tolerance or number of iterations. Typically, algorithms with a large number of parameters require the most trial and error to find a suitable combination. Although having many parameters typically provides greater flexibility, the

training time and accuracy of the algorithm can sometimes be quite sensitive to getting just the right settings.

8. **Number of features:** The number of features in some datasets can be very large compared to the number of data points. This is often the case with genetics or textual data. The large number of features can bog down some learning algorithms, making training time unfeasibly long. Some algorithms, such as Support Vector Machines, discussed later, are particularly well suited to this case.
9. **Understanding system constraints:** It is essential to be aware of the system's data storage capacity. Depending on the storage capacity of the system, one may not be able to store gigabytes of classification or regression models, or gigabytes of data to be clustered. This is the case, for instance, for embedded systems. Does the learning have to be fast? In some situation, training models quickly is necessary. Sometimes, one needs to rapidly update even on the fly.
10. **Identify the available algorithms**: Once we have a better understanding of the problem and data, we can determine the algorithms that are applicable and practical to implement using the available tools. Some of the factors affecting the choice of an algorithm are:
    - Whether the model meets the business goals.
    - How much preprocessing does the model need?
    - How accurate the model is.
    - How explainable the model is.
    - How fast the model is: How long does it take to build a model, and how long does the model take to make predictions?
    - How scalable the model is.

A crucial criterion influencing the selection of an algorithm is model complexity. Generally speaking, a model is more complex if:

- It relies on more features to learn and predict (e.g., using two features vs ten features to predict a target).
- It relies on more complex feature engineering (e.g., using polynomial terms, interactions, or principal components).
- It has more computational overhead (e.g., a single decision tree vs. a random forest of 100 trees).

Additionally, the same machine learning algorithm can be made more complex by adjusting the number of parameters or selecting specific hyperparameters. In ML, a hyperparameter is a parameter whose value is set before the learning process begins. For example,

- A regression model can have more features, or polynomial terms and interaction terms.
- A decision tree can have varying depths.

Making the same algorithm more complex increases the chance of overfitting, so one has to be careful with the complexity of the model.

3.6 Performance Metrics of ML Algorithms

Typically, after completing the Feature Engineering steps, we select and implement a model to obtain the desired output. The next step is to determine the model's effectiveness based on a specific metric using a test dataset. Evaluating a machine learning model is an essential part of every data science project. There are various metrics that can be used to evaluate the performance of ML algorithms, such as classification and regression algorithms. Model evaluation metrics are used to assess the model's goodness of fit with the data. These metrics can also be used to compare different models and select the most suitable one. Such an evaluation helps to predict how good these models are. The metrics chosen to evaluate a machine learning model are crucial. The choice of metrics influences how the performance of machine learning algorithms is measured and compared. So, we should carefully choose the metrics for evaluating ML performance for the following reasons:

- How the performance of ML algorithms is measured and compared will be dependent entirely on the metric chosen.
- How the importance of various characteristics weigh in the result will be influenced completely by the choice of metric.

We will discuss this in detail in Chap. 4.

3.7 What Are the Most Common and Popular Machine Learning Algorithms?

Following the general introduction to the machine learning algorithm types, let us now discuss some key machine learning algorithms. In this section, we will cover a brief introduction to popular ML algorithms. Details of these algorithms are beyond the scope of this book; readers are referred to [7–9].

3.7.1 Linear Regression

Linear regression is perhaps one of the most well-known algorithms in statistics and machine learning. Commonly used in predictive modeling, it is primarily concerned with minimizing the error of a machine learning model or making the most accurate predictions possible, at the expense of explainability. Linear regression is a very simple approach to supervised learning. Linear regression has been in use for over 200 years and is a widely studied technique. It was developed in the field of statistics. It has been extensively studied as a model for understanding the relationship between input and output variables and has been borrowed by machine learning researchers. Simple linear regression allows us to understand the relationships between two variables. It is used to estimate real values (house price, temperature, sales etc.) based on continuous variable. The relationship is established between the independent and dependent variable by fitting the best line/hyper plane (for 2-dimesional/multi-dimensional instances). The algorithm illustrates the impact of changing the independent variable on the dependent variable. The independent variables are referred to as explanatory variables, as they explain the factors that impact the dependent variables. The dependent variable is often referred to as the predictor. Dependent variables are also known as response variables. It is a statistical technique to predict the response variable based on the input of the predictor variable.

Linear regression is the simplest form of regression due to two key characteristics. First, linear regression only captures linear relationships. Second, linear regression only handles relationships between two variables. If you have more than two variables in your dataset, you should consider multiple regression, which is discussed below.

Regression analysis is implemented to do for the following:

- With it, shown in Fig. 3.2, we can establish a linear relationship between the independent and the dependent variables.
- The input variables $x1, x2 \ldots xn$ are responsible for predicting the value of y.
- In order to explain the dependent variable precisely, we need to identify the independent variables carefully. This will allow us to establish a more accurate causal relationship between these two variables.

3.7.1.1 Linear Regression Is of the Following Two Types

- **Simple Linear Regression**—Based on the value of the single explanatory variable, the value of the response variable changes, The equation of a simple linear regression model to calculate the value of the dependent variable, Y based on the predictor X is as follows:

$$y_i = \beta_0 + \beta_i x + \varepsilon_i$$

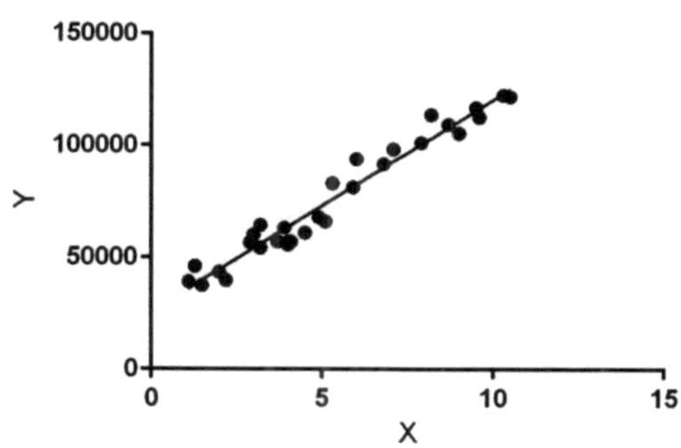

Fig. 3.2 Linear regression example

- where the value of y_i is calculated with input variable x_i for every i^{th} observation. β's are known as regression coefficients. The ith value of x has ε_i as its error in the measurement. These coefficients determine the goodness of predictions of our model. The goal is to find statistically significant values of the parameters β's that minimize the difference between y and y_i (predicted value). If we are able to determine the optimum values of these parameters, then we will have the line of best fit that can be used to predict the value of the output given a new value of input. The best fit line is the one for which the total prediction error is minimum. This method is known as ordinary least squares (OLS). The OLS technique tries to reduce the sum of squared error defined in the following equation by finding the best possible values of regression coefficients. OLS uses squared error that has nice mathematical properties, thereby making it easier to differentiate and compute gradient descent. The residual is calculated as:

$$e_i = y_i - \hat{y}_i$$

- and squared error is defined as:

$$J(\beta) = \frac{1}{n}\sum_{i=1}^{n} e_i^2$$

- **Multiple Linear Regression**—The value of the response variable is dependent upon more than one explanatory variable. Therefore, the simple linear regression model cannot be utilized as there is a need to undertake multiple linear regression for analyzing the predictor variables. The equation of multiple linear regression is as follows:

$$y = \beta_0 + \beta_1 x_1 + \beta_2 x_2 + \ldots + \beta_p x_p + \varepsilon$$

- The x's are explanatory variables and determine y.

Let's understand the algorithm using an example of predicting house prices based on various factors. Below is how the Linear Regression Algorithm will help in this scenario:

1. Collect a dataset of houses with known sales prices, including features such as size, number of bedrooms and bathrooms, lot size, and age of the house, etc.
2. Determine which features might influence the price.
3. Use the dataset to train the model, finding the best-fitted line/hyperplane that minimizes the difference between the predicted and actual prices.
4. Use the trained model to predict prices for new houses based on the features.
5. Test the model on a separate set of houses with known prices to assess the model's accuracy.

Using the Linear Regression algorithm, one can effectively estimate house prices, providing valuable insights for buyers and sellers.

3.7.1.2 Assumptions of Linear Regression

Using linear regression requires that the model conform to the following assumptions:

- The regression model is linear in parameters (which are coefficients and the error term).
- Linear regression requires that residuals be normally distributed. If the maximum likelihood (not Ordinary Least Squares is used to compute the estimates, then this implies that the dependent and independent variables are also normally distributed.
- The mean of the residuals is zero. The error term refers to the variance present in the response variable that the independent variables fail to explain. The model is said to be unbiased if the mean of the error variable is zero.
- Variance of the residuals is constant, i.e., residuals are evenly distributed around the mean, or residuals are approximately equal for all predicted dependent variable values. This condition is known as *homoscedasticity*. If the variance changes, it is referred to as *heteroscedasticity*.
- Independent variables should not be perfectly correlated (i.e., there should be no multicollinearity). A perfect correlation between two variables suggests that they contain the same information. In other words, both variables are different forms of the same variable. If variables are correlated, it becomes extremely difficult for the model to determine the true effects of independent variables on the dependent variable.
- Residuals should not be correlated with each other. This problem is also known as autocorrelation. This is particularly applicable to time series data. Autocorrelation is the correlation of a time series with lags of itself. When the residuals are correlated, it means that the current value is dependent on the previous values and that there is a definite, unexplained pattern in the dependent variable that appears in the disturbances.
- Residuals should not be correlated with the independent variables. If residuals are correlated with the independent variable, one can use the independent variables to predict the error. This correlation between error terms and independent variables is known as endogeneity.

3.7.1.3 Advantages of Linear Regression Machine Learning Algorithm

- It is one of the most interpretable machine learning algorithms, making it easy to explain to others.
- It is easy to use as it requires minimal tuning.
- It is the most widely used machine learning technique.

3.7.1.4 Disadvantages

- The model makes strong assumptions about the data.
- It works only with numeric features, so categorical data requires extra processing.
- It does not do well with missing values and in the presence of outliers.

Some good rules of thumb when using this technique are to remove variables that are very similar (correlated) and to remove noise from your data, if possible. It is a fast and simple technique and a good first algorithm to try.

3.7.2 K-Nearest Neighbors (KNN)

The KNN is a very simple and very effective machine learning algorithm. It is a non-parametric, lazy-learning algorithm, which means that there is no explicit training phase before classification. The KNN algorithm uses the entire dataset as the training set, rather than splitting the dataset into a training and test set. The KNN algorithm assumes that similar entities exist in close proximity. In other words, similar entities share features that are closely related to each other. Figure 3.3 displays identical data points that are closely grouped. KNN can be used to predict loan approval, calculate credit ratings, speech recognition, handwriting detection, Image recognition, or intrusion detection etc. KNNs are used in real-life scenarios where non-parametric algorithms are required. These algorithms do not make any assumptions about how the data is distributed. Not having to worry about distribution is a significant advantage. This means that KNN can be applied to a wide range of datasets. The KNN divides the data points into different classes based on a similar

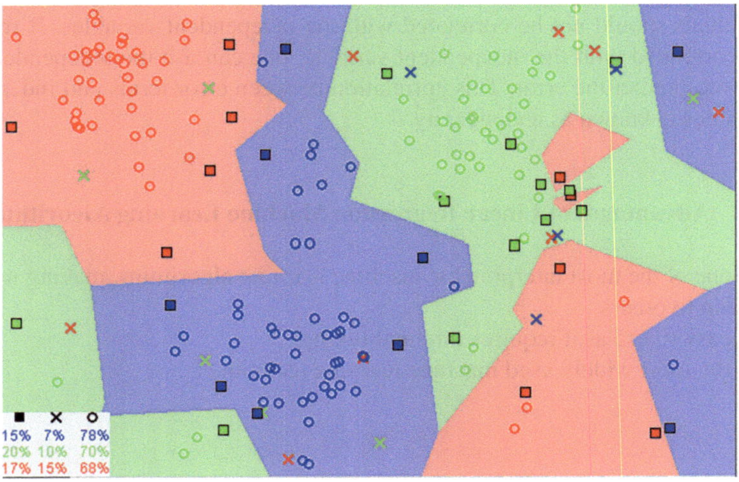

Fig. 3.3 Similar data points exist close to each other

measure, such as the distance function. Then a prediction is made for a new data point by searching through the entire data set for K most similar instances (the neighbors) and summarizing the output variable for K instances.

The main purpose of an algorithm is to classify the data into several classes to predict the class of a new data point. The K-nearest-neighbor algorithm estimates the likelihood that a data point belongs to one group or another. It essentially examines the data points surrounding a single data point to determine which group it actually belongs to. For example, if one point is on a grid and the algorithm is trying to determine which group that data point is in (Group A or Group B, for example), it would look at the data points nearby to see which group the majority of the points are in. In KNN algorithm, the predictions are made for a new data set by searching through the entire training set for the K similar instances, the neighbors, and summarizing the output variable for those *K* instances.

In the classification setting, the KNN algorithm essentially boils down to forming a majority vote between the *K* most similar instances to a given "unseen" observation. Similarity is defined according to a distance metric between two data points. A popular choice is the Euclidean distance given by

$$d(x,x') = \sqrt{(x_1 - x_1')^2 + \cdots + (x_n - x_n')^2}$$

where x and x' are two vectors and d is the distance between them but other measures can be more suitable for a given setting and include the Manhattan, Chebyshev, and Hamming distance.

Given a positive integer *K*, a new observation *x*, and a similarity metric *d*, the KNN classifier performs the following two steps:

- It runs through the whole dataset computing *d* between *x* and each training observation. We'll call the K points in the training data that are closest to *x* the set *A*. Note that K is usually odd to prevent a tie situation.
- It then estimates the conditional probability for each class, that is, the fraction of points in *A* with that given class label. (Note ***I(x)*** is the indicator function which evaluates to 1 when the argument *x* is true and) otherwise.

$$P(y = j|X = x) = \frac{1}{k}\sum_{i \varepsilon A} I(y^i = j)$$

Finally, *x* gets assigned to the class with the largest probability.

KNN is memory-intensive, performs poorly for high-dimensional data, and requires a meaningful distance function to calculate similarity.

Let's understand the algorithm using a popular data set (Iris) for classifying flowers into species. Below is how the KNN algorithm will help in this scenario:

1. Get the data that is available at various sites.
2. Determine key features for classification, such as petal/sepal dimensions.
3. Normalize features for accurate distance calculations.

4. For a new flower, calculate distances to all flowers, identify the K-nearest neighbors, and classify based on the majority species.
5. Test the model on a separate dataset to assess accuracy and adjust K as needed.

Using the KNN algorithm, one can classify flowers effectively, similar to how a botanist identifies species through comparison.

3.7.2.1 Assumptions

- KNN assumes that the data is in some *identifiable feature space*. More precisely, the data points are in a metric space. The data can be scalars or could be from a multidimensional vector space. Since the points are in feature space, they have a notion of distance. This need not necessarily be Euclidean distance, although it is the one commonly used.
- Each of the training data consists of a set of vectors and a class label associated with each vector. In the simplest case, it will be either + or −(for positive or negative classes). But KNN works equally well with an arbitrary number of classes.

We work with a number "*K*." This number determines how many neighbors (defined based on the distance metric) influence the classification. This is usually an odd number. If $K = 1$, then the algorithm is referred to as the nearest neighbor algorithm.

3.7.2.2 How Does the KNN Algorithm Work?

1. Select the *K* where *K* is the number of neighbors.
2. Compute the test point's distance from each point.
3. Sort the distances in ascending (or descending) order.
4. Use the sorted distances to select the *K* nearest neighbors.
5. Use majority rule (for classification) or averaging (for regression) to assign the class or value, respectively.

3.7.2.3 Advantages of the KNN Algorithm

- Simple to understand and easy to implement.
- Zero to little training time.
- Works just as easily with multiclass data sets.

One of the obvious drawbacks of the KNN algorithm is that it has a computationally expensive testing phase that is impractical in industry settings. Furthermore, KNN may sometimes suffer from a skewed class distribution. Finally, the accuracy of KNN can be severely degraded with high-dimensional data because there is little difference between the nearest and farthest neighbors.

3.7.2.4 K-Nearest Neighbor with Scikit-Learn

We will see how Python's Scikit-Learn library can be used to implement the KNN algorithm for classification.

3.7.3 Logistic Regression

Logistic regression is a classic predictive modeling technique and still remains a popular choice for modeling binary categorical variables. Logistic regression is the classification counterpart of linear regression. The logistic regression predicts the discrete variable, whereas the linear regression algorithm handles predictions with continuous values. It is used to predict the categorical dependent variable using a given set of independent variables. Logistic regression is an efficient and robust method for analyzing the effects of a group of independent variables on binary outcomes. It works by quantifying each independent variable's unique contribution to predict the output of a categorical dependent variable. Therefore, the outcomes must be categorical or should have discrete values. Each can be either yes or no, 0 or 1, true or false, etc. but instead of giving the exact value as 0 or 1, it gives the probabilistic values which lie between 0 and 1. This method can be applied in spam detection, credit card fraud detection, and predicting whether a given mass of tissue is benign or malignant, among other uses.

The name of this algorithm can be a bit confusing, as it is often associated with logistic regression, which is typically used for classification tasks. In fact, logistic regression is very similar to linear regression, except in its application. Linear regression is used for regression problems, whereas logistic regression is used for solving classification problems. The name "Regression" implies that a linear model is to fit into the linear space. Predictions are mapped to be between 0 and 1 through the logistic function shown in Fig. 3.4, which is a linear combination of features [10]. This means that predictions can be interpreted as class probabilities. The odds or probabilities that describe the outcome of a single trial are modeled as a function of explanatory variables. As the models themselves are still "linear", they work well when the classes are linearly separable (i.e., they can be separated by a single decision surface). In logistic regression, instead of fitting a regression line, we fit an "S" shaped logistic function, shown in Fig. 3.4, which predicts two maximum values (0 or 1). The curve from the logistic function indicates the likelihood of something, such as whether the email is spam or not, etc.

Using the components of linear regression reflected in the logit scale, logistic regression iteratively identifies the strongest linear combination of variables with the highest probability of detecting the observed outcome. The output is generated by a logarithmic transformation of the input values, defined by the logistic function $f(x) = 1/(1+ e^\wedge (-x))$. A threshold is then applied to force this probability into a binary classification. The goal of logistic regression is to train the model to find the values of coefficients such that it will minimize the error between the predicted

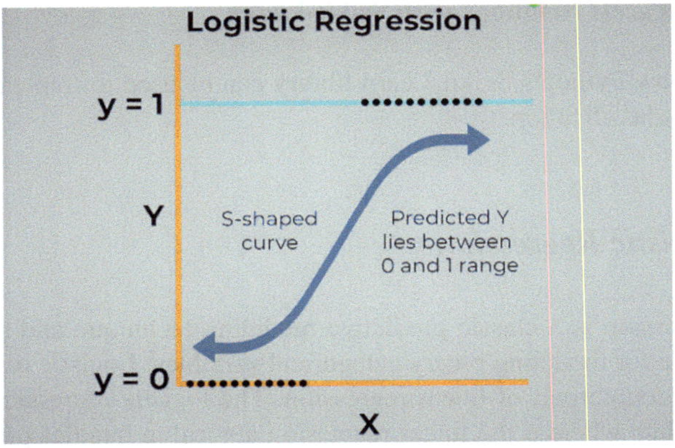

Fig. 3.4 Logistic function

outcome and the actual outcome. These coefficients are estimated using maximum likelihood estimation.

Predictions are mapped to be between 0 and 1 through the logistic function to a linear combination of features, which means that predictions can be interpreted as class probabilities. The odds or probabilities that describe the outcome of a single trial are modeled as a function of explanatory variables. Logistic regression algorithms help estimate the probability of falling into a specific level of the categorical dependent variable, given the predictor variables. The models themselves are still "linear," so they work well when the classes are linearly separable (i.e., they can be separated by a single decision surface).

3.7.3.1 Type of Logistic Regression

On the basis of categories, Logistic regression can be classified into three types:

- **Binomial**: In binomial logistic regression, there can be only two possible types of the dependent variables, such as 0 or 1, Yes or No, etc.
- **Multinomial**: In multinomial logistic regression, there can be 3 or more types of dependent (categorical) variables. Multinomial logistic regression can be used to distinguish between cats and dogs, or sheep.
- **Ordinal**: In ordinal logistic regression, there can be 3 or more possible ordered types of dependent variables, such as small, medium, or large.

3.7.3.2 Assumptions

Logistic regression does not make many of the key assumptions of linear regression and general linear models based on the ordinary least squares algorithms. In particular, regarding linearity, normality, homoscedasticity, and measurement level, these assumptions are often set aside. Firstly, it does not need a linear relationship between the dependent and independent variables. The primary reason why logistic regression can handle various relationships is that it applies a non-linear log transformation to the predicted odds ratio. Secondly, the independent variables do not need to be multivariate normal, although multivariate normality yields a more stable solution. Also, the error terms (i.e., the residuals) do not need to be multivariate normally distributed. Logistic regression does not need variances to be heteroscedastic for each level of the independent variables. Finally, it can handle ordinal and nominal data as independent variables. The independent variables do not need to be metric (interval or ratio scaled). However, some other assumptions still apply, which are described below:

- The outcome is a binary or dichotomous variable, such as true or false, 1 or 0. Reducing an ordinal or even metric variable to a dichotomous level loses a significant amount of information, which makes this test inferior to ordinal logistic regression in these cases.
- Logistic regression assumes linearity of independent variables and log odds. While it does not require the dependent and independent variables to be related linearly, it requires that the independent variables are linearly related to the log odds. In other words, there is a linear relationship between the logit of the outcome and each predictor variable.
- There are no influential values (extreme values or outliers).
- The model should be fitted correctly. Neither over-fitting nor under-fitting should occur. Only the meaningful variables should be included. A good approach to ensure this is to use a stepwise method to estimate the logistic regression.
- It requires quite a large sample size. Because maximum likelihood estimates are less powerful than ordinary least squares (e.g., simple linear regression, multiple linear regression). It has been observed that OLS requires 5 cases per independent variable in the analysis, while ML needs at least 10 cases per independent variable. Some statisticians recommend at least 30 cases for each parameter to be estimated.
- There is no high intercorrelation (i.e., multicollinearity) among the predictors.

3.7.3.3 Advantages

- It is easier to inspect and less complex.
- It is a robust algorithm, as the independent variables do not need to have equal variance or a normal distribution.
- The algorithm does not assume a linear relationship between the dependent and independent variables; hence, it can also handle non-linear effects.
- The algorithm can be regularized to avoid overfitting.
- The model can be easily updated with new data using stochastic gradient descent.

3.7.3.4 Disadvantages

- Logistic regression tends to underperform when there are multiple or non-linear decision boundaries.
- The model is not flexible enough to naturally capture complex relationships.
- Logistic models tend to overfit the data when the data is sparse and high-dimensional. It requires more data to achieve stability and meaningful results.
- It is not robust to outliers and missing values.

Let's understand this algorithm using an example of predicting whether a customer will churn (leave) a subscription service. Below is how the Logistic Regression Algorithm will help in this scenario:

1. Gather customer data, including age, subscription plan, usage frequency, and payment history.
2. Assess factors influencing churn, such as engagement and satisfaction levels.
3. Train the model using the dataset to estimate coefficients for features related to churn probability.
4. Input new customer features to calculate churn probability and classify as likely to churn (1) or not (0).
5. Test on a separate dataset to measure accuracy, precision, and recall, adjusting as needed.

Using the Logistic Regression Algorithm, businesses can effectively predict customer churn, allowing for proactive measures to retain customers—much like a keen manager identifying at-risk clients to enhance satisfaction and loyalty.

3.7.4 Naïve Bayes Classifier Algorithm

The Naïve Bayes Algorithm is a supervised machine learning algorithm used for classification tasks, and it operates on the principle of Bayes' theorem. It calculates the probability of a data point based on its features belonging to a particular category. Naïve Bayes (NB) classifier is among the most popular learning method grouped by similarities that works on the popular Bayes' theorem of probability with an assumption of independence between predictors—to build ML models particularly document classification, etc. Along with simplicity, Naïve Bayes is known to outperform even highly sophisticated classification methods. Naïve Bayes is called naïve because it assumes that each input variable is independent. This is a strong assumption and unrealistic for real data; nevertheless, the technique is very effective on a large range of complex problems. If the NB conditional independence assumption actually holds, a Naïve Bayes classifier will converge quickly than discriminative models like logistic regression, so one needs less data to train the model. Even if the NB assumption of independence of features doesn't hold, it still often does a great job in practice. Despite its simplicity, the classifier does surprisingly well and is often used due to the fact that it outperforms more sophisticated classification methods. It is called naïve because it assumes that the occurrence of a certain feature is independent of the occurrence of other features. Such as if the fruit is to be identified on the basis of color, shape, and taste, then a yellow, spherical, and sweet fruit is recognized as an orange. Hence, each feature individually contributes to identifying that it is an orange without depending on each other. It is called Bayes' because it depends on the principle of Bayes' Theorem.

3.7.4.1 Bayes' Theorem

- Bayes' theorem is also known as Bayes' rule or Bayes' law, which is used to determine the probability of a hypothesis with prior knowledge. It depends on the conditional probability.
- The formula for Bayes' theorem is given as:

$$P(A|B) = \frac{P(B|A)P(A)}{P(B)}$$

where

$P(A|B)$ is the Posterior probability.: Probability of hypothesis A on the observed event B.

$P(B|A)$ is the Likelihood probability.: Probability of the evidence given that the probability of a hypothesis is true.

$P(A)$ is the Prior probability: Probability of the hypothesis before observing the evidence.

$P(B)$ is the Marginal probability: Probability of evidence.

Using Bayes' theorem, it is possible to build a learning system that predicts the probability of the response variable belonging to some class, given a new set of attributes.

An NB model comprises two types of probabilities that can be calculated directly from the training data:

1. The probability of each class
2. The conditional probability for each class given each input value

A naïve Bayes model multiplies several different calculated probabilities together to identify the posterior probability for each class and select the class with the highest probability. This is called the maximum a posteriori probability. Conditional independence of the features reduces the complexity of model.

Naive Bayes *classifiers* are linear classifiers that are known for being simple yet very efficient. Naïve Bayes (NB) classifier is amongst the most popular learning methods grouped by similarities that works on the popular Bayes' theorem of probability with an assumption of independence between predictors—to build ML models. It is a probabilistic classifier, which predicts on the basis of conditional probability for its likely classification. Along with simplicity, Naïve Bayes is known to outperform even highly sophisticated classification methods. The method has a strong assumption of independent input variables, which is not always realistic for real data. Nevertheless, the technique is very effective on a large range of complex problems. If the NB conditional independence assumption actually holds, a Naïve Bayes classifier will converge quickly than the discriminative models such as logistic regression. Also, one needs less data to train the model. Even if the NB assumption of independence of features doesn't hold, it still often does a good job in practice. NB algorithm finds usage in text classification, spam filtering, sentiment analysis, recommendation systems, etc. However, strong violations of the independence assumptions and non-linear classification problems may lead to poor performances of naive Bayes classifiers.

In this model, we calculate the posterior probability for each class and select the class with the highest probability. This is called the maximum a posteriori probability. Conditional independence of the features reduces the complexity of the model.

3.7.4.2 Additive Smoothing

If during testing time, we come across a feature that we encountered during training time, then the individual conditional probability of that particular feature will become zero, thus making class-conditional probabilities equal to zero. So, we have to modify the formula for calculating individual conditional probability. In order to avoid the problem of zero probabilities, an additional smoothing term can be added to the multinomial Bayes model.

3.7.4.3 Types of Naïve Bayes Model

- **Gaussian**: The Gaussian model assumes that features follow a normal distribution. This means that if predictors take continuous values instead of discrete ones, then the model assumes that these values are sampled from the Gaussian distribution.
- **Multinomial**: The multinomial Naïve Bayes classifier is used when the data is multinomial distributed. It is primarily used for document classification problems. It means a particular document belongs to which category, such as politics, sports, entertainment, etc. The features/predictors used by the classifier are the frequency of the words present in the document.
- **Bernoulli**: The Bernoulli classifier works similarly to the Multinomial classifier, but the predictor variables are the independent Boolean variables. For example, whether a particular word is present or not in a document.

3.7.4.4 Assumptions

- The fundamental Naive Bayes assumption is that each feature makes an independent and equal (i.e., identical) contribution to the outcome.

3.7.4.5 How the Naïve Bayes Algorithm Works?

The following steps are followed while working with this algorithm:

1. Calculate a priori probability for the given class labels.
2. Calculate the conditional probability for each attribute and class.
3. Multiply the same class conditional probability.
4. Multiply the prior probability by the probability obtained in step 3.
5. Determine which class has a higher probability. The higher probability class belongs to the given input set.

3.7.4.6 Advantages

NB models perform well in practice, even though the conditional independence assumption rarely holds true.
Easy to implement and can scale with data.
Good choice when CPU and memory resources are a limiting factor.
Performs well when the input variables are categorical.
It is easier to predict the class of the test dataset.

3.7.4.7 Disadvantages

- NB assumes that all features are independent or unrelated, so it cannot learn the relationship between features.

Let's understand this algorithm using an example of Gmail, where you want to classify incoming emails as "spam" or "not spam" to reduce clutter in your inbox. Below is how the Naïve Bayes Classifier Algorithm will help this scenario:

1. Gather a labeled dataset of emails, including both spam and legitimate messages.
2. Identify key features, such as specific words (e.g., "free," "winner") and the number of links in an email.
3. Train the Naïve Bayes Classifier using the dataset, calculating the probabilities of each email being spam.
4. For new emails, analyze their features to compute probabilities and classify them based on the highest likelihood.
5. Test the classifier on a separate set of emails to measure accuracy and make adjustments as needed.

Using the Naïve Bayes Classifier, you can efficiently filter spam, saving time and ensuring you focus on important emails, much like a skilled bouncer at a club.

3.7.5 Support Vector Machine Algorithm

Support Vector Machines (SVMs) are supervised machine learning algorithms that can be used for classification and regression problems, such as support vector classification (SVC) and support vector regression (SVR), which are also applicable for regression. SVM is mostly used for classification. SVM algorithm can be used for face detection, image classification, text categorization, etc. It can solve linear and non-linear problems and works well for many practical challenges. They were extremely popular around the time of their development in the 1990s and continue to be the go-to methods for high-performing algorithms with some tuning. The Support Vector Machine Algorithm tries to find the hyperplane that maximizes the distance between the classes (known as margin maximization). A hyperplane is a line that splits the input variable space and, in SVM, it's selected to best separate the points in the input variable space by their class, either class 0 or class 1. In two dimensions, you can visualize this as a line. Let's assume that this line can completely separate all of our input points. The SVM learning algorithm determines the coefficients that yield the optimal separation of the classes by the hyperplane. The distance between the hyperplane and the closest data points is referred to as the margin. The best or optimal hyperplane that can separate the two classes is the line that has the largest margin. Only these points are relevant in defining the hyperplane and constructing the classifier. These points are referred to as the support vectors. They support or define the hyperplane. In practice, an optimization algorithm is used to find the values for the coefficients that maximize the margin. SVM might be one of the most powerful out-of-the-box algorithms and is worth trying.

SVM is based on the idea of finding a hyperplane that best separates the features into different domains. SVM calculates the distance between two observations. The objective of the SVM is to find a hyperplane in N-dimensional space (where N is the number of features) that distinctly classifies the data. In other words, the idea behind SVM is to find decision planes that define decision boundaries. A decision plane is one that separates a set of objects having different class memberships. As there are many such linear hyperplanes, the SVM algorithm attempts to maximize the distance between the various classes involved, a process referred to as margin maximization. If the line that maximizes the distance between the classes is identified, then the probability of generalizing well to new data is enhanced. An example of binary classes is shown in Fig. 3.5. The points closest to the hyperplane are called support vectors (SV), and the distance of the vectors from the hyperplane is called the margin. The basic intuition to develop here is that the farther support vector points are from the hyper-plane, the higher is the probability of correctly classifying points in their respective regions or classes. SV points are very critical in determining the hyperplane because if the position of vectors changes, then the hyperplane's

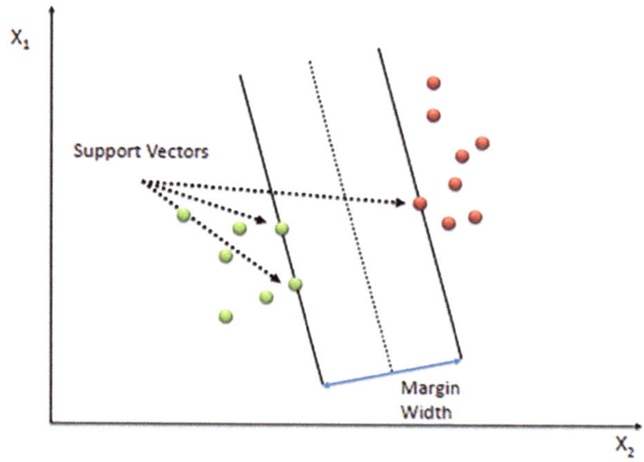

Fig. 3.5 Support vector machines

position is altered. Technically, this hyperplane can also be referred to as a maximizing margin hyperplane. If an SVM is given a data point closer to the classification boundary than the support vectors, then the SVM declares that data point to be too close for accurate classification. This defines a "no-man's land" for all points within the margin of the classification boundary. Since the support vectors are the data points closest to this "no-man's land" without being in it, intuitively, they are also the points most likely to be misclassified.

While it sounds simple, not all datasets are linearly separable. In fact, in the real world, almost all the data is randomly distributed, which makes it hard to separate different classes linearly. SVM uses a kernel trick to solve such types of problems. The kernel trick performs data transformations to determine an optimal boundary for separating the data. Based on the labels or outputs defined. Kernel trick is a method of using a linear classifier to solve non-linear problems. It entails transforming seemingly linearly inseparable data. The Kernel function maps the original non-linear observations into a higher-dimensional space, where they become separable, as shown in Fig. 3.6. In this figure, we find a way to map the data from 2-dimensional to 3-dimensional space and are able to find a decision boundary that clearly divides between different classes. The kernel trick enables us to operate in the original feature space without computing the coordinates of the data in a higher dimension [11]. There are many kernel functions, e.g., Gaussian/RBF kernel, polynomial kernel, sigmoid, etc. Care must be taken when choosing the kernel function to prevent overfitting the model. Thus, choosing the right kernel function (including the right parameters) and regularization (discussed in other part of the book) are of great importance. The most used type of kernel function in RBF is because it has a localized and finite response [12].

Kernel tricks are used to map a non-linearly separable function into a higher-dimensional linearly separable function. A support vector machine training algorithm finds the classifier represented by the normal vector and bias of the hyperplane.

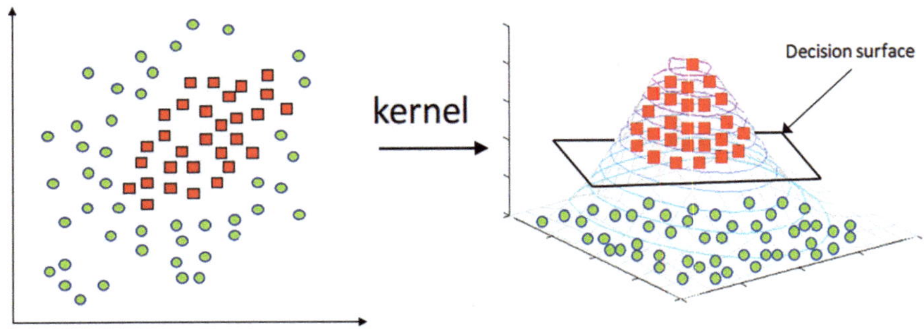

Fig. 3.6 Kernel method

This hyperplane (boundary) separates different classes by as wide a margin as possible.

An SVM finds the classifier represented by the normal vector and bias.

3.7.5.1 Assumptions

SVMs can be defined as linear classifiers under the following two assumptions:

- The margin should be as large as possible.
- The support vectors are the most useful data points because they are the ones most likely to be incorrectly classified.

The second assumption leads to a desirable property of SVMs. After training, the SVM can throw away all other data points and just perform classification using the support vectors. For example, in Fig. 3.6, we need only three points. This means that once classification is done, an SVM can predict a data point's class very efficiently.

3.7.5.2 Types of SVM

SVM can be of two types.

- **Linear SVM**: Linear SVM is used for linearly separable data, which means that if a dataset can be classified into two classes using a single straight line, then such data is termed linearly separable, and the classifier used is called a Linear SVM classifier.
- **Non-linear SVM**: A non-linear SVM is used to separate datasets that cannot be classified by using a straight line. Such data is termed non-linear data, and the classifier is called a non-linear SVM classifier.

3.7.5.3 Advantages

- It works well on a smaller and cleaner dataset.
- It is more efficient as it uses a subset of training points.
- It can model non-linear boundaries and there are many kernels to choose from.
- It is robust against over-fitting, especially in high-dimensional spaces.

3.7.5.4 Disadvantages

- It is not suitable for larger datasets due to the training time.
- It is less effective on noisier datasets with overlapping classes.
- It is memory-intensive.
- It is trickier to tune due to the importance of picking the right kernel.
- It does not scale well with large datasets.

Let's understand this algorithm using an example of classifying images of cats and dogs. Below is how the SVM algorithm will help in this scenario:

1. Collect a dataset of images, including features such as pixel values.
2. Extract relevant features that distinguish the classes, such as shape, color, and texture.
3. Use the dataset to train the SVM and find the optimal hyperplane.
4. Test the model on a separate set of images to measure accuracy.

3.7.6 Decision Trees

A Decision Tree (DT) algorithm is a tree-like structure of decision rules that are applied to the input features to predict the possible outcomes. DT belongs to supervised machine learning algorithms used for both classification and regression tasks. It creates a model that predicts the value of a target variable by learning simple decision rules inferred from the data. The tree can be explained by two entities, namely decision nodes and leaves. The leaves represent the final outcomes, and each node (decision node or internal node) within the tree represents a test on a specific feature.

These methods are used for regression as well as classification problems. Using the DT with a given set of inputs, one can map the various outcomes that result from the consequences or decisions. These trees are used to provide graphical outputs to the user based on several independent variables. DT is capable of handling both heterogeneous and missing data. DT algorithms are further capable of producing comprehensible rules. Classification can be performed with minimal computational effort. DTs are used in marketing and sales, reducing churn rate, anomaly and fraud detection, medical diagnosis, etc. (Fig. 3.7)

A decision tree is a flowchart-like structure that uses a branching method to illustrate all possible outcomes of a decision. Each node within the tree represents a test

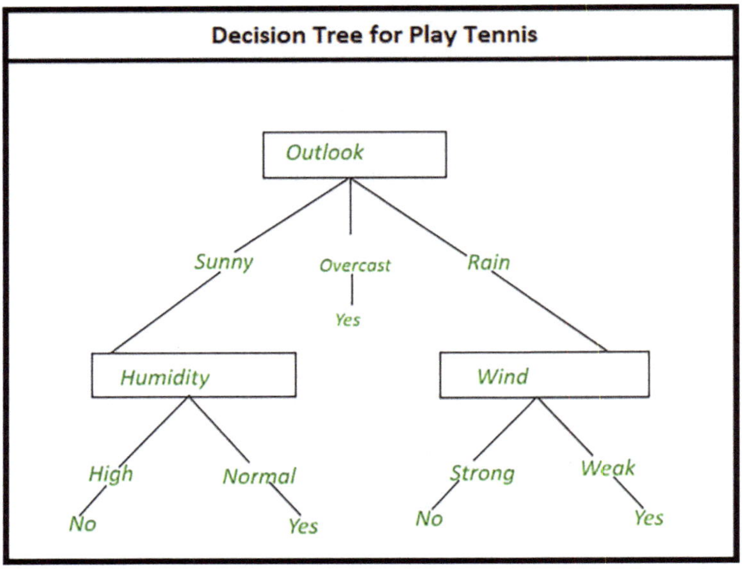

Fig. 3.7 Decision tree

on a specific variable, and each branch is the outcome of that test. It is a graphical representation of possible solutions to a decision based on certain conditions. It's called a decision tree because it starts with a single node known as the root node, which then branches off into a number of solutions, just like a tree. The decision tree algorithm can be used for both regression and classification. The decision tree method is capable of handling both heterogeneous and missing data. Trees are further capable of producing understandable rules. The tree can be explained by two entities, namely decision nodes and leaves. The leaves represent the outcomes, and each node (decision node or internal node) within the tree represents a test on a specific feature.

In a decision tree building algorithm, first, the best feature is placed at the root, and then the training dataset is split into subsets. Splitting of data depends on the features of the data. This process is repeated until the entire dataset is classified, and a leaf node is reached at each branch. A DT is largely used for non-parametric machine learning modeling for regression and classification problems. To find solutions, a DT makes sequential, hierarchical decisions about the outcome variable based on the predictor data. So, what does all that mean?

Hierarchical means that the model is defined by a series of questions that lead to a class label or a value when applied to any observation. Once set up, the model operates like a protocol, following a series of "if this occurs, then that follows" conditions that produce a specific result from the input data.

A non-parametric method means that there are no underlying assumptions about the distribution of the errors or the data. It basically means that the model is constructed based on the observed data.

The understanding level of the DT algorithm is easily comparable to that of other classification algorithms. The DT algorithm can be applied to both regression and classification tasks. DT method is capable of handling heterogeneous as well as missing data. Trees are further capable of producing understandable rules. DTs often mimic human-level thinking, making it simpler to understand the data and draw some good interpretations. DTs make it easy to see the logic behind the data interpretation (unlike SVM, ANN, etc.).

This algorithm is known as Hunt's algorithm, which is both greedy and recursive. Greedy means that at each step, it evaluates to maximize (or minimize) an objective function locally. Similarly, recursive means that it splits the larger question into smaller questions and resolves them using the same argument. The decision to split at each node is made based on a metric called purity. A node is 100% impure when a node is split evenly 50/50 and 100% pure when all of its data belongs to a single class. To optimize our model, we need to achieve maximum purity and minimize impurity.

DT may use multiple criteria to decide to split a node into two or more sub-nodes. With each split, the purity of the node is enhanced with respect to the target variable. DT splits the nodes on all the available attributes and selects the split that results in the most homogeneous sub-nodes. Given a data table that contains attributes and classes of the attributes, we can measure homogeneity (or heterogeneity) of the table based on the classes. We say a table is pure or homogeneous if it contains only a single class. If a data table contains several classes, then we say that the table is impure or heterogeneous. Several indices are used to measure the degree of impurity quantitatively. Some well-known indices to measure the degree of impurity are entropy, the Gini index, and classification error.

The popular measures used are:

- Information gain
- Gini index

3.7.6.1 Information Gain

Information gain is the measurement of changes in entropy after the segmentation of a dataset based on an attribute. It calculates how much information a feature provides about a class. According to the value of information gain, we split the node and build the DT. A DT algorithm always tries to maximize the value of information gain, and a node/attribute having the highest information gain is split first. It can be calculated using the formula given below:

$$\text{Information Gain} = \text{Entropy}(\text{S}) - (\text{weighted avg})^{*}\ \text{Entropy}(\text{for each attribute})$$

Entropy is a metric to measure the impurity in a given feature. It specifies randomness in data. Entropy can be calculated as:

$$\text{Entropy}(s) = \sum_j -p_j \log_2 p_j$$

where p_j is the probability of class j and s is the total number of data points.

3.7.6.2 Gini Index

The Gini index is a measure of impurity or purity used while creating a decision boundary. An attribute with a low Gini index should be preferred as compared to the high Gini index. Gini index can be calculated using the formula below:

$$\text{GiniIndex} = 1 - \sum_j p_j^2$$

These measures will calculate values for every attribute. The values are sorted, and attributes are placed in the tree by following the order, i.e., the attribute with a high value (in case of information gain) is placed at the root. While using information gain as an algorithm, we assume attributes to be categorical, and for the Gini index, attributes are assumed to be continuous.

3.7.6.3 Decision Tree Terminology

- **Root Node:** The Root node is from where the DT starts. It represents the entire dataset, which further gets divided into two or more homogeneous sets.
- **Leaf Node:** Leaf nodes are the final output node, and the tree cannot be segregated further after getting a leaf node.
- **Splitting:** Splitting is the process of dividing the decision node/root node into sub-nodes according to the given conditions.
- **Branch/Sub Tree:** A tree formed by splitting the tree.
- **Pruning:** Pruning is the process of removing the unwanted branches from the tree.
- **Parent/Child node:** The root node of the tree is called the parent node, and other nodes are called the child nodes.

3.7.6.4 Assumptions

Below are some of the assumptions we make while using DT:

- Initially, the entire training set is considered the root.
- Feature values are preferred to be categorical. If the values are continuous, then they are discretized prior to building the model.
- Records are distributed recursively on the basis of attribute values.
- In order to place attributes as root or internal node of the tree, some statistical approaches are used, which are mentioned above.

3.7.6.5 How Does the Decision Tree Classifier Work?

The complete process can be better understood using the algorithm given below:

1. The tree is constructed in a top-down recursive manner.
2. At the start, all the training examples are at the root.
3. Examples are partitioned recursively based on selected attributes.
4. Attributes are selected on the basis of an impurity function (e.g., information gain).

Condition for Stopping Partitioning

- All examples for a given node belong to the same class.
- There are no remaining attributes for further partitioning—majority class is the leaf.
- There are no examples left.

3.7.6.6 Advantages

- Robust to errors and if the training data contains error-decision tree algorithms will be best suited to address such problems.
- Very instinctual and can be explained to anyone with ease.
- Data type is not a constraint, as they can handle both categorical and numerical variables.
- Do not require any assumption about the linearity in the data and hence can be used where the parameters are non-linearly related.
- No assumption about the structure and space distribution.
- Save data preparation time, as they are not sensitive to missing values and outliers.

3.7.6.7 Disadvantages

- The greater number of decisions in a tree, the less is the accuracy of any expected outcome.
- Decision trees do not fit well for continuous variables and result in instability and a classification plateau.
- Decision trees are easy to use when compared to other decision-making models, but creating large decision trees that contain several branches is a complex and time-consuming task.
- Decision tree machine learning algorithms consider only one attribute at a time and might not be best suited for actual data in the decision space.
- Unconstrained, individual trees are prone to overfitting, but this can be alleviated by ensemble methods (discussed next).

Let's understand this algorithm using an example of predicting whether a customer will purchase a product. Below is how the Decision Trees Algorithm will help in this scenario:

1. Collect customer features (age, income, browsing behavior) and purchase labels (yes/no).
2. Determine key factors influencing purchases, like income and product interest.
3. Create a tree structure that splits data based on feature values.
4. Use the model to predict whether new customers will purchase.
5. Test on a separate dataset to measure accuracy and adjust as needed.

Single decision trees are used very rarely, but in combination with many others they build very efficient algorithms such as Random Forest or Gradient Tree Boost. There are a couple of algorithms to build DTs, e.g., CART (Classification and Regression Trees, ID3 (Iterative Dichotomiser 3), etc.

3.7.7 Ensemble Learning

An Ensemble method is a technique that combines the predictions of several base estimators built with a given learning algorithm to improve generalizability and robustness over a single estimator [13]. A model that comprises many models is called an Ensemble model. The general framework is shown in Fig. 3.8.

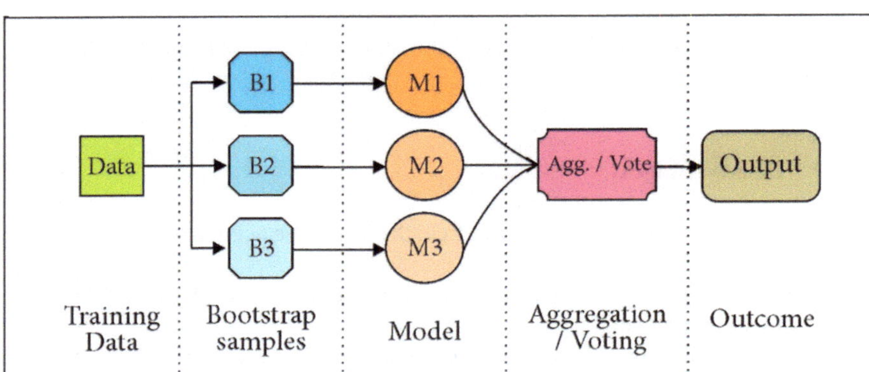

Fig. 3.8 Ensemble learning method

3.7.7.1 Types of Ensemble Learning

- Boosting
- Bootstrap aggregation (bagging).

We will discuss each method now.

Boosting

Boosting refers to a group of algorithms that utilize weighted averages to fine-tune weak learners into stronger learners. It is all about "teamwork." Each model that runs dictates what features the next model will focus on. In boosting, one learns from others, which in turn enhances the learning.

Bootstrap Aggregation (Bagging)

Bootstrap refers to random sampling with replacement. Bootstrap allows us to better understand the bias and the variance of the dataset. Bootstrap involves random sampling of a small subset of data from the dataset.

It is a general procedure to reduce the variance for those algorithms that have high variance. Bagging allows each model to run independently and then aggregates the outputs at the end, without favoring any model.

3.7.8 Random Forests

DTs are sensitive to the specific data on which they are trained. If the training data is changed, the resulting DT will be quite different, and in turn, the predictions will be quite different. Additionally, DTs are computationally expensive to train, carry a significant risk of overfitting, and tend to find local optima because they can't trace back, i.e., there is no backtracking after they have split and advanced. To address these weaknesses, we turn to Random Forest which illustrates the power of combining multiple DTs into a single model.

Random forests, or "random decision forests," are an ensemble learning method called bootstrap aggregation or bagging, combining multiple algorithms to generate better results for classification, regression, and other tasks. Each individual decision tree classifier is weak, but when combined with others, it can produce excellent results. As the name of the algorithm suggests, this ML algorithm creates a forest and makes it somewhat random.

Random Forest is a popular method for utilizing tree algorithms to achieve high accuracy while overcoming the overfitting problem encountered in single decision tree algorithms. It also helps to identify the most significant features. Random

Forest is highly scalable to any number of dimensions and has generally quite acceptable performance. With Random Forest, however, learning may be slow (depending on the parameterization), and it is not possible to iteratively improve the generated models.

The algorithm begins with a "DT" (a tree-like graph or model of decisions) and a node at the top that receives input. It then traverses down the tree, with data being segmented into smaller and smaller sets, based on specific variables. The forest that it builds is an ensemble of DT and most of the time it is trained with the "bagging" method. The basic concept behind the bagging method is that combining multiple learning models enhances the overall result. Bagging can be used to reduce the variance for those algorithms that have high variance, typically DTs. Bagging allows each model to run independently and then aggregates the outputs at the end, without favoring any specific model. There is no interaction between trees during the build-ups. A random forest is, therefore, considered to be a meta-estimator (i.e., it combines the results of many predictions), which aggregates many DTs.

For classifying a new observation, each tree gives a classification, and the forest chooses the classification with the most votes (over all the trees in the forest). For regression, it is the average of all the trees' outputs.

Each tree is planted and grown as follows: if the number of cases in the training set is N, then a sample of N cases is taken at random, with replacement. This sample will be the training set for growing the tree. Whereas if there are M input features/variables, then a number $m < M$ is specified such that at each node, m variables are selected at random out of the M, and the best split on this m is used to split the node. The value of m is held constant during the forest's growth. Each tree is grown to the maximum extent possible. Moreover, each tree is grown on a different sample of different data. Since random forest has the features to calculate out-of-bag (OOB) error internally, a cross-validation does not make sense in random forest. The number of features that can be used to split on at each node is limited to a certain percentage of the total (which is known as a hyperparameter). This ensures that the ensemble model does not rely too heavily on some individual feature and makes fair use of all potentially predictive features. Each tree draws a random sample from the original dataset when generating its splits, adding a further element of randomness that prevents overfitting. By reducing the features to a random subset that may be considered at each split point, it forces each DT in the ensemble to be more different. The effect is that the predictions, and in turn, prediction errors made by each tree in the ensemble are relatively less correlated. When the predictions from these less correlated trees are averaged to make a prediction, it often results in better performance than the bagged DTs. Various hyperparameters: number of samples, number of features, number of trees, and tree depth. Figure 3.9 explains the working of the Random Forest algorithm:

Random forest is a bagging technique and not a boosting technique. The trees in random forests are run in parallel. There is no interaction between these trees while building them. Figure 3.10 shows the two types.

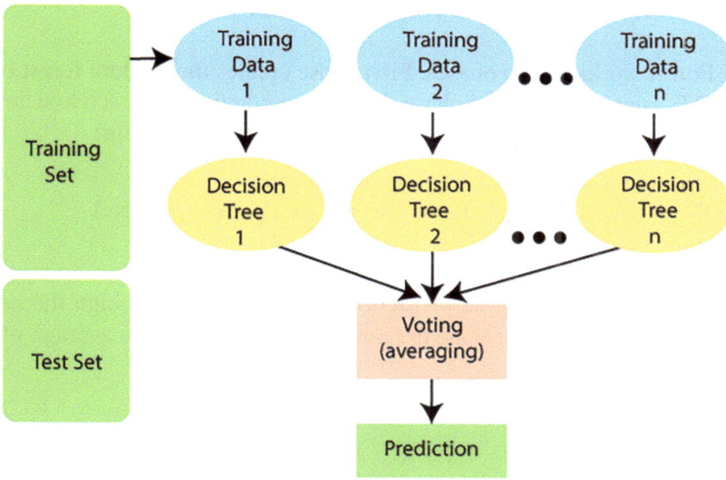

Fig. 3.9 Random forest algorithm

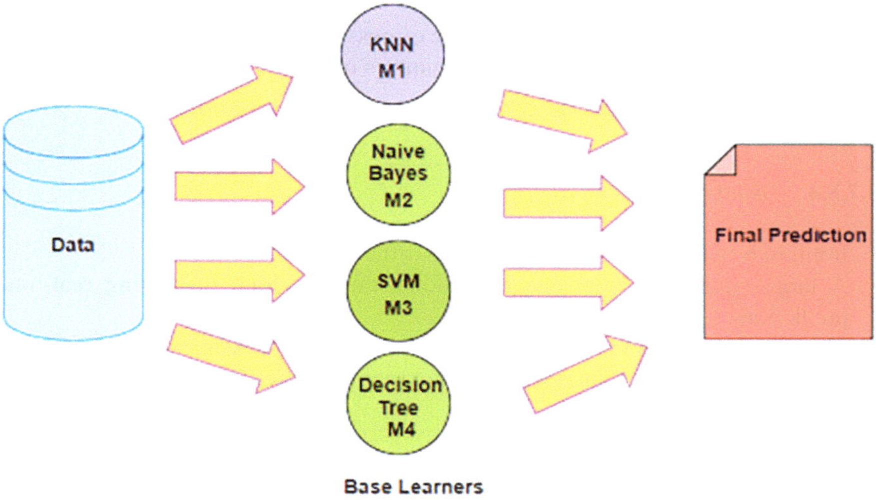

Fig. 3.10 Ensemble methods

3.7.8.1 Assumptions

Given below are the two assumptions for a better Random Forest classifier:

- There should be some actual values in the feature variable of the dataset for the classifier to predict more accurate results.
- The predictions from each tree should have very low correlations.

3.7.8.2 How Does the Random Forest Algorithm Work?

Random Forest works in two phases. First phase creates the random forest by combining *N* DTs, and in the second phase, it makes predictions for each tree created in the first phase. The process can be explained in the following steps:

1. Select random *K* data points from the training set.
2. Build the DTs associated with the selected data points (Subsets).
3. Choose the number *N* for DTs that we want to build.
4. Repeat Steps 1 & 2.
5. For new data points, find the predictions of each DT, and assign the new data points to the category that wins the majority of votes and the average of all the outputs for regression.

3.7.8.3 Advantages

- No over-fitting problem encountered.
- Can be used for both classification and regression tasks.
- Very little preprocessing of data.
- Accuracy is maintained in the presence of missing data and is robust to outliers.
- Implicit feature selection as it gives estimates on what variables are important.
- Algorithm can be grown in parallel.

3.7.8.4 Disadvantages

- It might be easy to use, but analyzing is difficult.
- A large number of trees can slow down the algorithm in making real-time predictions.
- The algorithm tends to favor features with more categories or levels. In such situations, variable importance scores do not seem to be reliable.

3.7.9 AdaBoost

When nothing works, boosting does. Nowadays, many people use boosting algorithms such as XGBoost or LightGBM. AdaBoost is the first stepping stone in the world of boosting [14]. It is called Adaptive Boosting because the weights are reassigned to each instance, with higher weights assigned to cases that are incorrectly classified. AdaBoost (Adaptive Boosting) is a powerful ensemble learning technique that combines multiple weak classifiers to create a strong classifier. Proposed by Yoav Freund and Robert Schapire in 1996, it has since been widely used in machine learning. It is one of the earliest implementations of the boosting algorithm. It forms the basis of other boosting algorithms, such as gradient boosting and XGBoost. It is mainly used for classification, and the based learner (the ML algorithm that is boosted) is usually a decision tree with only one level, also called stumps. It utilizes weighted errors to construct a robust classifier from a collection of weak classifiers. It adjusts the weights of incorrectly classified instances to focus learning on more challenging data points. AdaBoost combines multiple models to produce more accurate results, and this is done in two phases:

1. Multiple weak learners are allowed to learn on the training data.
2. Combining these models to generate a meta-model, this meta-model aims to resolve the errors as predicted by the individual weak learners.

The main idea behind AdaBoost is to iteratively train the weak classifier on the training dataset, with each successive classifier giving more weightage to the data points that are misclassified. The final AdaBoost model is decided by combining all the weak classifiers that have been used for training, with the weightage given to the models according to their accuracies. The weak model, which has the highest accuracy, is given the highest weight, while the model with the lowest accuracy is given a lower weight. It works in the following steps:

1. Initialize weights: Assign equal weights to all training examples.
2. Train weak classifiers: Train a weak classifier (e.g., decision stump, a simple decision tree with only one split) on the training data using the current weights.
3. Compute classifier weight: Calculate the weight of the weak classifier based on its classification error. The lower the error, the higher the weight.
4. Update example weights: Adjust the weights of the training examples. Increase the weights of misclassified examples, making them more important in the next iteration.
5. Normalize weights: Normalize the example weights so that they sum up to one.
6. Repeat steps 2–5: Repeat steps 2 to 5 for a predefined number of iterations or until a certain condition is met.
7. Combine weak classifiers: Combine the weak classifiers by assigning weights to them based on their classification performance.
8. Make predictions: Use the combined model to make predictions on new, unseen data.

3.7.10 Apriori Algorithm

Apriori Algorithm is a foundational method in data mining used for discovering frequent item sets and generating association rules. Its significance lies in its ability to identify relationships between items in large datasets, which is particularly valuable in market basket analysis. For example, if a grocery store finds that customers who frequently purchase bread also tend to buy butter, it can utilize this information to optimize product placement or marketing strategies. The Apriori Algorithm generates association rules using the IF_THEN format. This means that if event A occurs, then event B also occurs with a certain probability. The Apriori Algorithm is used for mining frequent item sets and generating association rules in transactional databases. Market basket analysis employs this algorithm to help behemoths like Amazon and Netflix translate the heaps of information about their users into simple rules of product recommendations. It analyses the associations between millions of products and uncovers insightful rules.

3.7.10.1 How the Apriori Algorithm Works?

The Apriori Algorithm operates through a systematic process that involves several key steps:

1. **Identifying Frequent Itemsets**: The algorithm begins by scanning the dataset to identify individual items (1-item) and their frequencies. It then establishes a **minimum support threshold**, which determines whether an itemset is considered frequent.
2. **Creating Possible item group**: Once frequent 1-itemgroup(single items) are identified, the algorithm generates candidate 2-itemgroup by combining frequent items. This process continues iteratively, forming larger itemsets (k-itemgroup) until no more frequent itemgroup can be found.
3. **Removing Infrequent Item groups**: The algorithm employs a pruning technique based on the **Apriori Property**, which states that if an itemset is infrequent, all its supersets must also be infrequent. This significantly reduces the number of combinations that need to be evaluated.
4. **Generating Association Rules**: After identifying frequent itemsets, the algorithm generates **association rules** that illustrate how items relate to one another, using metrics like **support, confidence,** and **lift** to evaluate the strength of these relationships.

3.7.10.2 Key Metrics of Apriori Algorithm

- **Support**: This metric measures the frequency with which an item appears in the dataset, relative to the total number of transactions. A higher support indicates a more significant presence of the itemset in the dataset. Support tells us how often a particular item or combination of items appears in all the transactions ("Bread is bought in 20% of all transactions.")
- **Confidence**: Confidence assesses the likelihood that an item Y is purchased when item X is purchased. It provides insight into the strength of the association between two items.
- Confidence indicates how often items occur together. ("If bread is bought, butter is bought 75% of the time.")

- **Lift**: Lift measures the likelihood that two items are purchased together compared to being purchased independently. A lift greater than 1 suggests a strong positive association. Lift shows how strong the connection is between items. ("Bread and butter are much more likely to be bought together than by chance.")

3.7.10.3 Applications of Apriori Algorithm

Below are some applications of the Apriori algorithm used in today's companies and startups

1. **E-commerce:** Used to recommend products that are often bought together, like a **laptop + a laptop bag**, increasing sales.
2. **Food Delivery Services:** Identifies popular combos, such as **burger + fries**, to offer **combo deals** to customers.
3. **Streaming Services:** Recommends related movies or shows based on what users often watch together, such as **action and superhero movies**.
4. **Financial Services:** Analyzes spending habits to suggest personalized offers, such as **credit card deals** based on frequent purchases.
5. **Travel & Hospitality:** Creates travel packages (e.g., **flight + hotel**) by finding commonly purchased services together.

3.7.11 K Means Clustering Algorithm (Unsupervised Learning—Clustering)

Sometimes the goal is to assign labels based on features, with no initial information about the labels. This task is achieved by creating clusters and assigning labels to each cluster. For example, clustering can be used when there is a large group of users and there is a need to divide them into groups based on identifiable attributes. In other words, we attempt to identify homogeneous subgroups within the data, ensuring that data points in each cluster are as similar as possible. Euclidean-based distance or correlation-based distance is usually used as a similarity measure. The decision of which similarity measure to use is application-specific.

K-means is probably one of the better-known and frequently used algorithms. K-means uses an iterative refinement method to produce its final clustering based on the number of clusters defined by the user (represented by the variable K) and the dataset. For example, if you set K equal to 3, then the dataset will be grouped into 3 clusters.

K-means is a general-purpose algorithm that makes clusters based on geometric distances between points. K-means is a centroid-based algorithm that means points are grouped in a cluster according to the distance (mostly Euclidean) from the centroid. K-means is the most widely used centroid-based clustering algorithm. Centroid-based algorithms are efficient but sensitive to initial conditions and outliers. K-means is an efficient, effective, and simple clustering algorithm. This algorithm is widely employed in various fields, such as marketing, image processing, and social network analysis, to discover inherent patterns within datasets

K-means is a non-deterministic iterative method. K-means starts off with arbitrarily chosen data points as proposed means of the data groups and iteratively recalculates new means in order to converge to a final clustering of the data points. The clusters are grouped around centroids, resulting in a globular shape and similar sizes. Typically mean is taken to define a centroid representing the center of the cluster. The centroid might not necessarily be a member of the dataset.

The initial result of running this algorithm may not be the best possible outcome and rerunning it with different randomized starting centroids might provide a better performance (different initial objects may produce different clustering results). For this reason, it is a common practice to run the algorithm multiple times with different starting points and evaluate different initiation methods. But another question arises: how does one know the correct value of *K*, or how many centroids to create? There is no universal answer. Although the optimal number of centroids or clusters is not known a priori, various approaches exist to estimate it. One commonly used approach is testing different numbers of clusters and measuring the resulting sum of squared errors, choosing the *K* value at which an increase will cause a very small decrease in the error sum, while a decrease will sharply increase the error sum. The point that defines the optimal number of clusters is known as the elbow point, and can be used as a visual measure to find the best pick for the value of *K*.

Because clustering is unsupervised (i.e., there is no "right answer"), data visualization is usually used to evaluate results. If there exists a "right answer" (i.e., we have pre-labeled groups in the training data), then classification algorithms are typically more appropriate.

K-means can be applied to data that has a smaller number of dimensions, is numeric, and is continuous. For example, it has been successfully used for document clustering, identifying crime-prone areas, customer segmentation, insurance fraud detection, public transport data analysis, clustering of IT alerts, etc.

3.7.11.1 Assumptions

- K-means assumes that the variance of the distribution of each attribute (variable) is spherical.
- All the constituent variables have similar variance.
- The prior probability for all *K* clusters is the same, i.e., each cluster has roughly an equal number of observations.

If any of these three assumptions is violated, then K-means will fail.

3.7.11.2 How Does K-means Algorithm Work?

Given *K* (the number of clusters), the K-means algorithm works as follows:

1. Randomly choose *K* data points (seeds) to be the initial centroids, cluster centers.
2. Compute the sum of the squared distances between data points and all centroids.
3. Assign each data point to the centroid that is closest to it.
4. Re-compute the centroids for the clusters by taking the average of all data points that belong to each cluster.
5. If the convergence criterion is not met, proceed to step 3.

3.7.11.3 Convergence Criterion

Most of the convergence happens in the first few iterations. Essentially, there are three stopping criteria that can be adopted to stop the K-means algorithm:

- Centroids of newly formed clusters do not change.
- Points remain in the same cluster.
- The maximum number of iterations is reached.

We can stop the algorithm if the centroids of newly formed clusters are not changing. Even after multiple iterations, if we obtain the same centroids for all the clusters, we can conclude that the algorithm is not learning any new patterns, and this is a sign to stop the training.

Another clear sign is that we should stop the training process if the points remain in the same cluster even after training the algorithm for multiple iterations.

Finally, we can stop the training if the maximum number of iterations is reached. Suppose we have set the number of iterations to 100. The process will repeat for 100 iterations before stopping.

Because clustering is unsupervised (i.e., there's no "right answer"), data visualization is usually used to evaluate results. If there is a "right answer" (i.e., you have pre-labeled groups in the training data), then classification algorithms are typically more appropriate.

3.7.11.4 Advantages

- Fast, simple, and flexible.

Given a smaller value of *K*, *K*-means clustering computes faster than hierarchical clustering for a large number of variables.

3.7.11.5 Disadvantages

- Requires specification of the number of clusters in advance, which may not be easy to do.
- Produces poor clusters if the data does not have globular clusters.

There are other clustering algorithms also (Hierarchical, DBSCAN, etc.).

3.7.12 Deep Learning

During recent years, Deep Learning (DL) has acquired the status of a buzzword in the technical community. Every month, we see new developments and trends being reported in the field of AI and ML applications that transform businesses. Machine learning is embedded in the majority of business operations and has proven to be an integral part of decision-making. However, it is Artificial Intelligence with DL, which amplifies the overall capability in specific domains. The benefits of using models based on these technologies in businesses have led to a significant shift in the way companies invest in and adopt these technologies. In this chapter, we will be discussing the concept of DL to provide an intuition of how it works. Deep Learning is transforming the way machines understand, learn, and interact with complex data. Deep learning mimics neural networks of the human brain, enabling computers to autonomously uncover patterns and make informed decisions from vast amounts of unstructured data.

Deep learning has evolved into a significant subset of artificial intelligence (AI) in recent years, revolutionizing several sectors by enabling machines to perform tasks that were previously thought to be beyond their capabilities. Deep learning has achieved impressive results in various fields, from image recognition to natural language processing, making it a cornerstone of modern AI systems.

Currently, AI is advancing at a great pace, and DL is one of the main contributors to that. With unlimited applications like prediction, speech, image processing, and recognition, natural language processing, gene mapping, and more, DL is being extensively used by companies. Incorporating AI and ML will be a paradigm shift for companies that have, to date, depended on rule-based engines to drive their businesses. This helps in establishing an autonomous and self-healing process in organizational growth. Deep learning has become the primary driver of numerous new applications, such as self-driving cars, and it is time to examine why this is the case. It is beneficial to understand the basics of DL, as they are transforming the world we live in.

DL is a particular kind of machine learning that achieves great power and flexibility by learning to represent the world as a nested hierarchy of concepts, where each concept is defined in relation to simpler concepts and more abstract representations are computed in terms of less abstract ones. The algorithm has a unique feature, i.e., automatic feature extraction. It means that this algorithm automatically identifies the relevant features required for solving the problem. It reduces the burden on the programmer to explicitly select the features. It forms a hierarchy of low-level features. This enables DL algorithms to solve more complex problems consisting of a vast number of nonlinear transformational layers. In a DL neural network, each hidden layer is responsible for training a unique set of features based on the output of the previous layer. As the number of hidden layers increases, the complexity and abstraction of data also increase. It can be used to solve supervised, unsupervised, or semi-supervised learning.

In DL, we don't need to explicitly program everything. They can automatically learn representations from data such as images, video, or text, without introducing hand-coded rules. Their highly flexible architectures can learn directly from raw data and improve their predictive performance when provided with additional data. For example, in face recognition, how pixels in an image create lines and shapes, how those lines and shapes create facial features, and how these facial features are arranged into a face. For example, a DL model known as a Convolutional Neural Network can be trained using a large number (as in millions) of images, such as those containing cats. This type of neural network typically learns from the pixels contained in the images it acquires. For example, it can classify groups of pixels that are representative of a cat's features, such as paws, ears, and eyes, indicating the presence of a cat in an image.

Conventional machine learning algorithms and models have always been very powerful in processing structured data and have been widely used by various businesses for customer segmentation, credit scoring, fraud detection, churn prediction, and other applications. The success of these models highly depends on the performance of the feature engineering phase: the more we work closely with the business to extract relevant knowledge from the structured data, the more powerful the model will be.

When it comes to unstructured data (images, text, voice, videos), hand-engineered features are time-consuming, brittle, and not scalable in practice. That is where Neural Networks-based DL scores. This is due to their ability to automatically discover the representations needed for feature detection or classification from raw data. This replaces manual feature engineering, allowing a machine to both learn the features and utilize them to perform specific tasks. Improvements in Hardware (GPUs) and Software (advanced models/research related to AI) also contributed to deepening the learning from data using DL (Fig. 3.11).

The primary difference between deep learning and machine learning lies in the structure of the underlying neural network architecture. "Nondeep," traditional machine learning models use simple neural networks with one or two computational layers. Deep learning models use three or more layers—but typically hundreds or thousands of layers—to train the models.

While supervised learning models require structured, labeled input data to make accurate outputs, deep learning models can use unsupervised learning. With unsupervised learning, deep learning models can extract the characteristics, features, and relationships necessary to produce accurate outputs from raw, unstructured data. Additionally, these models can even evaluate and refine their outputs for increased precision.

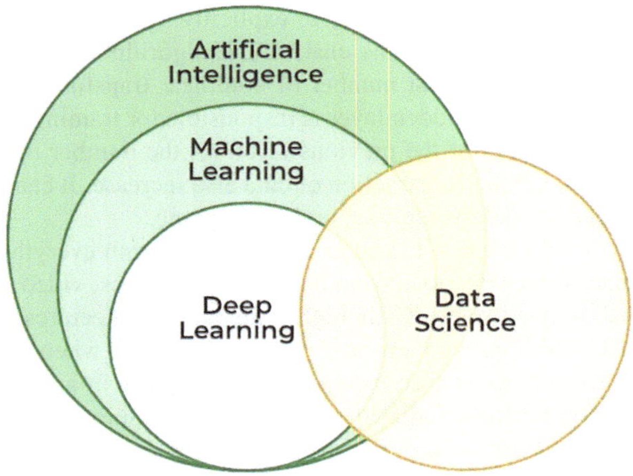

Fig. 3.11 Deep learning

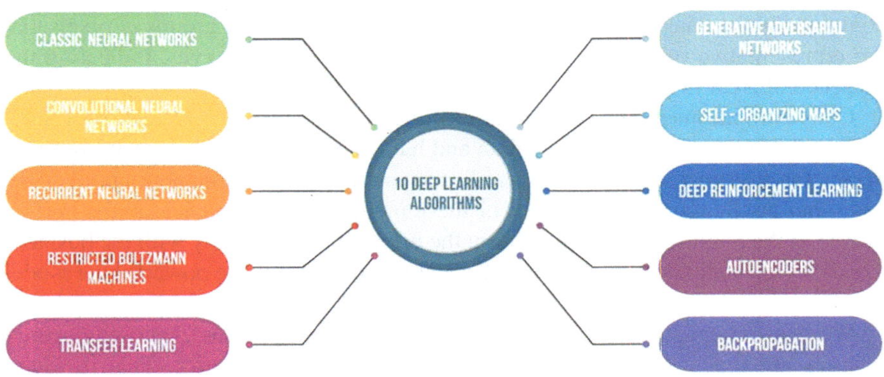

Fig. 3.12 Deep learning algorithms [15]

3.7.12.1 Type of Deep Learning Models

Deep learning algorithms are incredibly complex, and there are different types of neural networks to address specific problems or datasets as shown in Fig. 3.12. We will discuss a few of them

Artificial Neural Networks

Artificial neural networks (ANNs) are inspired by biological systems, such as the human brain, and how it processes information. Neural networks flourished in the mid-1980s due to their inherent parallel and distributed processing capability. However, research in this field was hindered by the ineffectiveness of the back-propagation training algorithm, which was widely used to optimize the parameters of neural networks. In recent years, new and improved training techniques such as unsupervised pre-training and layer-wise greedy training have led to a resurgence of interest in neural networks. Increasingly powerful computational capabilities, such as graphical processing units (GPUs) and massively parallel processing (MPPs), have also spurred the revival of neural networks. The resurgence of research in neural networks has given rise to the invention of models with thousands of layers; shallow neural networks have evolved into deep learning neural networks. Deep neural networks have been very successful for supervised learning. When used for speech and image recognition, deep learning algorithms perform exceptionally well, often surpassing even human capabilities. Applied to unsupervised learning tasks, such as feature extraction, deep learning also extracts features from raw images or speech with less or no human intervention.

With ANN, extremely complex models can be modeled and can be utilized as a kind of black box, without performing unpredictable complex feature engineering before training the model. Joined with the "deep approach," even more unpredictable models can be picked up to realize new possibilities, e.g., Object recognition has been as of late enormously enhanced utilizing Deep Neural Networks. Applied to unsupervised learning tasks, such as feature extraction, deep learning also extracts from raw images or speech with much less human intervention. ANNs also learn by example and through experience, and they are extremely useful for modeling nonlinear relationships in high-dimensional data or where the relationship amongst the input variables is difficult to understand. ANNs can be used for both regression and classification.

ANNs have interconnected processing elements that work in unison to solve specific problems. A neural network consists of three parts: input layer, hidden layer, and output layer, as shown in Fig. 3.13. The training samples define the input and output layers. Hidden layers between inputs and outputs are used in order to

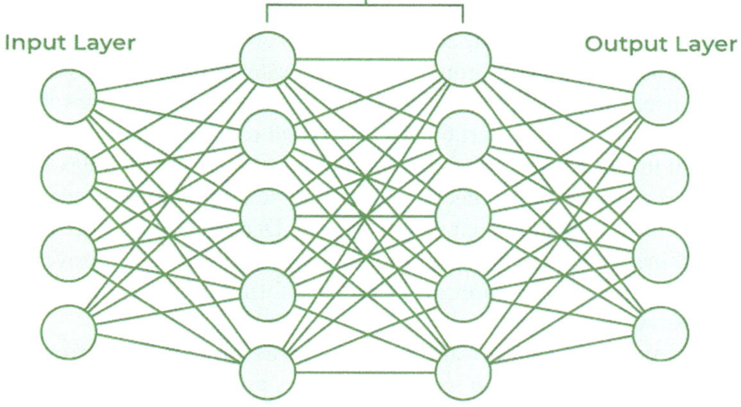

Fig. 3.13 A typical neural networks architecture

model an intermediary representation of the data that other algorithms cannot easily learn. The number of hidden layers determines the model's complexity and capacity.

ANN comprises "units" arranged in a series of layers, each of which connects to layers on either side. With ANN, extremely complex models can be modeled and can be utilized as a kind of black box, without playing out unpredictable complex feature engineering before training the model. Joined with the "deep approach," even more unpredictable models can be picked up to realize new possibilities. For example, object recognition based on geometry has been enhanced enormously by utilizing Deep Neural Networks. Applied to unsupervised learning tasks, such as feature extraction. Deep learning has also been used to extract from raw images or speech with much less human intervention. ANNs also learn by example and through experience, and they are extremely useful for modeling non-linear relationships in high-dimensional data or where the relationship amongst the input variables is difficult to understand. ANNs can be used for both regression as well as classification.

Deep Neural Network (DNN)

For the past decade, deep neural networks have been used in image recognition, speech, and even play games with high accuracy. The name Deep Neural evolved from the use of many layers making it a 'deep' network to learn more complex problems. The success stories of DL have only emerged in the last few years because training a network is computationally intensive and requires large amounts of data. The success of DL was facilitated by the increasing availability and affordability of faster computation and massive data storage.

A DNN is an ANN with multiple hidden layers between the input and output layers. DNN is a neural network with a certain level of complexity (having multiple hidden layers in between the input and output layers). They are capable of modeling and processing non-linear relationships through the activation function described in the earlier chapter. The DNN determines the correct mathematical operation to transform the input into the output, whether it involves a linear or non-linear relationship. The network moves through the layers, calculating the probability of each output. The user can review the results and select which probabilities the network should display (e.g., above a certain threshold) and return the proposed label. Each mathematical manipulation, as such is considered a layer, and complex DNNs have many layers, hence the name "deep" networks. The system must process layers of data between the input and output to solve a task. Deep neural network represents a type of machine learning where the system uses many layers to derive high-level functions from input information. It means transforming the data into a more creative and abstract component.

DNNs are typically feedforward networks in which data flows from the input layer to the output layer without looping back. A simplified version of DNN is represented as a hierarchical (layered) organization of neurons (similar to the neurons in the brain) with connections to other neurons. These neurons pass a signal to other

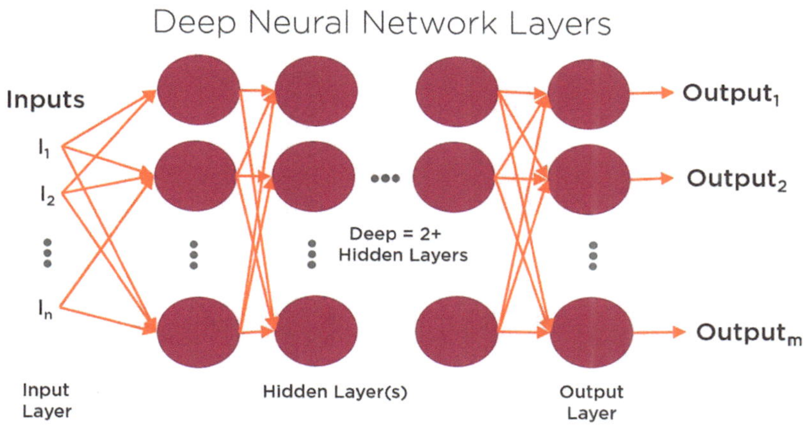

Fig. 3.14 Deep neural network architecture

neurons based on the received input and form a complex network that learns with a feedback mechanism. The input data is given to the neurons in the first layer also known as input layer which then provides its output to the neurons within next layer and so on till the final layer or the output layer. Each layer can have one or many neurons, and each of them will compute a small function, i.e., an activation function. If the incoming neurons result in a value greater than some threshold, the output is passed else ignored. The connection between two neurons of successive layers has an associated weight. The weight defined the strength of the input to the output for the next neuron and ultimately, to the overall final output. Initially weights would be all random, but during training, these weights are updated iteratively to learn to predict a correct output. Iterating the process several times step-by-step, with more and more data, helps the network update the weights appropriately to create a system where it can make a decision for predicting the output based on the rules. The prediction accuracy of a network depends on its weights and biases. A typical DNN can be defined by building blocks like a neuron, layers (input, hidden, and output), weights, an activation function, and a learning mechanism (back-propagation). Figure 3.14 represents a typical deep neural network.

DNNs are good at classification prediction problems using labeled data. They are flexible networks that can be applied to various scenarios.

Recurrent Neural Network (RNN)

A recurrent neural network is a class of artificial neural networks where connections between nodes form a directed graph along a temporal sequence. This allows it to exhibit temporal dynamic behavior. Derived from feedforward neural networks, RNNs can use their internal state (memory) to process variable-length sequences of

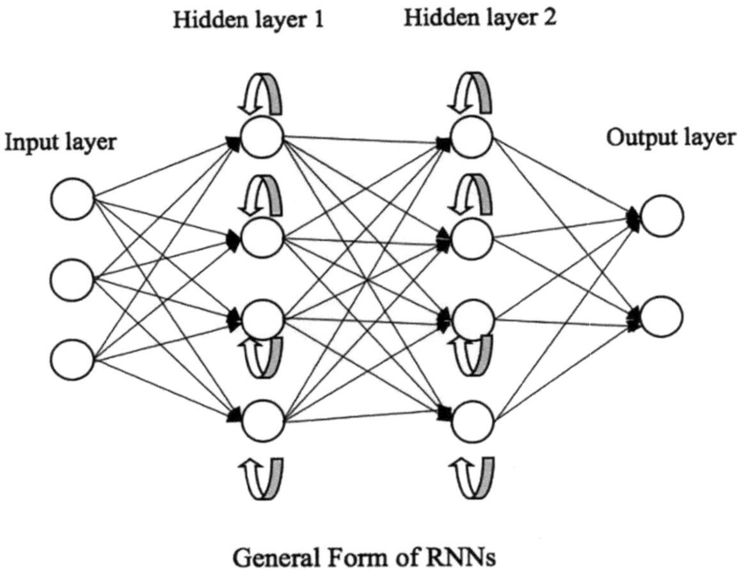

Fig. 3.15 Recurrent neural network

inputs [15]. This makes them applicable to tasks such as unsegmented, connected handwriting recognition [16] or speech recognition [17].

RNNs are neural networks in which data can flow in any direction. The basic concept is to utilize sequential information. In a normal neural network, it is assumed that all inputs and outputs are independent of each other. If we want to predict the next word in a sentence, we have to know which words come before it. RNNs are called recurrent as they repeat the same task for every element of a sequence, with the output being based on the previous computations. RNNs have a memory that captures information about what has been previously calculated. In theory, RNNs can utilize information from very long sequences, but in practice, they can only look back a few steps. A typical RNN is shown in Fig. 3.15.

A Recurrent Neural Network remembers the past, and its decisions are influenced by what it has learnt from the past. Note: Basic feedforward networks "remember" things, too, but they remember things they learned during training. For example, an image classifier learns what a "1" looks like during training and then uses that knowledge to classify things in production.

RNNs learn similarly while training; in addition, they remember things learnt from prior input(s) while generating output(s). RNNs can take one or more input vectors and produce one or more output vectors, and the output(s) are influenced not just by weights applied on inputs like a regular NN, but also by a "hidden" state vector representing the context based on prior input(s)/output(s). Therefore, the same input can produce a different output depending on the previous inputs in the series. Long short-term memory networks (LSTMs) are the most commonly used RNNs. The readers are referred to [17–19] for more details.

Recurrent neural networks are the mathematical engines to parse language patterns and sequenced data. They are used in time series data, e.g., natural language processing and speech recognition. They are also used in language translation, stock predictions, and algorithmic trading.

Long Short-Term Memory (LSTM)

Long Short-Term Memory (LSTM) is a type of artificial recurrent neural network (RNN) architecture used in the field of deep learning. Unlike standard feedforward neural networks, LSTMs have feedback connections, allowing them to exploit temporal dependencies across sequences of data. LSTM is designed to handle the issue of vanishing or exploding gradients, which can occur when training traditional RNNs on sequences of data. This makes them well-suited for tasks involving sequential data, such as natural language processing (NLP), speech recognition, and time series forecasting. LSTM networks improve on recurrent neural networks by selectively forgetting irrelevant information while retaining important details, making them practical for tasks requiring long-term context retention.

LSTM networks introduce memory cells, which can retain information over long sequences. Each memory cell has three main components: an input gate, a forget gate, and an output gate. These gates help regulate the flow of information in and out of the memory cell.

The input gate determines how much of the new input to store in the memory cell. It takes the current input and the previous hidden state as inputs, and outputs a value between 0 and 1 for each element of the memory cell.

The forget gate decides which information to discard from the memory cell. It takes the current input and the previous hidden state as inputs, and outputs a value between 0 and 1 for each element of the memory cell. A value of 0 indicates that the information is ignored, while a value of 1 indicates that it is retained.

The output gate controls how much of the memory cell's content should be used to compute the hidden state. It takes the current input and the previous hidden state as inputs, and outputs a value between 0 and 1 for each element of the memory cell.

By using these gates, LSTM networks can selectively store, update, and retrieve information over long sequences. This makes them particularly effective for tasks that require modeling long-term dependencies, such as speech recognition, language translation, and sentiment analysis.

LSTM models have demonstrated state-of-the-art performance across various tasks, including machine translation, speech recognition, and text summarization.

Convolution Neural Networks (CNN)

Convolutional Neural Networks, also known as CNNs or ConvNets, are a class of deep neural networks, most commonly applied to analyzing visual imagery. These networks have been some of the most influential innovations in the field of computer vision. 2012 was the first year that neural nets grew to prominence as Alex Krizhevsky used them to win that year's ImageNet competition, dropping the classification error record from 26% to 15%, an astounding improvement at the time [20]. Ever since then, a host of companies have been using the core of their services. They have applications in image and video recognition, recommender systems, medical image analysis, natural language processing, and time series. It is mostly applied to images because there is no need to check each pixel individually. CNN checks an image by blocks, starting from the upper left corner and moving pixel by pixel until it is successfully completed. Then the result of every verification is passed through a convolution layer, where data elements have connections while others don't. Based on this data, the system can produce the verification result and conclude what is in the picture.

Convolutional networks are simply neural networks that use convolution in place of general matrix multiplication in at least one of their layers. CNNs use relatively little pre-processing compared to other image classification algorithms. This means that the network learns from the filters that were previously hand-engineered in traditional algorithms. This independence from prior knowledge and human effort in feature design is a major advantage. CNNs are composed of five basic blocks:

- An input layer
- Convolution layer
- Nonlinear layer
- Pooling layer
- Fully connected layer

Convolution Layers extract features, patterns, structures, and characteristics from the input data by applying filters that detect specific patterns or characteristics.

Pooling Layers summarize and condense the information within specific regions of the input, helping to make the network more computationally efficient and robust to variations in the input.

Nonlinear Layers: Activation functions introduce non-linearity, the property of a function or system where the output is not directly proportional to the input, into the network, allowing it to learn complex relationships between the input and output. Every activation function takes a single number and performs a certain fixed mathematical operation on it. Figure 3.16 shows various activation functions (Table 3.1):

Fully Connected Layers make the final predictions based on the layered features, connecting every neuron in one layer to every neuron in the next layer.

By combining these components and stacking multiple layers together, CNNs can learn hierarchical representations of data and complex patterns.

The readers are referred to [21] for more details (Fig. 3.17).

Convolution neural networks are the image crunchers to identify objects. CNN image recognition is better in some scenarios than humans, and that ranges from cats to identifying vehicles, and fusion energy research.

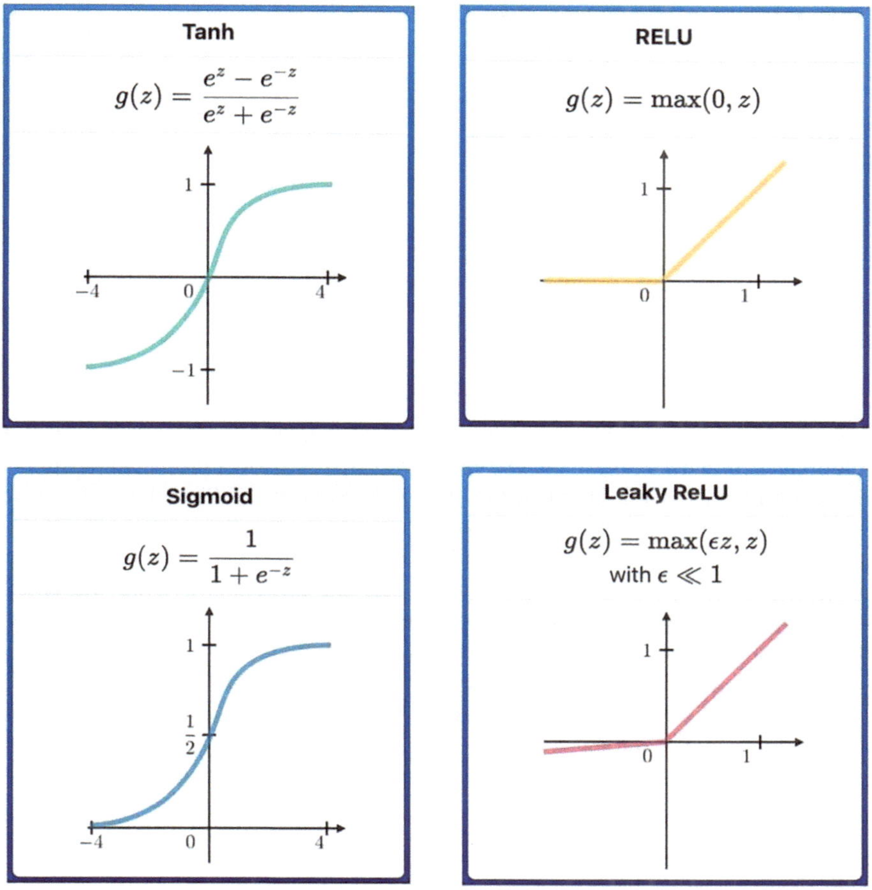

Fig. 3.16 Activation functions

Table 3.1 Activation functions

Activation function	Pros	Cons
Sigmoid	Used in the output layer for binary classification	Output ranges from 0 to 1
Tanh	Better than sigmoid	Updates parameters slowly when points are at extreme ends
ReLu	Updates parameters faster as slope is 1 when $x > 0$	Zero slope when $x < 0$

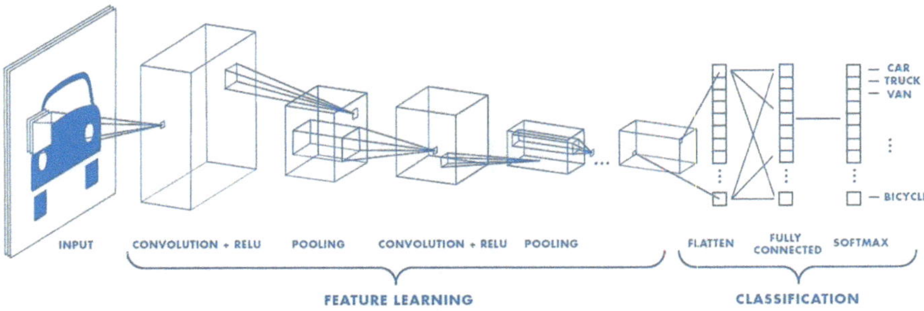

Fig. 3.17 Convolution neural network architecture

Types of Convolutional Neural Networks

Kunihiko Fukushima and Yann LeCun laid the foundation of research around convolutional neural networks in their work in 1980 [20] and "Backpropagation Applied to Handwritten Zip Code Recognition" in 1989, respectively. More famously, Yann LeCun successfully applied backpropagation to train neural networks to identify and recognize patterns within a series of handwritten zip codes. He would continue his research with his team throughout the 1990s, culminating with "LeNet-5", which applied the same principles of prior research to document recognition. Since then, several variant CNN architectures have emerged with the introduction of new datasets, such as MNIST and CIFAR-10, as well as competitions, including the ImageNet Large Scale Visual Recognition Challenge (ILSVRC). Some of these other architectures include:

- AlexNet
- VGGNet
- GoogLeNet
- ResNet

The reader is referred to the references [22–25] for details, as they are outside the scope of this book.

How to Choose a Network?

How to choose an appropriate network? One has to decide depending upon the problem being solved e.g., either building a classifier or trying to find patterns in the data. The following points should be considered while selecting a network [16]:

- For text processing, sentiment analysis, any language model that operates at the character level, parsing, and named entity recognition, we use a recurrent network.
- For image recognition, use a convolutional network.
- For speech recognition, we use a recurrent network.

In general, multilayer perceptrons or DNN with rectified linear activation function or ReLU are good choices for classification. For time series analysis, it is always recommended to use recurrent network.

Autoencoders and Variational Autoencoders

Deep learning made it possible to move beyond the analysis of numerical data, by adding the analysis of images, speech, and other complex data types. Among the first class of models to achieve this were variational autoencoders (VAEs). They were the first deep-learning models to be widely used for generating realistic images and speech, which empowered deep generative modeling by making models easier to scale, a cornerstone of what we think of as generative AI.

Autoencoders work by encoding unlabeled data into a compressed representation and then decoding the data back into its original form. *Plain* autoencoders were utilized for various purposes, including reconstructing corrupted or blurry images. *Variational* autoencoders added the critical ability not just to reconstruct data, but also to output variations on the original data.

This ability to generate novel data sparked a rapid succession of new technologies, from generative adversarial networks (GANs) to diffusion models, capable of producing increasingly realistic—but fake—images. In this way, VAEs set the stage for today's generative AI.

Autoencoders are built out of blocks of encoders and decoders, an architecture that also underpins today's large language models. Encoders compress a dataset into a dense representation, arranging similar data points closer together in an abstract space. Decoders sample from this space to create something new while preserving the dataset's most important features.

The biggest advantage of autoencoders is the ability to handle large batches of data and show input data in a compressed form, so the most significant aspects stand out—enabling anomaly detection and classification tasks. This also speeds transmission and reduces storage requirements. Autoencoders can be trained on unlabeled data so they might be used where labeled data is not available. When unsupervised training is used, there is a time savings advantage: deep learning algorithms learn automatically and gain accuracy without needing manual feature engineering. In addition, VAEs can generate new sample data for text or image generation.

There are disadvantages to autoencoders. The training of deep or intricate structures can be a drain on computational resources. And during unsupervised training, the model might overlook the needed properties and instead simply replicate the input data. Autoencoders might also overlook complex data linkages in structured data, so that it does not correctly identify complex relationships. The reader is referred to [26].

Generative Adversarial Networks (GANs)

Generative adversarial networks (GANs) are neural networks that are used both in and outside of artificial intelligence (AI) to create new data resembling the original training data. These can include images appearing to be human faces—but are generated, not taken of real people. The "adversarial" part of the name comes from the back-and-forth between the two portions of the GAN: a generator and a discriminator.

- The generator creates *something*: images, video, or audio, and then produces an output with a twist. For example, a horse can be transformed into a zebra with some degree of accuracy. The result depends on the input and how well-trained the layers are in the generative model for this use case.
- The discriminator is the adversary, where the *generative* result (fake image) is compared against the *real* images in the dataset. The discriminator tries to distinguish between the real and fake images, videos, or audio.

GANs train themselves. The generator creates fakes while the discriminator learns to spot the differences between the generator's fakes and the true examples. When the discriminator is able to flag the fake, then the generator is penalized. The feedback loop continues until the generator succeeds in producing output that the discriminator cannot distinguish.

The primary benefit of GAN is creating realistic output that can be difficult to distinguish from the originals, which in turn can be used to train machine learning models further. Setting up a GAN to learn is straightforward, since they are trained using unlabeled data or with minimal labeling. However, the potential disadvantage is that the generator and discriminator might go back and forth in competition for a long time, creating a significant system drain. One training limitation is that a considerable amount of input data might be required to obtain a satisfactory output. Another potential problem is "mode collapse," when the generator produces a limited set of outputs rather than a wider variety. The reader is referred to [26].

Transformer Models

Transformer models, which combine an encoder-decoder architecture with a text-processing mechanism, have revolutionized the training of language models. An encoder converts raw, unannotated text into representations known as embeddings; the decoder takes these embeddings together with previous outputs of the model and successively predicts each word in a sentence.

Using fill-in-the-blank guessing, the encoder learns how words and sentences relate to each other, building up a powerful representation of language without having to label parts of speech and other grammatical features. Transformers can be pretrained at the outset without a particular task in mind. After these powerful representations are learned, the models can later be specialized—with much less data—to perform a requested task.

Several innovations make this possible. Transformers process words in a sentence simultaneously, enabling parallel text processing and speeding up training. Earlier techniques, including recurrent neural networks (RNNs), processed words sequentially. Transformers also learned the positions of words and their relationships—this context enables them to infer meaning and disambiguate words such as "it" in long sentences.

By eliminating the need to define a task upfront, transformers made it practical to pretrain language models on vast amounts of raw text, enabling them to grow dramatically in size. Previously, labeled data was collected to train a single model on a specific task. With transformers, one model trained on a massive amount of data can be adapted to multiple tasks by fine-tuning it on a small amount of labeled task-specific data.

Language transformers are currently used for both non-generative tasks, such as classification and entity extraction, and generative tasks, including machine translation, summarization, and question answering. Transformers have surprised many people with their ability to generate convincing dialogue, essays, and other content.

Natural language processing (NLP) transformers offer remarkable power, as they can run in parallel, processing multiple portions of a sequence simultaneously, which significantly speeds up training. Transformers also track long-term dependencies in text, enabling them to understand the overall context more clearly and produce superior output. In addition, transformers are more scalable and flexible in order to be customized by task.

Regarding limitations, due to their complexity, transformers require substantial computational resources and a considerable amount of training time. Also, the training data must be accurate, on-target, unbiased, and plentiful to produce accurate results. The reader is referred to [26].

3.8 Summary

This chapter provided a comprehensive exploration of machine learning algorithm selection, focusing on regression and classification algorithms and their applicability to specific problems. It offered a balanced analysis of various machine learning algorithms, discussing their advantages and limitations. The chapter then delved into deep learning algorithms, covering both basic architectures and more complex models, such as Convolutional Neural Networks (CNN), Recurrent Neural Networks (RNN), and Long Short-Term Memory (LSTM) networks. It addressed the intricacies and challenges associated with these advanced algorithms. Furthermore, the chapter outlined strategies for developing algorithm architectures that aligned with the entire machine learning lifecycle, including data collection, algorithm development, integration, deployment, and MLOps activities. This comprehensive approach ensured that readers gained a thorough understanding of not only the technical aspects of machine learning algorithms but also their practical implementation and optimization in real-world scenarios. By mastering these skills, one can be at the forefront of developing powerful models with a wide range of real-world problems using machine learning algorithms.

3.9 Points to Ponder

1. When should you use classification over regression?
2. What is the difference between supervised and unsupervised learning?
3. What are some supervised machine learning applications?
4. What is the difference between a discriminative and a generative model?
5. How would you prune a decision tree?
6. How would you choose a machine learning algorithm for a particular dataset?

3.10 Answers

1. When should you use classification over regression?
   One should choose classification over regression if the output is expected to belong to specific categories.
2. What is the difference between supervised and unsupervised learning?
   The difference between supervised and unsupervised learning is the way labeled data is treated. Unsupervised learning does not require labeled data, whereas supervised learning does.
3. What are some supervised machine learning applications?
   Some examples are:
   - Fraud detection
   - Sentiment analysis
   - Stock price prediction
   - Health diagnostics
4. What is the difference between a discriminative and a generative model?
   A discriminative model learns the difference between data categories while a generative model learns data categories.
5. How would you prune a decision tree?
   Pruning a decision tree refers to the process of removing branches with weak predictive power. This simplifies the model, increases predictive accuracy, and avoids overfitting.
6. How would you choose a machine learning algorithm for a particular dataset?
   There is no perfect algorithm that works for every situation. One will choose an algorithm using the following question:
   - What is the organization's goal?
   - Is the data labeled, unlabeled, or mixed?
   - Does the problem relate to clustering, regression, or classification?
   - Is the data categorical or continuous?

References

1. Chappell D (2015) Introducing azure machine learning: a guide for technical professionals
2. https://www.knowledgehut.com/blog/data-science/machine-learning-algorithms
3. https://medium.com/@Zelros/a-brief-history-of-machine-learning-models-explainability-f1c3301be9dc
4. https://hackernoon.com/choosing-the-right-machine-learning-algorithm-68126944ce1f
5. https://medium.com/@aravanshad/how-to-choose-machine-learning-algorithms-9a92a448e0df
6. http://blog.echen.me/2011/04/27/choosing-a-machine-learning-classifier/
7. Mitchell TM (1997) Machine learning. McGraw-Hill International
8. Bishop CM (2006) Pattern recognition and machine learning. Springer
9. James G, Witten D, Hastie T, Tibshirani R (2013) An introduction to statistical learning, with application in R. Springer, New York
10. https://www.cxtoday.com/data-analytics/gartner-magic-quadrant-for-analytics-and-business-intelligence-platforms-2023/
11. https://medium.com/@zxr.nju/what-is-the-kernel-trick-why-it-is-imporant-98a98db0961d
12. https://data-flair.training/blogs/svm-kernel/functions/
13. https://www.snowflake.com/guides/machine-learning-platforms
14. https://www.analyticsvidhya.com/blog/2021/09/adaboost-algorithm-a-complete-guide-for-beginners/
15. https://en.wikipedia.org/wiki/Deep_learning
16. Alex G, Liwicki M, Fernandez S, Bertolami R, Bunke H, Schmidhuber J (2009) A novel connectionist system for improved unconstrained handwriting recognition. IEEE Trans Pattern Anal Mach Intell 31(5):855–868
17. Li X, Wu X (2014) Constructing long short-term memory based deep recurrent neural networks for large vocabulary speech recognition
18. Bengio Y, LeCun Y, Hinton G (2015) Deep learning. Nature 521(7553):436–444
19. Goodfellow I, Bengio Y, Cournville A (2016) Deep learning
20. Fukushima K (1980) Neocognitron: a self-organizing neural network model for a mechanism of pattern recognition unaffected by shift in position. Biol Cybern 36:193–202
21. https://miro.medium.com/max/875/1*NGPAHYYqs6yRUhMj2BbLaw.jpeg
22. Krizhevsky A, Sutskever I, Hinton GE (2012) Imagenet classification with deep convolution neural networks. In: Advances in neural information processing systems, pp 1097–1105
23. Simonyan K, Zisserman A. Very deep convolutional networks for large-scale image recognition, 1025, ICLR. https://arxiv.org/pdf/1409.1556
24. Christian et al (2015) Going deeper with convolutions. CVPR2015
25. He K et al. Deep residual learning for image recognition. https://arxiv.org/pdf/1512.03385v1
26. https://www.ibm.com/think/topics/deep-learning

Chapter 4
Performance Metrics and Training of AI/ML Clinical Model

Abstract This chapter provides a comprehensive exploration of performance metrics for various machine learning algorithms, including supervised, unsupervised, generative AI, NLP, and reinforcement learning. It delves into the selection of appropriate metrics and addresses critical issues such as overfitting, underfitting, outlier handling, confusion metrics, and monotonicity. The chapter also covers best practices for dataset splitting, cross-validation techniques, and balancing bias and variance. Recent developments in RAG for LLMs and the Hallucination Index Report for foundation models are discussed. An in-depth examination of training methodologies for different machine learning models is presented, along with hyperparameter tuning strategies. The chapter explores techniques such as early stopping and regularization to enhance model generalizability, as well as model pruning and optimization of computational cost.

Keywords Supervised · Unsupervised · Generative AI · Large language models (LLMs) · Natural language processing (NLP) · Reinforcement learning · Classification · Regression · Deep learning · Bias · Variance · Cross validation · RAG · Hallucination rate · Hallucination index report · Dataset splitting · Overfitting · Underfitting · Early stopping · Droput · Regularization · Computational cost · AI/ML-enabled medical devices · AI/ML-enabled medical device performance · Training · Evaluation · Performance metrics · Evaluation metrics · Segmentation quality · Dimensionality reducation · Deep learning · multimodal

4.1 Machine Learning Algorithms

Machine learning algorithms can be broadly categorized into three main types.

A. Pandey et al., *Artificial Intelligence in Medical Software*,
https://doi.org/10.1007/978-3-032-10808-1_4

4.1.1 Supervised Machine Learning

Supervised learning [1] is a fundamental approach in machine learning where algorithms are trained using labeled data to predict outcomes or classify information. These algorithms work by mapping input data to known output labels during the training process. The key characteristics of supervised learning include the use of labeled training data, where each input is paired with its corresponding correct output. This allows the algorithm to learn from these input-output pairs, gradually adjusting its parameters to minimize prediction errors. The primary goal of supervised learning is to develop models that can accurately predict outcomes on new, unseen data. This approach is efficient in scenarios with a clear relationship between inputs and outputs, and a large amount of labeled data is available for training. Supervised learning forms the basis for many practical applications in fields such as image recognition, natural language processing, and predictive analytics.

Types of Supervised Machine Learning

- Classification (categorical predictions)
- Regression (continuous value predictions)

Common Supervised Algorithms

- Linear regression
- Logistic regression
- Decision trees
- Random forest
- Support vector machines
- Neural networks
- K-nearest neighbors

4.1.2 Unsupervised Machine Learning

Unsupervised learning [2] is a sophisticated machine learning approach that discovers hidden patterns and structures within unlabeled datasets without predefined outputs or target variables. Unlike supervised learning, these algorithms work directly with raw, untagged data, seeking to uncover inherent relationships, groupings, and underlying organizational principles within complex datasets. The key characteristics of unsupervised learning include its ability to process unlabeled data autonomously and identify intrinsic structures and connections that might not be immediately apparent to human observers. By exploring data without explicit guidance, unsupervised learning algorithms can reveal unexpected insights, cluster similar data points, reduce dimensionality, and transform complex datasets into more manageable and interpretable representations. This approach is particularly valuable when ground truth labels are unavailable or expensive. It is a powerful technique for exploratory data analysis, anomaly detection, and understanding complex systems across various domains, including customer segmentation, scientific research, and pattern recognition.

However, unsupervised learning techniques can be used for tasks that are related to both regression and classification:

- **Clustering:** This is one of the primary tasks in unsupervised learning, where algorithms group similar data points. While not a classification in the traditional sense, clustering can be seen as an automatic categorization [3].
- **Dimensionality Reduction:** This technique simplifies data while preserving meaningful relationships, which can be helpful as a preprocessing step for regression and classification tasks [4].
- **Association:** This involves discovering relationships between variables in large datasets, which can inform regression and classification problems [5].

It is important to note that while unsupervised learning doesn't directly perform regression or classification, its outputs can be used to inform or enhance supervised learning tasks. For example, a supervised classification model could use cluster assignments from an unsupervised algorithm as features. In summary, unsupervised learning is distinct from regression and classification, which are typically supervised learning tasks. However, unsupervised techniques can complement and support regression and classification problems in various ways.

Common Unsupervised Algorithms

- K-means
- Hierarchical clustering
- Principal component analysis
- Self-organizing maps

4.1.3 Reinforcement Machine Learning

Reinforcement learning [6] is a dynamic and innovative branch of machine learning that focuses on training algorithms through interaction with an environment. In this approach, an agent learns to make decisions by performing actions and receiving feedback through rewards or penalties. The key characteristics of reinforcement learning include its emphasis on learning through trial and error, where the agent explores different strategies to maximize cumulative rewards over time. This method is particularly adept at developing sophisticated decision-making strategies, as the agent must balance immediate rewards with long-term benefits. Unlike supervised or unsupervised learning, reinforcement learning simulates a more natural learning process, akin to how humans and animals learn from experience. This makes it especially suitable for complex, sequential decision-making tasks in uncertain environments. Reinforcement learning has found significant applications in robotics, game playing, autonomous systems, and resource management, where adaptive and optimized decision-making is crucial.

Applications of Reinforcement Learning

- Robotics
- Game playing
- Autonomous systems
- Resource management

4.1.4 Semi-Supervised Machine Learning

Semi-supervised learning [7] is an innovative approach that bridges the gap between supervised and unsupervised learning techniques. This method leverages both labeled and unlabeled data during training, providing a powerful solution for scenarios where obtaining fully labeled datasets is challenging or costly. By combining the strengths of both approaches, semi-supervised learning can improve model performance and generalization, especially when labeled data is scarce. It allows algorithms to learn from the structure of unlabeled data while still benefiting from the guidance provided by labeled examples.

4.1.5 Deep Learning

Deep learning [8], on the other hand, represents a cutting-edge subset of machine learning based on artificial neural networks with multiple layers. These advanced algorithms can learn complex patterns and representations across various levels of abstraction. Deep learning models, such as convolutional neural networks (CNNs) and recurrent neural networks (RNNs), can automatically extract features from raw data, thereby eliminating the need for manual feature engineering. This ability to learn hierarchical representations makes deep learning particularly effective in handling high-dimensional data, such as images, speech, and text. The multi-layered structure of deep learning models enables them to capture intricate relationships and nuances in data, resulting in breakthrough performances in various domains, including computer vision, natural language processing, and speech recognition.

4.1.6 Ensemble Machine Learning

Ensemble machine learning [9] models combine multiple individual models to produce more accurate and robust predictions than any single model alone. This approach leverages the strengths of various algorithms while mitigating their weaknesses. Ensemble methods typically employ bagging, boosting, or stacking techniques to aggregate predictions from diverse base models, often referred to as weak learners. By combining these models, ensembles can reduce overfitting, decrease variance, and improve generalization on unseen data. Popular ensemble methods include random forests, gradient-boosting machines, and voting classifiers. These techniques have proven highly effective in various machine learning tasks, including classification, regression, and anomaly detection, often outperforming single models in competitions and real-world applications.

Each type of machine learning algorithm has its strengths and is suited for different problems and datasets. The choice of algorithm depends on the specific task, available data, and desired outcomes.

4.2 AI/ML Algorithm for Medical Device Intended Use

The FDA's approach to accepting machine learning model results for medical device intended use is multifaceted and evolving, emphasizing a risk-based evaluation that considers the specific context and potential impact of each AI/ML device. Rather than imposing a universal accuracy threshold, the FDA assesses whether performance shortfalls could adversely affect patients, potentially requiring higher accuracy for devices with more severe consequences [10, 11]. This evaluation extends beyond simple accuracy metrics to include sensitivity, specificity, and predictive values, ensuring comprehensive performance assessment across relevant metrics for the intended use [12].

For 510(k) clearance, AI/ML devices must demonstrate "substantial equivalence" to legally marketed devices, with performance comparable to or surpassing existing solutions [13]. Clinical validation in representative patient populations is crucial, as is the implementation of continuous monitoring and update plans, such as Predetermined Change Control Plans (PCCPs), to manage the device's evolution in real-world settings [13]. The FDA emphasizes transparency, explainability, and adherence to Good Machine Learning Practices (GMLP) throughout the product lifecycle [10].

Medical device manufacturers are encouraged to engage with the FDA early in the development process, thoroughly document their work, and prepare comprehensive regulatory submissions addressing all relevant guidance [11, 13]. As the field rapidly advances, staying informed about the latest FDA guidelines and best practices is crucial for successfully navigating the regulatory landscape and ensuring the safe and effective integration of AI/ML devices into healthcare [14, 15].

4.3 Performance Metrics for AI/ML-Enabled Medical Devices

Developing standardized performance metrics for AI/ML-enabled Medical Devices requires a comprehensive and thoughtful approach. Best practices in this domain emphasize tailoring metrics to specific problems and dataset characteristics, recognizing that a one-size-fits-all approach is often inadequate [15, 16]. Using multiple complementary metrics is crucial for obtaining a holistic view of performance, particularly in tasks involving regression or classification [17].

Assessing generalizability across diverse patient populations and clinical settings is crucial, as is utilizing external validation datasets to evaluate real-world performance. Measuring bias and fairness ensures equitable performance across all relevant subpopulations [18]. Incorporating clinical relevance by collaborating with domain experts to define meaningful performance thresholds is crucial for aligning metrics with patient outcomes and informed clinical decision-making. Evaluating reliability, consistency, and data quality using specialized metrics helps ensure robust performance over time and across different environments [19].

Consideration of model interpretability is increasingly important, with metrics needed to quantify the explainability of model decisions. Leveraging standardized datasets and benchmarks, when available, enables meaningful comparisons across different models [20, 21]. Alignment with regulatory guidelines, such as those from the FDA, is crucial for compliance. Finally, enabling continuous monitoring through automated performance tracking systems supports ongoing assessment of real-world performance [13, 22]. By adhering to these practices, Medical Device manufacturers can create more robust, meaningful, and compliant performance metrics for AI/ML-enabled medical devices, facilitating better evaluation and comparison of these innovative medical devices.

4.4 Performance Metrics for Supervised Machine Learning Models

When evaluating supervised machine learning models, using multiple metrics is often beneficial as each provides different insights into model performance. The choice of metrics depends on the specific problem, the nature of the data, and the relative importance of different types of errors in your application. Here are some key performance metrics for supervised machine learning models:

4.4.1 *Classification Metrics*

Multiple metrics are often beneficial when evaluating classification models, as each provides different insights into model performance. The choice of metrics can depend on the specific problem, the class distribution, and the relative importance of various errors in your application.

4.4.1.1 Accuracy

Accuracy measures the overall correctness of the model's predictions. While easy to interpret, accuracy can be misleading for imbalanced datasets.

$$\text{Accuracy} = \{\text{TP} + \text{TN}\} / \{\text{TP} + \text{TN} + \text{FP} + \text{FN}\}$$

where

- TP = True Positives
- TN = True Negatives
- FP = False Positives
- FN = False Negatives

4.4.1.2 Precision (Positive Predictive Value)

Precision measures the accuracy of positive predictions. It's useful when the cost of false positives is high. How many of all the instances that were predicted as positive were positive? High Precision indicates that the model has a low rate of false positives. Precision is critical in scenarios where false positives carry a high cost (e.g., spam detection) [23, 24].

$$\text{Precision} = \left(\begin{matrix}\text{Correctly classified actual positives /}\\ \text{everything classified positive}\end{matrix}\right) = \{\text{TP}\} / \{\text{TP} + \text{FP}\}$$

4.4.1.3 Recall (Sensitivity, True Positive Rate)

Recall measures the proportion of actual positives correctly identified. It's important when the cost of false negatives is high. Recall (also known as sensitivity, true positive rate, or probability of detection) measures the ability of a model to identify all relevant instances. It is defined mathematically as:

$$\begin{aligned}\text{Recall} &= (\text{Correctly classified actual positives / All actual positives})\\ &= \{\text{TP}\} / \{\text{TP} + \text{FN}\}\end{aligned}$$

Sensitivity is often used in medical contexts to reflect how effectively a test can identify patients with a specific condition. In this sense, sensitivity emphasizes the importance of detecting true positives, which aligns directly with the definition of recall in machine learning [24, 25]. Precision improves as false positives decrease, while recall improves when false negatives decrease. However, as seen in the previous section, increasing the classification threshold tends to reduce the number of false positives and increase the number of false negatives, while decreasing the threshold has the opposite effect. As a result, Precision and recall often show an inverse relationship, where improving one of them worsens the other [24].

4.4.1.4 Specificity (False Positive Rate, Probability of False Alarm)

A false positive rate, also known as specificity, corresponds to the proportion of negative data points that are mistakenly considered positive among all negative data points. Please note that both FPR and TPR have values from 0 to 1. Specificity is crucial in contexts where identifying true negatives is essential (e.g., medical tests) [26].

$$\begin{aligned}\text{Specificity} &= (\text{incorrectly classified actual negatives} / \text{all actual negatives}) \\ &= \text{FP} / \text{FP} + \text{TN}\end{aligned}$$

A perfect model would have zero false positives and, therefore, an FPR of 0.0, which is to say, a 0% false alarm rate. In an imbalanced dataset where the number of actual negatives is very low, say 1–2 examples, FPR is less meaningful and less useful as a metric [24].

4.4.1.5 F1 Score

The F1 score is the harmonic mean of Precision and recall. It provides a single score that balances both Precision and recall. It is used to measure the accuracy of tests and directly indicates the model's performance. The range of the F1 score is between 0 and 1, with the goal being to get as close as possible to 1. It can be a better measure if we seek a balance between Precision and recall if there is a class imbalance (many Actual Negatives and fewer Actual positives). Class imbalance is always present in real-life situations; hence, it is always better to use an F1 score over accuracy [26].

$$\text{F1} = 2^{*}\{\text{Precision}^{*}\text{Recall}\} / \{\text{Precision} + \text{Recall}\}$$

4.4.1.6 Confusion Matrix

While not a single metric, the confusion matrix for a binary classification provides a tabular summary of the model's performance, as shown in Fig. 4.1.

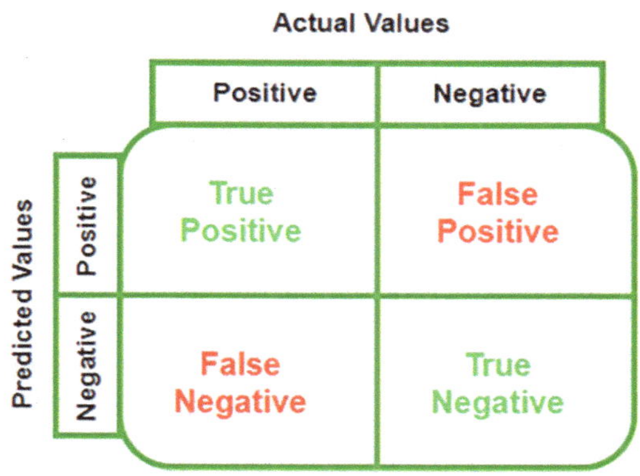

Fig. 4.1 Confusion matrix [27]

A confusion matrix for a multiclass problem is a powerful tool for evaluating the performance of a classification model. It provides a detailed breakdown of how well the model's predictions match the class labels. Here's an overview of constructing and interpreting a multiclass confusion matrix, examples, and calculations for Precision, recall, and F1 score.

A multiclass confusion matrix is structured as a square matrix where:

- Rows represent the actual classes.
- Columns represent the predicted classes.
- Diagonal elements: correctly classified instances.
- Off-diagonal elements: misclassifications.

Each cell in the matrix indicates the number of instances that belong to a specific actual class and were predicted to belong to a particular predicted class. The diagonal elements represent correct predictions (True Positives), while off-diagonal elements indicate misclassifications.

4.4.1.7 Log Loss

For probabilistic classifiers, log loss (also called cross-entropy loss) [28] measures the performance of a model where the prediction is a probability value between 0 and 1. The Log Loss formula penalizes incorrect predictions heavily, primarily when a low probability is assigned to the true class.

For a binary classification:

$$\log \text{Loss} = -\left(y \cdot \log(p) + (1-y) \cdot \log(1-p)\right)$$

where

- y: actual label (0 or 1)
- p: predicted probability of the positive class

For multiclass classification problems with C classes, where y_i is a binary indicator (0 or 1) if class label i is the correct classification for observation j, and p_{ij} is the predicted probability that observation j belongs to class i, the log loss is extended as follows:

$$\text{logLoss} = -\frac{1}{N}\sum_{j=1}^{N}\sum_{i=1}^{C} y_{ij} \cdot \log\left(p_{ij}\right)$$

where

- N: Total number of observations
- C: Total number of classes
- y_{ij}: Binary indicator (1 if observation j belongs to class i, otherwise 0)
- p_{ij}: The predicted probability that observation jj belongs to class ii

Lower log loss indicates better performance. It functions by penalizing all false classifications. When working with this metric, the classifier must assign a specific probability to each class for every sample. In a nutshell, the range of log loss varies from 0 to infinity (∞). The closer it is to 0, the higher the prediction accuracy. Minimizing Log Loss is a top priority.

4.4.1.8 Detection Rate

This metric represents the proportion of correct positive class predictions among all predictions.

$$\text{Detection Rate} = \text{TP} / \text{TP} + \text{FP} + \text{FN} + \text{TN}$$

4.4.1.9 Receiver Operating Characteristic Curve (ROC)/Area Under Curve (AUC) Score

The ROC curve [29] is a graph that displays the performance of the classification model at all thresholds. As the name suggests, the AUC, as shown in Fig. 4.2, is the entire area below the two-dimensional area below the ROC curve. This curve generates two important metrics: sensitivity and specificity.

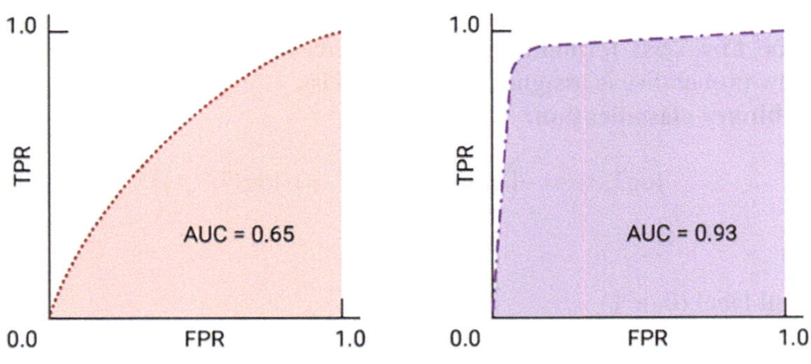

Fig. 4.2 ROC and AUC of two hypothetical models. The curve on the right, with a greater AUC, represents the better of the two models [29]

AUC-ROC measures the model's ability to distinguish between classes across all possible thresholds. A higher AUC indicates better performance, with a value of 1.0 being perfect and 0.5 indicating no better than random guessing.

4.4.2 *Regression Metrics*

Multiple metrics are often beneficial when evaluating regression models, as each provides different insights into model performance. The choice of metrics can depend on the specific problem, the scale of the target variable, and the importance of various types of errors in your application [30–32].

4.4.2.1 Mean Absolute Error (MAE)

This is calculated by taking the absolute difference between the predicted and actual values and then the average across all samples. MAE is less sensitive to outliers than **Mean Squared** and **Root Mean Squared Error**. This metric is popular among data scientists and machine learning engineers because it's easy to interpret—how of a regression model is on average. Below is the mathematical formula:

$$\text{MAE} = \frac{1}{n}\sum_{i=1}^{n}\left|y_i - \hat{y}_i\right|$$

4.4.2.2 Mean Squared Error (MSE)

Mean Squared Error is one of the most commonly used metrics for regression problems. The average of squared differences between predicted and actual values. MSE penalizes larger errors more heavily due to the squaring. A lower MSE indicates better model performance. For this reason, MSE is the most commonly used loss function in machine learning. Remember that MSE returns the error in units squared, so it's a bit trickier to interpret. Below is the mathematical formula:

$$\text{MSE} = \frac{1}{n}\sum_{i=1}^{n}\left(y_i - \hat{y}_i\right)^2$$

4.4.2.3 Root Mean Squared Error (RMSE)

The square root of MSE provides a metric in the same unit as the target variable. Like MSE, a lower RMSE indicates better model performance. The Root Mean Squared Error metric does everything MSE does, making the error more interpretable by bringing the value to the original units. RMSE still penalizes larger errors like MSE, but takes the additional step of returning the error values to the original unit. Here is the mathematical formula:

$$\text{RMSE} = \sqrt{\frac{1}{n}\sum_{i=1}^{n}\left(y_i - \hat{y}_i\right)^2}$$

4.4.2.4 Root Mean Squared Log Error (RMSLE)

The Root Mean Squared Log Error metric is similar to RMSE, but it takes the logarithm of the predicted and actual values before calculating the squared difference. It is often used when the target variable has a wide range of values. The RMSLE function first takes the natural logarithm of the predicted and actual values and then calculates the squared differences between them. Once done, the RMSE is calculated based on the logarithmic difference. Here is the mathematical formula:

$$\text{RMSLE} = \sqrt{\frac{1}{n}\sum_{i=1}^{n}\left(\log\left(\hat{y}_i + 1\right) - \log\left(y_i + 1\right)\right)^2}$$

4.4.2.5 R-Squared (Coefficient of Determination)

Measures the proportion of variance in the dependent variable explained by the independent variables. R^2 ranges from 0 to 1, with 1 indicating perfect prediction and zero indicating that the model predicts no better than a horizontal line. The *R*-squared metric, also known as the coefficient of determination, measures the extent to which the machine learning model explains the target variable. *R*-squared is calculated by dividing the variance of the predicted values by the variance of the actual values. Here is the mathematical formula:

$$R^2 = 1 - \frac{\sum_{i=1}^{n}\left(y_i - \hat{y}i\right)^2}{\sum_{i=1}^{n}\left(y_i - \overline{y}\right)^2}$$

4.4.2.6 Mean Absolute Percentage Error (MAPE)

The average of absolute percentage differences between predicted and actual values is known as MAPE. MAPE is scale-independent, making it helpful in comparing models across different scales. MAPE is calculated by taking the absolute difference between the predicted and actual values, dividing these by the actual values, and finally, taking the average across all samples. The MAPE metric is often used when the target variable is expressed as a percentage with a non-zero mean. Below is the mathematical formula:

$$\text{MAPE} = \frac{1}{n}\sum_{i=1}^{n}\left|\frac{y_i - \hat{y}_i}{y_i}\right| \times 100\%$$

where

- n is the number of samples
- y_i is the actual value
- $\hat{y}_i$ is the predicted value

4.4.2.7 Symmetric Mean Absolute Percentage Error (SMAPE)

The Symmetric Mean Absolute Percentage Error is quite like MAPE but can handle zero values. SMAPE is calculated by first taking the absolute value of the difference between the predicted and actual values, then dividing the sum of the predicted and actual values, and finally multiplying it by 2 and taking the average across all samples. This metric is helpful for all cases where MAPE is used and when the target variable can be zero. Below is the mathematical formula:

$$\text{SMAPE} = \frac{100\%}{n}\sum_{i=1}^{n}\frac{\left|y_i - \hat{y}_i\right|}{\left(\left|y_i\right| + \left|\hat{y}_i\right|\right)/2}$$

4.4.2.8 Mean Directional Accuracy (MDA)

The Mean Directional Accuracy metric is unique, as it measures the percentage of correct predictions regarding the direction of change (up or down). MDA is calculated by counting the times the predicted value has the same sign as the actual value, divided by the total number of samples. Below is the mathematical formula:

$$\text{MDA} = \frac{1}{n}\sum_{i=1}^{n}\left[sign\left(y_i - y_{i-1}\right) = sign\left(\hat{y}_i - y_{i-1}\right)\right]$$

4.4.2.9 Median Absolute Error (MedAE)

MedAE is identical to MAE but calculates the median over the mean. For this reason, MedAE is less sensitive to outliers. Below is the formula:

$$\text{MedAE} = \text{median}\left(\left|y_1 - \hat{y}_1\right|, \left|y_2 - \hat{y}_2\right|, \ldots, \left|y_n - \hat{y}_n\right|\right)$$

4.4.2.10 Explained Variance Score (EVS)

This metric measures the proportion of variance in the target variable that the model explains. EVS is calculated by taking the difference between the variance of the actual values and the variance of the residuals and then dividing it by the variance of the actual values. The values for this metric range from 0 to 1, where 1 indicates a perfect fit to the data. Below is the formula:

$$\text{EVS} = 1 - \frac{\text{Var}(y - \hat{y})}{\text{Var}(y)}$$

4.4.3 Selection of Regression Performance Metrics

When evaluating regression models, several common pitfalls should be considered when using performance metrics. Understanding these can help data scientists and analysts interpret results more accurately and make better model selection and improvement decisions. It is crucial to adopt a multifaceted approach when evaluating model performance. This involves using a combination of metrics to gain a well-rounded view and understanding of each metric's assumptions and limitations. Additionally, it is essential to consider the specific requirements and constraints of your problem domain. When comparing models, caution is necessary, especially when dealing with different datasets or scales. Always interpret the results within the context of the data and the problem you are trying to solve. By being aware of these common pitfalls and taking a thoughtful, comprehensive approach to model evaluation, data scientists can make more informed decisions and develop more robust and reliable regression models.

- **Overreliance on a Single Metric:** One of the most common pitfalls is relying too heavily on a single metric to evaluate model performance. Each metric has strengths and weaknesses, and using multiple metrics provides a more comprehensive view of model performance [32]. R-squared alone can be misleading, particularly when comparing models with different numbers of predictors, as it does not account for the model's complexity. Additionally, metrics such as Root Mean Square Error (RMSE) and Mean Square Error (MSE) are sensitive to outliers, which can distort their values and may not accurately reflect the model's overall performance. Therefore, relying solely on these metrics can lead to incorrect conclusions about a model's effectiveness.
- **Misinterpretation of *R*-squared:** *R*-squared is a widely used statistic, but is often misunderstood. A high *R*-squared doesn't necessarily mean the model is good or accurate. It only indicates how much variance in the dependent variable is explained by the independent variables [33]. *R*-squared will always increase when adding more variables, even if they don't improve the model's predictive power. This can lead to overfitting if not carefully monitored.

- **Sensitivity to Outliers:** Some metrics are more sensitive to outliers than others. MSE and RMSE heavily penalize significant errors due to squaring, making them particularly sensitive to outliers [32]. MAE is less affected by outliers, which can be both an advantage and a disadvantage depending on the specific problem and dataset.
- **Scale Dependence:** Not all metrics are scale-independent. MSE, RMSE, and MAE are scale-dependent, meaning their values are in the same units as the target variable. This makes it difficult to compare across different datasets or target variables with varying scales [32]. MAPE, being a percentage, is scale-independent; however, it can be problematic when dealing with values close to or equal to zero in the actual data [32].
- **Assumption of Error Distribution**: Some metrics make assumptions about the distribution of errors. RMSE assumes that errors are normally distributed. RMSE may not be the most appropriate metric if this assumption is violated. R-squared assumes a linear relationship between variables, which may not hold for all datasets [32].
- **Lack of Interpretability:** Certain metrics can be challenging to interpret, especially for non-technical stakeholders. While MSE is mathematically useful, its squared units make it less intuitive to explain than RMSE or MAE. Complex metrics, such as adjusted R-squared, may be more challenging to interpret than simpler ones, particularly for non-technical audiences [33].
- **Ignoring Domain-Specific Considerations:** Generic metrics may not always align with domain-specific needs. In some fields, underestimation might be more problematic than overestimation (or vice versa), but symmetric metrics like MSE or MAE don't account for this. The practical significance of errors may vary depending on the specific application, which is not captured by standard metrics.

4.4.4 Overfitting in Machine Learning

Good performance evaluation metrics are just one part of the equation; monitoring for overfitting issues is also crucial, as they can often go unnoticed. Overfitting occurs when a model is excessively tailored to its training dataset, including the random noise inherent in that data, leading to poor predictive performance on new data points. To mitigate overfitting, a common strategy is to split the data into training and test sets, typically using an 80/20 ratio, where 80 percent of the data is used for training and 20 percent for testing. This allows you to build the model with the training set and evaluate its performance with the test set. Conversely, underfitting happens when the model fails to capture the underlying correlations in the training data, and it is generally easier to detect; performance metrics will immediately reflect this issue. Ultimately, monitoring your classification models is essential to ensure you are not falling into the traps of overfitting or underfitting [34, 35]. Figure 4.3 contains a pictorial representation of overfitting and underfitting.

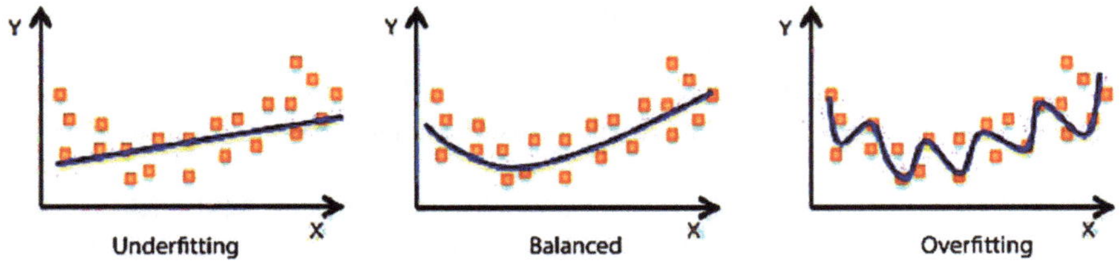

Fig. 4.3 Pictorial representation of overfitting and underfitting [34]

4.4.5 Outliers Affect the Performance Metrics in Regression Models

Outliers can significantly impact the performance metrics used to evaluate regression models. Understanding these effects is crucial for accurately assessing model performance and making informed decisions.

Mean Squared Error (MSE) and Root Mean Squared Error (RMSE) are highly sensitive to outliers because of the squaring of errors in their calculations. Outliers can disproportionately inflate the values of these metrics, potentially giving a misleading impression of the overall model performance. Even a few extreme outliers can significantly increase MSE and RMSE, which may obscure the model's effectiveness on most data points. This sensitivity often leads to overestimating model error, making it difficult to compare models across different datasets and interpret the practical significance of the reported error. Mean Absolute Error (MAE) is generally considered more robust to outliers compared to Mean Squared Error (MSE) and Root Mean Squared Error (RMSE). This robustness stems from MAE's approach of treating all errors equally, regardless of their magnitude, which makes it less sensitive to extreme values. However, it's important to note that MAE is not entirely immune to the influence of outliers. In cases where a dataset contains numerous extreme values, MAE can still be affected, albeit to a lesser extent than MSE or RMSE. This characteristic makes MAE a valuable metric in scenarios where outliers are present. However, it should still be used with an awareness of its limitations in handling datasets with a high proportion of extreme values [36].

Outliers can artificially inflate or deflate *R*-squared values, leading to an overestimation or underestimation of the model's explanatory power [36]. In some cases, outliers can cause the *R*-squared to increase even when the model's overall fit is poor for most of the data points. For datasets with values close to zero, outliers can cause Mean Absolute Percentage Error (MAPE) to become extremely large or undefined. MAPE can be skewed by outliers, mainly when they occur in conjunction with small actual values. While not a performance metric per se, it's important to note that outliers can significantly affect the estimated model coefficients in regression models. Outliers can pull the regression line towards themselves, distorting the model's overall fit. This can lead to biased coefficient estimates, which in turn affect all performance metrics. Outliers can indeed lead to overfitting and underfitting in regression models, depending on how they are handled and the specific characteristics of the dataset [36, 37].

4.4.6 Strategies to Address Outlier-Induced Overfitting and Underfitting

By carefully considering the presence and impact of outliers and employing appropriate techniques to address them, data scientists can develop more robust and accurate regression models that avoid overfitting and underfitting. To mitigate the impact of outliers on model performance:

- Use robust regression techniques: Methods like Huber regression, RANSAC, or Theil-Sen regression are less sensitive to outliers and can help prevent overfitting and underfitting [38].
- Employ cross-validation: This technique helps assess how well the model generalizes to unseen data, which is particularly important when dealing with outliers [39].
- Conduct thorough outlier analysis: Identify and investigate outliers using partial regression plots or statistical tests before deciding how to handle them [37]. Additionally, relying on a combination of performance metrics rather than a single measure provides a more comprehensive view of model performance in the presence of outliers. Conducting thorough outlier analysis through visual inspection and statistical tests is crucial before calculating performance metrics.
- Consider domain-specific implications: In some cases, outliers may represent important rare events that should not be ignored or removed. Understanding the context of the data is crucial [37].
- Use regularization techniques: Methods such as Lasso or Ridge regression can help prevent overfitting by adding a penalty term to the loss function, which mitigates the impact of outliers [37, 39].
- Transform variables: In some cases, applying appropriate transformations (e.g., log transformation) to the dependent or independent variables can reduce the impact of outliers and improve model fit [37, 39].

4.4.7 Significance of the Confusion Matrix

The confusion matrix is a powerful tool for evaluating classification models, offering a comprehensive view of performance beyond simple accuracy metrics. It provides a detailed breakdown of predictions, revealing true positives, true negatives, false positives, and false negatives. This granular information distinguishes between different types of errors, which is crucial in applications where error costs vary. The confusion matrix is particularly valuable for handling class imbalance, a common challenge in real-world datasets where accuracy alone can be misleading. It enables the calculation of more informative metrics such as Precision, recall, and F1-score, providing a multifaceted evaluation of model performance [40].

The confusion matrix helps identify weaknesses and guides improvement efforts by highlighting areas where a model makes mistakes. It is versatile, supporting binary and multiclass classification problems and facilitating easy comparison between different models. Probabilistic classifiers facilitate threshold selection by enabling the examination of performance at various decision points. The confusion matrix also aligns model evaluation with business objectives by providing insights into the costs associated with different error types. Furthermore, it serves as the foundation for advanced evaluation techniques, such as ROC curves and AUC analysis.

The confusion matrix empowers data scientists and machine learning practitioners to make informed decisions about model selection, refinement, and deployment. Its ability to provide a nuanced and thorough evaluation makes it an indispensable tool in developing and assessing classification models, ensuring that performance is understood and optimized across multiple dimensions [41].

4.4.8 Binary and Multiclass Confusion Matrices

The differences between binary and multiclass confusion matrices are significant and impact various aspects of model evaluation. Binary confusion matrices are simple 2 × 2 tables, while multiclass matrices expand to NxN dimensions, where N represents the number of classes. This structural difference affects the interpretation of cells: binary matrices clearly show True Positives, True Negatives, False Positives, and False Negatives, whereas multiclass matrices display correct classifications on the diagonal and misclassifications in off-diagonal cells. The complexity of error analysis increases substantially in multiclass scenarios, revealing intricate patterns of confusion between classes [41, 42].

Metric calculation also differs, with binary matrices allowing direct computation of accuracy, Precision, and recall, while multiclass matrices require additional steps or averaging techniques. Multiclass problems often employ a One-vs-All approach for class-specific metrics, which is unnecessary in binary classification. Visualization techniques vary, with multiclass matrices benefiting from heatmap representations to highlight patterns. Assessing class imbalance and interpreting error costs become more complex in multiclass scenarios, requiring careful examination of class distributions and domain-specific knowledge.

Performance evaluation in multiclass problems demands a more nuanced approach, considering both per-class performance and overall trends. While binary matrices have a fixed structure, multiclass matrices can be extended to accommodate any number of classes, increasing in size and complexity. Understanding these differences is crucial for accurately evaluating and interpreting the performance of classification models across various problem types, enabling data scientists to make informed decisions in model development and deployment [43].

4.4.9 One-Vs-All Approach in Multiclass Classification

The one-vs-all approach, known as one-vs-rest, offers numerous advantages in multiclass classification scenarios. Its simplicity and interpretability make it an attractive option, as each binary classifier focuses on distinguishing one class from all others, resulting in an intuitive and easily understandable model. This approach is particularly valuable because it enables the use of binary classification algorithms for multiclass problems, making it compatible with popular algorithms such as logistic regression, support vector machines, and perceptrons [44].

One-vs-all is computationally efficient, requiring only n_classes classifiers, where n_classes is the number of classes. This linear scalability makes it well-suited for problems with many classes. The approach also offers flexibility in handling class imbalance, as each binary classifier can be independently tuned. Furthermore, it facilitates individual class analysis, parallel training, and easy updates when new classes are added.

Another significant advantage of the method is its compatibility with probabilistic outputs, as many binary classifiers can provide probability estimates useful for ranking or decision-making in multiclass contexts. While one-vs-all has these benefits, it's important to note that its performance can vary depending on the specific problem and dataset. In some cases, approaches such as one-vs-one or native multiclass algorithms may be more suitable. The choice often depends on the characteristics of the problem, the dataset, and the base classifiers being used [44, 45].

4.5 Performance Metrics for Unsupervised Machine Learning Models

Performance metrics for unsupervised machine learning models help evaluate the quality of clustering, dimensionality reduction, and pattern identification, even in the absence of labeled data. Common metrics include [46]:

4.5.1 Clustering Metrics

- **Silhouette Coefficient:** Measures how similar an object is to its cluster compared to others. Higher values indicate better-defined clusters.

- **Davies–Bouldin Index**: This index evaluates the average similarity between each cluster and its most similar cluster. Lower values suggest better clustering.
- **Calinski–Harabasz Index:** Compares the variance between clusters to the variance within clusters. Higher values indicate better-defined clusters.
- **Adjusted Rand Index:** This measures the similarity between two clusterings, adjusted for chance. It's useful when you have some ground truth labels to compare against.
- **Mutual Information:** This metric measures the mutual dependence between two clusterings. It can be used to compare a clustering result to ground truth classes.
- **Cophenetic Correlation Coefficient:** Used in hierarchical clustering, this measures how faithfully a dendrogram preserves the pairwise distances between the original unmodeled data points.

4.5.2 Dimensionality Reduction Metrics

- **Cumulative Explained Variance:** Measures the amount of information retained after reducing dimensions. Higher values indicate better preservation of the data structure.
- **Reconstruction Error:** Quantifies the difference between the original and reconstructed data after dimensionality reduction. Lower values suggest better performance.
- **Trustworthiness:** Evaluates how well the local structure of high-dimensional data is preserved in lower dimensions.

4.5.3 General Metrics

- Within-Cluster Sum of Squares (WCSS): Measures the compactness of clusters. Lower values indicate tighter, more distinct clusters.
- Elbow Method: This method uses WCSS to determine the optimal number of clusters by identifying the "elbow" point in the WCSS curve.

When ground truth labels are available, additional metrics such as homogeneity, completeness, and V-measure can be used to assess clustering performance [47]. It's important to note that the choice of metric often depends on the specific unsupervised learning task (e.g., clustering vs. dimensionality reduction) and the characteristics of your data. Usually, combining multiple metrics provides a more comprehensive evaluation of model performance.

4.5.4 Challenges in Evaluating Unsupervised Learning Models

Evaluating unsupervised learning models presents a unique set of challenges that stem from the inherent nature of working with unlabeled data. The primary obstacle is the lack of ground truth labels, making objectively assessing model performance and the meaningfulness of learned representations challenging. This absence of clear benchmarks leads to high subjectivity in interpreting results, as different experts may draw varying conclusions from the same output. Unlike supervised learning, quantifying performance becomes a complex task, where metrics like accuracy or F1-score are readily available. Unsupervised models are also susceptible to data quality and preprocessing choices, making it challenging to discern whether poor performance is due to model limitations or data issues [48].

Further complications arise in determining optimal parameters, such as the number of clusters or components in clustering and dimensionality reduction tasks. The lack of standardized evaluation metrics across all unsupervised learning tasks adds to the difficulty, as does assessing the stability and robustness of results across different initializations or data samples. Practitioners often face a trade-off between model interpretability and performance, with more complex models potentially yielding better results but at the cost of reduced interpretability. Evaluating how well-learned patterns generalize to new, unseen data is particularly challenging without labeled test sets. Lastly, the effectiveness of unsupervised models is often highly context-dependent, making it difficult to compare models across different domains or applications. These challenges underscore the complexity of evaluating unsupervised learning models and highlight the need for careful consideration and domain expertise in interpreting their results [49].

To address these challenges, researchers and practitioners often evaluate unsupervised learning models using a combination of quantitative metrics, qualitative assessments, and domain expertise. Methods such as utilizing learned representations for downstream supervised tasks, conducting human evaluations, and assessing model stability can provide valuable insights into model performance.

4.6 Performance Metrics for Gen AI Models

The performance metrics for Gen AI models (e.g., **long language models**) [50, 51] collectively provide a comprehensive framework for evaluating generative AI models across multiple dimensions, ranging from technical performance to user experience and ethical considerations. The choice of metrics depends on the specific use case, application domain, and desired outcomes of the generative AI system. The performance metrics for generative AI models, along with their definitions and methods, and the significance of each metric, are discussed below.

4.6.1 Perplexity

Perplexity is a crucial metric used to evaluate the performance of generative language models. It measures how effectively a model can predict a sequence of text by quantifying its uncertainty in generating the next word or token. The concept is intuitive: lower perplexity scores indicate that the model is more confident in its predictions, suggesting a better understanding of the underlying data distribution [52, 53]. This metric is particularly valuable for assessing and comparing language models like GPT, which plays a significant role during the training and evaluation phases. By providing insights into a model's predictive capabilities, Perplexity helps researchers and developers gauge the quality and effectiveness of their language models, ultimately contributing to the advancement of natural language processing technologies.

4.6.2 BLEU (Bilingual Evaluation Understudy)

BLEU (Bilingual Evaluation Understudy) is a widely adopted precision-based metric in natural language processing, primarily used to assess the quality of machine-generated text, such as translations. It aims to compare AI-generated content with human-written reference text, providing a quantitative measure of similarity. BLEU calculates the n-gram overlap between the generated text and the reference, assigning scores on a scale from 0 to 1, where 0 indicates no match and 1 represents a perfect match. Despite its widespread use in machine translation and text generation tasks, BLEU has notable limitations. It doesn't account for recall or capture semantic meaning, which can sometimes lead to scores that don't fully reflect the true quality of the generated text. Nevertheless, BLEU remains a valuable tool in developing and evaluating language models, offering a standardized method for comparing different systems and tracking improvements in text generation capabilities [53, 54].

4.6.3 ROUGE (Recall-Oriented Understudy for Gisting Evaluation)

ROUGE (Recall-Oriented Understudy for Gisting Evaluation) is a sophisticated metric designed to assess the quality of machine-generated text, particularly in summarization tasks. Unlike BLEU, which focuses on Precision, ROUGE emphasizes recall, evaluating how much reference content is captured in the generated output. This metric comes in several variants, each measuring different aspects of textual similarity [53, 54].

4.6.4 Human Evaluation

Human evaluation is crucial in assessing the quality of AI-generated content, providing insights that automated metrics often fail to capture. This method involves human raters who carefully assess generated text based on attributes such as Fluency, Coherence, relevance, and overall satisfaction. Raters typically provide qualitative feedback or assign scores according to predefined criteria, allowing for a nuanced understanding of the content's quality. While human evaluation offers the advantage of capturing subtle aspects of language and context that machines may miss, it also presents challenges. The process can be subjective, as different raters may interpret criteria differently, and it is often time-consuming and expensive to implement on a large scale [53]. Despite these drawbacks, human evaluation remains invaluable in natural language processing. It is usually considered the gold standard, complementing automated metrics to provide a more comprehensive assessment of AI-generated text quality.

4.6.5 Fluency

Fluency is a critical metric in evaluating the quality of machine-generated text, focusing on the output's naturalness, grammatical correctness, and overall readability. This metric assesses whether the generated content flows smoothly and adheres to linguistic conventions, making it comprehensible to human readers [53]. Evaluating Fluency can be conducted through two primary approaches: human judgment, where experts assess the text's linguistic quality, and automated tools that detect grammatical errors and structural inconsistencies.

4.6.6 Coherence

Coherence is a fundamental metric in assessing the quality of AI-generated text, focusing on the logical consistency and precise flow of ideas throughout the content. This crucial aspect evaluates whether the generated text maintains a cohesive structure, ensuring that ideas are connected logically and that the overall narrative or argument makes sense as a unified whole. The assessment of Coherence can be conducted through two primary methods: human evaluation, where experts analyze the text's logical progression, and automated approaches that utilize semantic similarity measures between sentences or paragraphs [53].

4.6.7 Diversity

Diversity is a crucial metric in evaluating the performance of generative AI models. It focuses on the variety and non-repetitiveness of outputs across multiple generations or within a single output. This metric assesses the model's ability to produce varied and creative content, avoiding monotonous or repetitive responses. To quantify diversity [53], researchers employ techniques such as measuring distinct n-grams or calculating self-BLEU scores, which compare the similarity of a model's outputs.

4.6.8 Factual Accuracy

Factual accuracy is a vital metric for evaluating the reliability of generative AI models, ensuring that the information they generate is truthful and aligns with verified facts. This metric assesses the model's ability to produce content that is not only coherent but also factually correct, which is particularly crucial in domains where misinformation can lead to serious consequences, such as healthcare, legal, or financial applications. Evaluating factual accuracy often involves comparing the generated content against trusted external knowledge bases or conducting human-led fact-checking to verify the correctness of the information [52, 53]. This metric is especially relevant in knowledge-based systems like chatbots or automated report generation, where users rely on AI to provide accurate and dependable information. By prioritizing factual accuracy, developers can enhance trust in AI systems and ensure safe and effective deployment in critical real-world scenarios.

4.6.9 Relevance

Relevance is a critical metric in evaluating the performance of generative AI models, focusing on how effectively the generated output addresses the given prompt or query. This metric ensures that the AI's responses are coherent, factually accurate, and directly applicable to the user's intent or question. Assessment of relevance typically involves either sophisticated semantic similarity metrics that compare the generated content to the input prompt or human evaluation to gauge the alignment between the response and the user's expectations [53].

4.6.10 Toxicity and Bias

Toxicity and bias are crucial metrics in evaluating generative AI models. They focus on detecting and preventing harmful, offensive, or biased content in AI-generated outputs. These metrics are designed to assess the ethical implications of AI-generated content, ensuring that the model does not produce text that could be considered discriminatory, inflammatory, or otherwise inappropriate [50, 52]. The evaluation process typically involves using specialized classifiers or carefully curated datasets trained to identify problematic language patterns or content.

4.6.11 Inference Time

Inference time is a critical performance metric for generative AI models, measuring the speed at which a model can produce outputs in response to an input prompt. This metric is crucial for assessing the practical usability of AI systems, particularly in real-time applications where rapid response is essential [55]. The inference time of a model is influenced by several factors, including the model's size and complexity, the availability of hardware resources, and the implementation of optimization techniques. In applications such as chatbots or virtual assistants, where user experience heavily depends on quick and seamless interactions, inference time plays a pivotal role in determining the system's effectiveness. Fast inference times contribute to more natural and engaging user experiences, making this metric particularly important for latency-sensitive systems.

4.6.12 Hallucination Rate

The hallucination rate [56, 57] is a crucial metric in evaluating the reliability of generative AI models, measuring the frequency with which a model produces false, nonsensical, or unsupported information. This metric is crucial for evaluating the trustworthiness of AI-generated content, especially in domains where factual accuracy is paramount, such as news generation or information retrieval systems. Detecting hallucinations typically involves rigorous fact-checking processes, comparing the generated content against reliable sources, or performing internal consistency checks within the output. Minimizing the hallucination rate is vital for maintaining the credibility and utility of AI systems, as frequent inaccuracies can erode user trust and potentially lead to misinformation. In practical applications, particularly those involving sensitive or high-stakes information, a low hallucination rate is crucial for ensuring that AI-generated content can be trusted for decision-making or public consumption. As AI models become more sophisticated and are deployed in increasingly diverse and critical roles, focusing on reducing hallucinations remains a key priority for developers and researchers aiming to create more dependable and trustworthy AI systems.

4.6.13 Zero-Shot and Few-Shot Performance

Zero-shot and few-shot performance metrics [58] are crucial indicators of a generative AI model's versatility and adaptability, assessing its ability to handle new tasks with minimal or no task-specific training data. Zero-shot performance evaluates how well a model can perform a new task without prior examples. In contrast, few-shot performance measures its capability when provided with only a handful of examples. These metrics are evaluated by comparing the model's performance on new tasks against established benchmarks without the benefit of fine-tuning task-specific datasets. The importance of these metrics lies in their demonstration of a model's generalization capabilities, particularly for large language models like GPT-4. Strong zero-shot and few-shot performance indicates a model's ability to transfer knowledge across domains and adapt to new scenarios quickly. This adaptability is especially valuable in real-world applications where labeled training data may be scarce or expensive, making these metrics essential for assessing advanced AI systems' practical utility and flexibility in diverse and evolving environments.

4.6.14 Fine-Tuning Efficiency

Fine-tuning efficiency is a critical metric in deploying and optimizing generative AI models, measuring how effectively a pre-trained model can be adapted to specific tasks using task-specific data. This metric evaluates the balance between performance improvements and resource utilization during the fine-tuning process, considering time, data requirements, and computational resources. By assessing the gains in performance metrics after fine-tuning against the resources invested, fine-tuning efficiency helps developers and researchers determine whether the process is worthwhile for specialized applications. This metric is particularly relevant when deploying domain-specific generative AI models where the goal is to achieve optimal performance in a targeted area without excessive resource expenditure. Strong fine-tuning efficiency indicates a model's adaptability and cost-effectiveness, making it a valuable consideration in implementing AI systems across various specialized domains [59]. As AI technologies evolve, fine-tuning efficiency remains a key factor in maximizing the utility and ROI of generative AI models in real-world applications.

4.6.15 RAG for Long Language Models

Retrieval-Augmented Generation (RAG) is an AI framework designed to enhance the capabilities of large language models (LLMs) by incorporating external knowledge sources. RAG combines the power of LLMs with the ability to retrieve and integrate relevant information from external databases or knowledge bases [60]. This approach enables LLMs to access up-to-date and accurate information beyond their initial training data, thereby enhancing the quality and reliability of their outputs.

Retrieval-augmented generation (RAG) greatly enhances the accuracy and reliability of large language models (LLMs) by addressing several key limitations. One of its primary benefits is improved accuracy, as RAG grounds LLM responses in verified external information, reducing hallucinations and inaccuracies [61]. By allowing access to up-to-date data, RAG overcomes the constraints of outdated training datasets, ensuring responses are based on current facts. This approach is also cost-effective, as it eliminates the need for frequent retraining of large models with new data. Additionally, RAG fosters trust by providing source attribution, which increases user confidence in AI-generated outputs.

RAG ensures enhanced factual grounding by enabling LLMs to retrieve relevant external information, minimizing the risk of generating outdated or incorrect responses. It significantly reduces hallucinations by supplying precise and contextually relevant data from external sources, resulting in more reliable outputs. Furthermore, RAG enhances contextual understanding by incorporating domain-specific knowledge and recent developments that may not have been incorporated into the model's initial training. This makes it particularly effective for specialized or rapidly evolving topics. Another advantage of RAG is its ability to integrate real-time knowledge, ensuring that LLMs reflect the latest developments and facts rather than relying solely on static training data. Source attribution adds transparency, allowing users to trace information to its origin and verify its accuracy. RAG enhances relevance by retrieving information specifically tailored to a user's query, resulting in more targeted and pertinent responses. Moreover, it addresses knowledge gaps in an LLM‘s original training data by incorporating external domain knowledge, resulting in more comprehensive and accurate outputs across a wide range of topics. By combining the generative power of LLMs with the ability to retrieve and integrate external knowledge, RAG significantly improves its accuracy, reliability, and usefulness across various applications. This hybrid approach ensures that LLMs deliver more trustworthy and contextually appropriate responses while staying current with evolving information needs [62, 63].

The RAG process typically involves the following steps:

- Query Processing: The user's input is analyzed and converted into a suitable format for information retrieval.
- Information Retrieval: Semantic search techniques fetch Relevant data from external sources.
- Context Augmentation: The retrieved information is combined with the original query.
- Generation: The LLM uses the augmented input to produce a response [60, 63].

Implementing RAG: To integrate RAG with LLMs, developers typically need to [62, 63]:

- Set up a vector database to store and efficiently retrieve relevant information.
- Implement semantic search capabilities for effective information retrieval.
- Design prompts that effectively combine retrieved information with user queries.
- Fine-tune the LLM to generate responses based on the augmented input.

4.6.16 RAG and Fine-Tuning

Retrieval-Augmented Generation (RAG) and fine-tuning are two approaches to enhancing the performance of large language models [59, 64, 65], each with its strengths. While fine-tuning excels at tailoring models for specific tasks by adjusting their internal parameters, RAG offers several distinct advantages. One of RAG's key benefits is its ability to access and utilize the most current information. In contrast, fine-tuned models are constrained by the data on which they were initially trained. This makes RAG particularly valuable in rapidly evolving fields or for queries requiring up-to-date knowledge. RAG also boasts a broader knowledge base, allowing it to handle a wider range of queries without needing model retraining. This versatility is instrumental in applications with a vast and diverse scope of potential questions. Regarding security and data privacy, RAG provides enhanced protection by keeping sensitive information separate from the model, a crucial feature in many enterprise and governmental applications. Additionally, maintaining an RAG system is often more straightforward and cost-effective, as updates typically involve modifying the external knowledge base rather than retraining the entire model.

In conclusion, while RAG and fine-tuning have their place in improving AI accuracy, RAG offers unique advantages in dynamic knowledge integration, scalability, and cost-effectiveness [59, 64]. It is particularly well-suited for applications that require access to extensive, constantly updated databases and for scenarios where maintaining current and accurate information is paramount. The choice between RAG and fine-tuning ultimately depends on the application's specific requirements, with RAG emerging as a powerful solution for many modern AI challenges.

4.6.17 Limitations of Using RAG for Real-Time Applications

Retrieval-augmented generation (RAG) faces several limitations when applied to real-time applications despite its benefits for improving AI accuracy [66]. One of the primary challenges is increased latency, as searching and retrieving relevant information from external sources can introduce delays, particularly when dealing with large datasets [65]. The computational overhead of combining retrieved information with the original query and generating a response can further slow down response times [64]. Performance constraints also pose significant challenges for RAG in real-time scenarios. Complex queries across multiple tables with millions of records can be time-consuming, especially for large organizations [66]. Some RAG implementations may lack advanced indexing capabilities, resulting in slower data retrieval and processing times for structured data queries [66].

Scalability becomes a critical issue as the number of concurrent users increases. Handling thousands or millions of users executing queries simultaneously can strain the system, especially for transactional workloads that require quick read-write operations [67]. The retrieval and integration processes can be resource-intensive, potentially leading to bottlenecks when scaling to meet real-time demands [7]. Ensuring data freshness and consistency in real-time scenarios presents another challenge. While many data warehouses and data lakes support real-time data ingestion, it often depends on the source system's streaming capabilities [66]. Maintaining consistency between the retrieval source and operational systems can be difficult in rapidly changing environments [66].

Cost considerations are also significant, as cloud-based data warehouses and lakes often operate on a pay-per-use model, where each query incurs charges based on the data volume and access frequency [66]. The dynamic nature of real-time queries can lead to unpredictable and potentially high costs, especially when dealing with large-scale operations [66]. Lastly, implementing RAG for real-time use cases can be technically challenging. Designing a system that can efficiently retrieve, process, and integrate information in real-time requires a sophisticated architecture [65]. Keeping the retrieval sources up to date and ensuring the RAG system remains effective over time adds to the operational burden [65].

Understanding these limitations is crucial for organizations to assess the suitability of RAG for their real-time applications and implement strategies to mitigate potential issues. While RAG offers powerful capabilities for enhancing AI accuracy, carefully considering these constraints is essential for successful deployment in time-sensitive scenarios.

4.6.18 Hallucination Index Report [68]

The Hallucination Index, developed by Galileo [56, 68], is a comprehensive framework designed to evaluate and rank large language models (LLMs) based on their tendency to hallucinate during common generative AI tasks. Its primary purpose is to assist teams in selecting the most suitable LLM for their specific projects and use cases. Galileo recently released its second annual Hallucination Index, which focuses on Retrieval Augmented Generation (RAG) performance and evaluates 22 leading LLMs, expanding its scope by adding 11 new models compared to the previous year.

The evaluation method employs a proprietary metric called "context adherence" to assess output inaccuracies, testing models with inputs ranging from 1000 to 100,000 tokens across short, medium, and long context lengths. Key findings from the report include Anthropic's Claude 3.5 Sonnet emerging as the best overall performing model, open-source models narrowing the performance gap with closed-source counterparts, and some smaller models outperforming larger ones in specific scenarios. Notably, many models maintained high performance even with extended context lengths, and emerging global players like Mistral and Alibaba made significant progress [68].

The report highlights several trends, including the improvement in open-source model performance, challenges to the assumption that larger models are always superior, strong performance across longer context lengths, and a growing international effort in LLM development. The Hallucination Index is a valuable resource for AI teams and leaders, enabling them to make informed decisions about balancing price and performance when choosing LLMs and providing real-world data on model performance for various tasks and context lengths [68]. This comprehensive evaluation tool reduces hallucinations in generative AI applications, contributing to more reliable and accurate AI systems.

4.7 Performance Metrics for Natural Language Processing

Natural Language Processing (NLP) machine learning models can be evaluated using various performance metrics [69], depending on the specific task and desired outcomes. When selecting performance metrics for NLP models, it is essential to consider several key factors to ensure accurate and meaningful evaluation. First, prioritize task-specific relevance by selecting metrics that closely align with the specific NLP task and its objectives. Second, a balanced assessment should be achieved using multiple metrics to understand the model's performance across different dimensions comprehensively. For natural language generation tasks, incorporating human evaluation is crucial to assess qualitative aspects such as Fluency and Coherence. Additionally, leveraging established benchmarks like GLUE (General Language Understanding Evaluation) or MMLU (Massive Multitask Language Understanding) can provide standardized comparisons and valuable context for performance assessment. By thoughtfully selecting and analyzing these metrics, you can uncover insights into the strengths and weaknesses of your NLP models, guiding improvements and ensuring their effectiveness in real-world applications [62, 70]. Here's an overview of some key performance metrics used for NLP models:

4.7.1 Classification Metrics

These metrics are commonly used for tasks like sentiment analysis, text classification, and named entity recognition [69, 70]:

- Accuracy: The proportion of correct predictions among the cases examined.
- Precision: The ratio of true positive predictions to the total positive predictions.
- Recall: The ratio of true positive predictions to actual positive cases.

- F1-Score: The harmonic mean of Precision and recall, providing a balanced measure of the model's performance.

4.7.2 *Text Generation Metrics*

For tasks involving text generation, such as machine translation and summarization:

- BLEU (Bilingual Evaluation Understudy): Measures the quality of machine-translated text by comparing it to reference translations [69].
- ROUGE (Recall-Oriented Understudy for Gisting Evaluation): Evaluates the quality of generated summaries by comparing them to reference summaries [69].
- METEOR (Metric for Evaluation of Translation with Explicit Ordering): Assesses machine translation quality by considering word order and synonyms [70].
- CIDEr (Consensus-based Image Description Evaluation): Often used for image captioning tasks, measuring how well a generated description matches a set of reference descriptions [69].

4.7.3 *Language Modeling Metrics*

For evaluating language models [50, 53, 69]:

- Perplexity: Measures how well a probability model predicts a sample, with lower values indicating better performance.
- Bit Per Character (BPC): Used to evaluate character-level language models.

4.7.4 *Sequence Labeling Metrics*

For tasks like part-of-speech tagging and dependency parsing [51, 52]:

- Unlabeled Attachment Score (UAS): Measures the accuracy of dependency arcs in dependency parsing.
- Labeled Attachment Score (LAS): Similar to UAS, but it also considers the correctness of dependency labels.

4.7.5 *Other Important Metrics*

- Word Error Rate (WER): Commonly used in speech recognition to measure the edit distance between the predicted and reference transcriptions.

- Exact Match: Used in question-answering tasks to measure the percentage of predictions that match the ground truth.

4.8 Performance Metrics for Reinforcement Learning

Evaluating reinforcement learning (RL) algorithms requires a comprehensive and statistically sound approach to ensure reliable conclusions. Given the inherent stochasticity in RL, conducting multiple trials for each algorithm is crucial, as it provides a more accurate representation of their performance across various scenarios. This multi-trial approach helps mitigate the impact of random fluctuations and provides a more robust assessment of an algorithm's capabilities [71, 72].

Resmeaningfully compare statistical tests, such as the Mann–Whitney U test, to draw meaningful comparisons between RL algorithms. These tests offer a quantitative basis for determining whether observed differences in performance are statistically significant or merely due to chance. This statistical rigor is crucial for making informed decisions about algorithm selection and improvement.

A thorough evaluation should consider the algorithm's final performance and learning dynamics throughout the training process. This dual focus provides insights into the algorithm's ultimate capabilities, efficiency, and stability during the learning phase. Understanding these learning dynamics can be particularly valuable for applications where rapid adaptation or consistent improvement is crucial.

To assess an algorithm's generalization capabilities, evaluating its performance across various environments or tasks is essential. This broad-spectrum evaluation helps identify highly versatile and robust algorithms that can adapt to multiple challenges, rather than being overly specialized for a specific scenario. Ultimately, the selection of evaluation metrics should be closely aligned with the particular goals of the RL task. Different applications may prioritize other aspects of performance, such as sample efficiency, stability, or asymptotic performance. By tailoring the evaluation approach to these specific requirements, researchers and practitioners can ensure that their assessments provide meaningful insights that directly inform the development and deployment of RL algorithms in real-world applications. Here are some key performance metrics used for evaluating reinforcement learning models [71, 72]:

- Cumulative Reward/Return:
  - The total reward accumulated over one or more episodes.
  - Measures how well the agent is maximizing its objective.
- Average Return:
  - The mean reward obtained per episode over multiple runs.
  - Provides a more stable measure of performance than individual episode returns.

- Learning Curve:
  - A plot showing how the performance (usually average return) changes over time or training iterations.
  - It helps visualize the learning progress and convergence of the algorithm.
- Success Rate:
  - The percentage of episodes where the agent completes the task or reaches a goal state.
  - Useful for tasks with clear success/failure conditions.
- Sample Efficiency:
  - How quickly the agent learns in terms of environment interactions or data samples.
  - It can be measured by the number of steps or episodes needed to reach a certain performance threshold.
- Stability:
  - How consistent is the agent's performance across different runs or environments?
  - It can be measured by the variance in returns or success rates.
- Policy Quality:
  - Measures of how good the learned policy is, such as state visitation frequencies or action probabilities.
- Exploration vs Exploitation Balance:
  - Metrics that show how well the agent balances exploring new actions versus exploiting known good actions.
- Convergence Speed:
  - How quickly the algorithm converges to a stable policy.
- Robustness:
  - How well the agent performs under different initial conditions or environment variations.

4.8.1 Pivotal Role of Reward Functions in Reinforcement Learning

The reward function is a cornerstone in reinforcement learning (RL), profoundly influencing both the learning process and the evaluation of agent performance. At its core, the reward function defines the fundamental objectives for the RL agent, shaping what constitutes success and guiding the agent's policy optimization. This direct link between rewards and policy updates means that the reward function's design can significantly impact learning efficiency, potentially leading to faster and more stable learning when well-crafted or, conversely, resulting in suboptimal or unintended behaviors if poorly designed [73, 74].

The influence of the reward function extends deeply into various performance metrics commonly used in RL. Metrics such as cumulative reward, average return per episode, and success rates are directly derived from or heavily influenced by the reward structure. Moreover, the reward function is crucial in balancing short-term gains against long-term value, which in turn affects sample efficiency and convergence speed metrics. It also shapes the agent's approach to exploration versus exploitation, impacting its ability to discover optimal solutions. Beyond immediate performance indicators, the design of the reward function has far-reaching effects on the agent's overall capabilities. It influences how well the learned policy generalizes to new situations, affecting robustness and transfer learning metrics. In many applications, the reward function also incorporates ethical and safety considerations, expanding the scope of performance evaluation beyond mere task completion to include broader societal impacts [75, 76].

Ultimately, the reward function serves as a lens through which we define, measure, and achieve performance in reinforcement learning. Its central role in shaping the agent's behavior, learning trajectory, and ultimate capabilities underscores the critical importance of thoughtful reward design in developing effective and responsible RL systems.

4.8.2 Visualizing the Learning Process in Reinforcement Learning

Visualizing the learning process in reinforcement learning (RL) is crucial for understanding agent behavior and identifying potential issues with the reward function. A comprehensive approach employs various techniques to provide insights into different aspects of the learning journey [77]. Learning curves, which plot cumulative or average reward over time, provide a high-level view of the agent's improvement and convergence, with sudden drops or plateaus potentially indicating problems with the reward function.

State visitation frequency heatmaps and action distribution histograms provide deeper insights into the agent's behavior patterns, revealing potential sticking points or action biases that might stem from reward imbalances. For Q-learning approaches, visualizing Q-values through heatmaps can illuminate how the agent's value estimates evolve. Analyzing reward distributions and episode lengths helps identify issues like reward sparsity or structural problems in the learning environment [74, 75].

In simpler environments, direct policy visualization can offer intuitive insights into the agent's learned behaviors. Tracking intermediate goal achievements and the balance between exploration and exploitation provides additional context for the learning process. Comparing performance against benchmark tasks and baseline algorithms offers a standardized measure of progress [71, 77].

These visualization techniques collectively form a powerful toolkit for RL practitioners. By leveraging these methods, researchers and developers can gain deep insights into their agents' learning processes, identify potential issues with reward functions, and refine their approaches iteratively. This visual analysis is instrumental in developing RL systems that perform well and exhibit desirable and interpretable behaviors, ultimately leading to more robust and reliable AI systems.

4.9 Monotonicity in Machine Learning Models

Monotonicity in machine learning is a crucial concept that describes a consistent relationship between an input feature and a model's output. This property exists when the output consistently increases or decreases as the input feature changes. The importance of monotonicity in machine learning models cannot be overstated. It significantly enhances model interpretability and explainability, making it easier for stakeholders to understand and trust the model's decisions [78–80]. Moreover, monotonicity often aligns with domain knowledge and business logic, ensuring that the model's behavior matches expert expectations. This alignment can lead to reduced overfitting and improved generalization, as the model is constrained to follow logical patterns. Monotonicity is particularly valuable in regulated industries such as finance and healthcare, where model decisions can have significant real-world impacts. It provides a level of predictability and reliability often required by regulatory bodies and can help defend model decisions if challenged. Understanding and implementing monotonicity in machine learning models can thus lead to more robust, trustworthy, and effective predictive systems.

Detecting Monotonicity [81, 82]

- Partial Dependence Plots (PDP):
- Visualize the relationship between a feature and the model's predictions.
- A monotonic relationship will show a consistently increasing or decreasing trend.
- Individual Conditional Expectation (ICE) plots:
  - Similar to PDPs, but show individual data point relationships.
  - Monotonic relationships will have lines that don't cross each other.
- Feature Importance Analysis:
  - Examine feature coefficients in linear models.
  - Analyze feature split directions in tree-based models.

- Correlation Analysis:
  - Calculate correlation coefficients between features and predictions.
  - Strong positive or negative correlations may indicate monotonicity.
- Model-specific techniques:
  - For tree-based models: Examine split directions in decision trees.
  - For neural networks: Analyze weights and activations of hidden layers.
- Monotonicity Constraints [82, 83]:
  - Some algorithms (e.g., XGBoost, LightGBM) allow specifying monotonicity constraints.
  - Check if the model respects these constraints after training.
- Testing with Synthetic Data:
  - Generate data with known monotonic relationships.
  - Verify if the model captures these relationships correctly.
- Domain Knowledge:
  - Consult with domain experts to identify expected monotonic relationships.
  - Compare model behavior with these expectations.

Implementing Monotonicity [83]

- Use algorithms that support monotonicity constraints (e.g., XGBoost, LightGBM).
- For neural networks, apply specialized techniques like Monotonic Neural Networks or constraints on weights.
- In tree-based models, enforce monotonicity by controlling split directions.

Example (Using LightGBM)

```
import lightgbm as lgb


# Specify monotonicity constraints
monotone_constraints = [1, 0, -1]  # 1: positive, 0: none, -1:
negative

model
= lgb.LGBMRegressor(monotone_constraints=monotone_constraints)
model.fit(X_train, y_train)
```

Monotonicity significantly reduces overfitting in machine learning models by enforcing constraints that help the model generalize better to unseen data. By aligning the model's behavior with domain knowledge and logical relationships between features and outcomes, monotonic constraints enable the model to capture accurate underlying patterns rather than fitting to noise in the training data. This alignment improves generalization and performance on new examples [81]. Additionally, limiting the model's flexibility to only monotonic relationships reduces overall complexity, preventing it from learning overly intricate, non-monotonic patterns that may not generalize well. Monotonic models also produce more stable and predictable outputs across the feature space, thereby mitigating erratic behavior caused by overfitting when slight variations in input data occur. In noisy datasets, these constraints enable the model to focus on overall trends rather than local fluctuations caused by noise. This is particularly beneficial in tree-based methods where monotonicity can be enforced during tree construction. Studies have shown that models with monotonic constraints often exhibit better performance on test data compared to their unconstrained counterparts, as they capture more generalizable patterns. Overall, incorporating monotonicity into machine learning models enhances interpretability and alignment with domain knowledge, leading to more robust and reliable predictions across various applications [82, 83].

4.10 Train Supervised AI/ML Model

Figure 4.4 shows the key steps to train a supervised machine learning model:

To develop a supervised machine learning model, prepare data, collect and clean the dataset, split it into features and target variables, and divide it into training and test sets. Next, choose an appropriate machine learning model and create an instance using a library such as scikit-learn, TensorFlow, or pytorch. Train the model on your training data using the fit() method, then make predictions on the test data with the predict() function. Evaluate the model's performance using the metrics discussed in previous sections of this chapter, and assess the features of importance. Tune hyperparameters using techniques such as grid search or random search, and perform cross-validation if necessary. Once satisfied with the model's performance, retrain it on the full dataset and save it for future use. Throughout this process, consider key factors such as feature scaling, handling missing data, checking for multicollinearity, applying regularization to prevent overfitting, selecting relevant features, monotonicity, and comparing multiple model types. The machine learning libraries offer

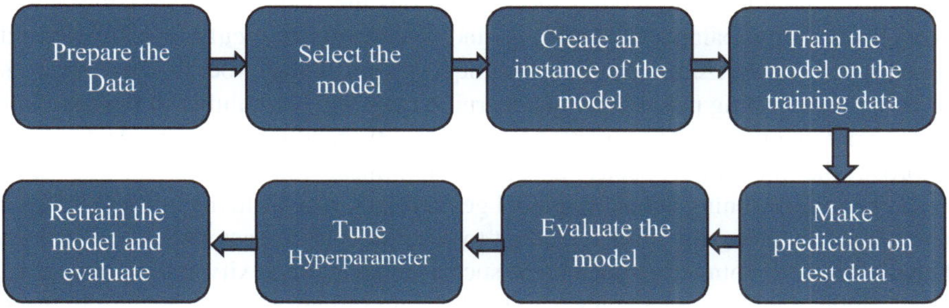

Fig. 4.4 Supervised machine learning model training

user-friendly implementations of various algorithms and evaluation metrics, making them an excellent tool for developing effective machine learning models [1, 7, 8, 84, 85].

4.10.1 Improving the Accuracy of AI/ML Model

Improving the accuracy of a machine learning model involves a multifaceted approach that combines various techniques and strategies. Feature engineering plays a crucial role, where new relevant features are created, existing ones are transformed, and irrelevant or redundant features are removed. Handling outliers is equally important, as they can significantly skew the model's performance. Experimenting with different algorithms can help identify the most suitable approach for the given dataset and develop a balance between accuracy and explainability. Hyperparameter tuning can optimize the model's performance using grid or random search methods. Increasing the training data can enhance the model's learning capabilities, provided that it is feasible. Addressing overfitting through regularization techniques and cross-validation is essential for creating a robust model. Non-linear relationships can be captured by incorporating polynomial features or other non-linear transformations. Feature scaling and addressing multicollinearity are necessary preprocessing steps that can improve model accuracy. Advanced techniques, such as boosting algorithms and deep learning models, can be employed to handle complex relationships. Analyzing residuals and incorporating domain expertise can provide valuable insights for further refinement. The key to developing an accurate machine learning model lies in iteratively testing different approaches, analyzing results, and continuously refining the model through proper data preparation, feature engineering, algorithm selection, and tuning [86].

4.10.2 Hyperparameter Tuning to Improve Model Accuracy

Hyperparameter tuning is a crucial process that can significantly enhance the accuracy of machine learning models through various mechanisms. Optimizing model complexity helps balance underfitting and overfitting, particularly in tree-based models and regularized regression techniques. The tuning process also enhances learning by adjusting parameters like learning rates and the number of base models in ensemble methods. Additionally, it can indirectly facilitate feature selection and improve the model's capacity to handle nonlinear relationships in the data. Hyperparameter tuning enhances model generalization by utilizing cross-validation techniques, enabling the model to adapt to specific data characteristics and strike a balance between competing objectives, such as model complexity and training time. In ensemble methods, it can enhance the diversity of base models, leading to more robust predictions. The process also optimizes regularization strategies and can help address challenges posed by imbalanced data. Researchers can identify configurations that substantially improve model accuracy compared to default settings by systematically exploring the hyperparameter space using techniques like grid search, random search, or Bayesian optimization. This tailored approach ensures that the machine learning model is well-suited to the specific task and dataset, ultimately leading to more accurate and reliable predictions [87, 88].

4.10.3 Best Practices for Splitting the Dataset into Training and Test

When splitting a dataset into training and testing sets, it's crucial to follow best practices to ensure reliable model development and evaluation [89]. A standard approach is to allocate 70–80% of the data for training and 20–30% for testing, with larger datasets potentially using an 80–10-10 split for training, validation (tuning), and testing. It's essential to ensure representative sampling, maintain the same distribution in both sets, and use stratified sampling for imbalanced datasets. To maintain consistency and reproducibility, use a fixed random seed when shuffling data and create a script or function for the splitting process. It's critical to keep the test set separate and use it only for final model evaluation, never for training or hyperparameter tuning. For smaller datasets or more robust evaluation, consider using K-fold cross-validation. Including a validation set for hyperparameter tuning is also recommended. To avoid data leakage, split the data before applying any preprocessing or transformations. Ensure that the test set is large enough to provide a reliable estimate of model performance. For time series data, use a chronological split rather than random sampling. By adhering to these practices, you can create a robust train-test split that supports effective model development and evaluation [89, 90].

Consider implementing various best practices to ensure your training, validation, and test sets have the same distribution. Start by using stratified sampling to maintain consistent proportions of classes or categories across all sets, which is particularly crucial for imbalanced datasets. Randomize your entire dataset before splitting to avoid any inherent ordering, and use a consistent, reproducible splitting method with a fixed random seed for experimental consistency. Regularly check for distribution similarity by comparing basic statistics and using visualization techniques. For time series data, maintain chronological order while ensuring each set spans the whole time range. When dealing with grouped data, keep all samples from one group in the same set. Use cross-validation techniques, such as K-fold, to ensure all data is used for training and validation. If multiple features are crucial, consider stratifying on combinations of these features. Be vigilant about data leakage, ensuring no information from validation or test sets influences the training process. Use large sample sizes for each set to accurately represent the overall distribution. Finally, if your training data comes from a different source, consider using a "bridge" set to help measure distribution mismatch. By adhering to these practices, you can help ensure that your data sets are representative of the same underlying distribution, leading to more reliable model development and evaluation [91, 92].

When splitting datasets for machine learning tasks, it's crucial to be aware of and avoid several common pitfalls that can compromise the reliability and generalizability of your models. Ensuring an adequate sample size across all sets is fundamental, as insufficient data can lead to unreliable performance metrics. Data leakage, where information from validation or test sets influences the training process, must be strictly prevented to avoid overly optimistic results. Proper shuffling or sorting of data is essential, particularly for time series data, where the temporal order is crucial. Maintaining similar distributions of classes across splits, especially for imbalanced datasets, is vital and can be achieved through stratified sampling. Considering temporal aspects and potential data drift is crucial for time-based data, which may necessitate chronological splits rather than random ones. Choosing appropriate split ratios based on dataset size and problem complexity is important, as is avoiding cross-contamination by performing preprocessing after splitting. Keeping related points in the same set prevents leakage when dealing with grouped data. Finally, maintaining consistent preprocessing across all datasets ensures comparability. By carefully navigating these potential issues, you can develop more robust and effective machine learning models that perform well on new, unseen data [93, 94].

4.10.4 Training Ensemble Machine Learning Models

Training ensemble machine learning models for regression and classification tasks involves several key steps. Choose an ensemble method such as bagging (e.g., Random Forest), boosting (e.g., Gradient Boosting, AdaBoost), or stacking. Select appropriate base models, which can be a single algorithm type for bagging and boosting or diverse algorithms for stacking. Prepare your data by splitting it into training and test sets, and perform the necessary preprocessing. For bagging, train multiple models on bootstrap samples of the data; in boosting, train models sequentially, with each model focusing on the errors of its predecessors. Stacking involves training base models that are independent of the same training data. Combine predictions using methods like averaging or majority voting for bagging, weighted sum for boosting, or training a meta-model for stacking [95, 96].

Evaluate the ensemble using cross-validation or a held-out test set, comparing its performance to individual base models. Tune hyperparameters such as the number of base models or learning rates using techniques like grid search. Finally, the model will be retrained on the full dataset once its performance is satisfactory. Throughout the process, diversity among base models will be ensured, model complexity will be balanced with performance, appropriate evaluation metrics will be used, and computational resources will be considered. Libraries like scikit-learn implement many ensemble methods, facilitating experimentation with different approaches [96, 97].

4.11 Training Large Language Model

Training large language models (LLMs) begins with meticulous data preparation, where researchers aggregate vast repositories of high-quality text from diverse sources like websites, academic publications, books, and digital archives. This raw textual data undergoes rigorous cleaning and preprocessing to ensure consistency, remove noise, and create a pristine training foundation. Central to this process is selecting an advanced transformer-based architecture, which has revolutionized natural language processing. Researchers carefully calibrate the model's size—balancing computational requirements with performance goals—by determining the optimal number of parameters that can effectively capture linguistic nuances without overwhelming available computational resources [98–100]. The primary training processes of LLMs are shown in Fig. 4.5.

The initial pre-training phase represents a critical stage where the model learns fundamental language understanding through self-supervised techniques [7]. By exposing the model to unannotated data and employing sophisticated methods, such as masked language modeling, researchers enable the system to develop comprehensive linguistic comprehension. This computationally demanding phase can span months, requiring significant infrastructure and potentially consuming thousands of GPU hours. Following pre-training, instruction tuning transforms the model from a generalized language learner to a more specialized, task-oriented assistant. Through supervised learning on curated instruction-response datasets, the model learns to interpret and execute specific commands with increasing precision and contextual awareness [100].

Reinforcement learning [6] introduces a sophisticated refinement layer, where human feedback guides the model's behavioral evolution. By training a reward model based on nuanced human annotations, researchers can encourage desirable output characteristics and discourage unintended responses, creating a more aligned and reliable AI system. Hyperparameter optimization represents another crucial dimension, where advanced techniques, such as Bayesian optimization and grid search, fine-tune learning rates, batch sizes, and other critical training parameters. This meticulous calibration ensures maximum model performance and efficiency.

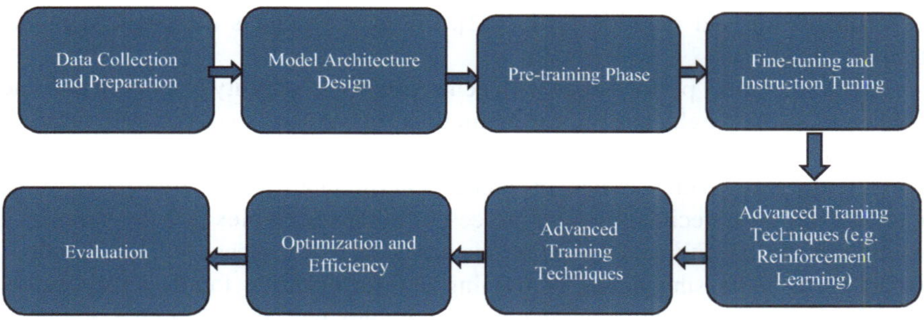

Fig. 4.5 Training large language model

Continuous evaluation remains paramount, with researchers employing comprehensive metrics, such as perplexity and task-specific assessments, to gauge the model's capabilities. This iterative improvement process allows for ongoing refinement based on empirical performance data and emerging training methodologies. Ethical considerations are increasingly central to LLM development. Researchers must proactively address potential biases in training data, implement robust content safeguards, and ensure the model generates responsible, unbiased outputs across diverse contexts [99–101].

The underlying computational infrastructure demands sophisticated distributed training architectures, typically leveraging multiple high-performance GPUs or specialized TPUs. Efficient data loading pipelines and optimization techniques, such as mixed-precision training and gradient accumulation, are essential for managing the immense computational complexity. Ultimately, training large language models represents a sophisticated intersection of machine learning, computational linguistics, and ethical technology development—an ongoing scientific endeavor pushing the boundaries of artificial intelligence's understanding and generation of human-like text [102].

4.11.1 Challenges of Training Large Language Models

Training large language models (LLMs) presents a complex array of challenges that span technical, ethical, and practical domains. The immense demand for computational resources is at the forefront, necessitating high-performance hardware and sophisticated distributed systems, which can be prohibitively expensive and complex to manage. This infrastructure challenge is closely tied to the environmental concerns arising from the substantial energy consumption required for training and operating these models. Data-related issues form another critical cluster of challenges. The need for vast, high-quality, and diverse datasets introduces significant hurdles in collection, cleaning, and preprocessing. Ensuring data quality while mitigating inherent biases is a labor-intensive process that directly impacts model performance and ethical considerations. The extended training periods, often stretching over weeks or months, further complicate the development process, limiting rapid experimentation and iteration [101, 102]. Figure 4.6 depicts a typical AI model training challenge.

Balancing model performance with generalization capability presents a delicate challenge. LLMs'high capacity for memorization can lead to overfitting, necessitating careful architecture design and hyperparameter tuning. This challenge intersects with the broader issues of model interpretability and transparency, as understanding the decision-making processes in these complex systems remains a significant hurdle. Ethical considerations permeate every aspect of LLM development. From addressing biases in training data to ensuring fair and responsible model outputs, developers must navigate a complex landscape of moral and societal implications. This extends to data privacy and security concerns, requiring

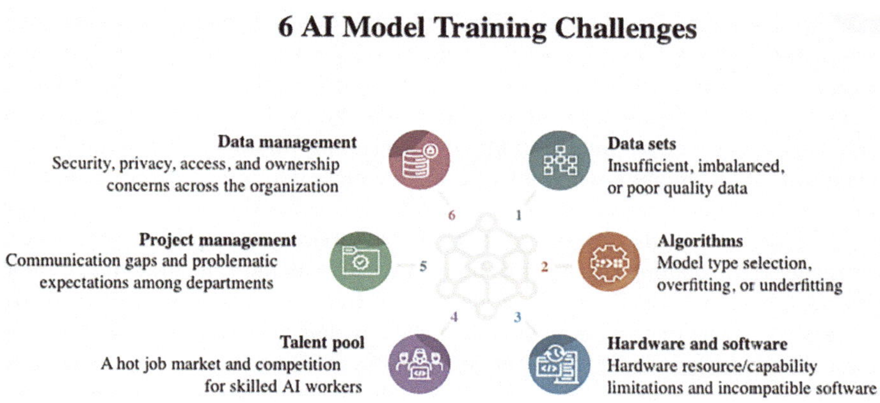

Fig. 4.6 AI model (e.g., LLMs) training challenges [104]

robust safeguards to protect sensitive information in training data and model outputs [101].

Practical challenges in deployment, such as managing inference latency and keeping models up to date with new information without full retraining, add another layer of complexity. These issues are particularly pertinent in real-time applications where response speed and up-to-date knowledge are crucial. Addressing this multifaceted set of challenges demands ongoing research, interdisciplinary collaboration, and a commitment to responsible development practices. As the field evolves, finding innovative solutions to these challenges will be key to unlocking the full potential of LLMs while mitigating associated risks and ethical concerns [102].

4.11.2 Techniques to Mitigate Overfitting in Large Language Models

Addressing overfitting in large language models requires a multifaceted strategy that combines various techniques to enhance generalization. The core of this approach is the expansion and diversification of training data, leveraging existing public datasets and innovative data augmentation techniques, such as back-translation and word replacement. These methods significantly broaden the model's exposure to linguistic variations, fostering a more robust language understanding [103].

Regularization plays a crucial role in preventing overfitting. Techniques like Dropout randomly deactivate neural connections during training, while L1 and L2 regularization penalize model complexity, encouraging the model to learn more generalizable features. Early stopping safeguards against overtraining by halting the process when validation performance declines, preserving the model's generalization ability [105].

Ensemble methods and cross-validation further enhance generalization by combining insights from multiple model variations and ensuring consistent performance across different data subsets. In some cases, simplifying the model architecture or focusing on the most salient features can prevent the model from learning noise or irrelevant patterns. Continual learning and fine-tuning strategies enable models to adapt to new information without requiring complete retraining, thereby maintaining relevance while preserving their generalization capabilities. Underpinning all these techniques is the critical process of hyperparameter optimization, where parameters like learning rate and batch size are carefully tuned to strike the optimal balance between model fit and generalization. Researchers and developers can significantly mitigate overfitting in large language models by integrating these diverse approaches and maintaining vigilant performance monitoring. This comprehensive strategy not only enhances the model's ability to generalize to unseen data but also improves its overall robustness and applicability across various linguistic tasks and domains [103–105].

4.12 Training Unsupervised Machine Learning Models

Training unsupervised machine learning models [2–5] is a nuanced process that begins with meticulous data collection and preparation. Researchers gather extensive unlabeled datasets relevant to their problem domain, carefully cleaning and preprocessing the information to ensure quality and consistency. This preparatory phase often involves handling missing values, addressing outliers, and normalizing or standardizing features to create a robust foundation for model training.

Selecting an appropriate unsupervised learning algorithm is crucial and depends on the project's specific goals. Whether opting for clustering techniques like K-means or DBSCAN, dimensionality reduction methods such as PCA, or sophisticated anomaly detection algorithms, this choice shapes the entire training process. Once selected, the model is initialized with carefully chosen parameters, setting the stage for the learning phase. The core of unsupervised learning lies in its ability to discern patterns and structures within data without explicit guidance. During training, the model iteratively refines its understanding, grouping similar data points in clustering tasks or discovering lower-dimensional representations in dimensionality reduction scenarios. This process often involves multiple iterations, with algorithms like K-means repeatedly adjusting cluster assignments and centers to optimize their representation of the data's inherent structure [106, 107].

Hyperparameter tuning plays a vital role in optimizing model performance. Techniques such as grid or random search are employed to fine-tune the model's configuration, ensuring it effectively captures the nuances of the data. Evaluation in unsupervised learning presents unique challenges, requiring metrics tailored to the specific task—such as silhouette scores for clustering or reconstruction error for dimensionality reduction. The final stages of the process involve carefully interpreting and validating the model's findings. Researchers analyze the uncovered patterns, leveraging domain expertise to contextualize and validate findings. This often leads to a refinement phase, where the model is adjusted or alternative algorithms are explored based on these insights. Ultimately, the validated model is applied to new, unseen data, unlocking valuable insights and patterns that drive innovation across various fields. This iterative and exploratory nature of unsupervised learning underscores its power in uncovering hidden structures within complex datasets. It offers a pathway to discover knowledge without the constraints of predefined categories or labels [106, 108].

4.13 Train a Deep Learning Neural Network

Training a deep learning neural network [8] is a multifaceted process that begins with meticulous data preparation. This crucial first step involves collecting a substantial, high-quality dataset relevant to the task, followed by thorough cleaning, handling missing values, and removing outliers. Feature engineering and selection may be necessary, along with appropriate scaling or normalization of features to ensure optimal model performance [109, 110].

The dataset is then strategically divided into training, validation, and test sets, typically in a 70/15/15 ratio, ensuring each subset accurately represents the overall data distribution. The network architecture is carefully chosen, considering the number of layers, neurons, and appropriate activation functions, with the output layer designed to match the specific task requirements. Initializing weights and biases, often using techniques such as Xavier or Glorot initialization, sets the stage for effective learning. Hyperparameter selection follows, encompassing crucial decisions on learning rate, batch size, number of epochs, and regularization techniques. The training process involves forward propagation, loss calculation, backpropagation for gradient computation, and weight updates using optimizers like Adam or SGD [111].

Throughout the training, close monitoring of performance metrics on both training and validation sets is essential, with early Stopping implemented to prevent overfitting. The process is inherently iterative, involving continuous evaluation and refinement of hyperparameters, architecture, and regularization techniques based on the performance of the validation set. The final model assessment is conducted on the held-out test set, followed by deployment to production environments with ongoing performance monitoring. Best practices such as mini-batch gradient descent, data shuffling between epochs, learning rate decay, batch normalization, and various regularization techniques are employed to enhance model robustness and generalization. This comprehensive approach to deep learning model training emphasizes the importance of experimentation and iteration. By systematically exploring different architectures and hyperparameters, practitioners can optimize their models for specific tasks, ultimately developing robust and effective neural networks capable of tackling complex real-world problems [111, 112].

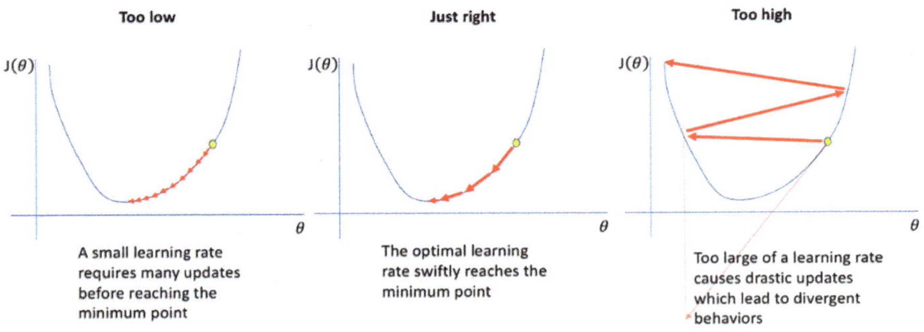

Fig. 4.7 Impact of learning rate on neural network training and optimization [114]

4.13.1 Learning Rate for Neural Networks

Selecting an optimal learning rate for neural networks is a critical yet nuanced process that combines systematic testing with adaptive strategies. The learning rate range test serves as a foundational technique, involving a gradual increase in learning rate during training and plotting the loss against it. As shown in Fig. 4.7, this method helps identify the sweet spot where learning accelerates and the upper bound where performance begins to degrade [113]. Adaptive learning rate methods, such as Adam, RMSProp, and Adagrad, offer an alternative approach by automatically adjusting the learning rate during training, potentially reducing the need for manual tuning. These can be complemented by learning rate schedules, which systematically decrease the rate over time, following patterns like step decay, exponential decay, or cosine annealing [111].

For a more exhaustive approach, grid search enables a comprehensive exploration of learning rates across a predefined range, typically from 0.1 to 10^-6, with 0.01 often serving as a reasonable starting point. Throughout this process, closely monitoring training dynamics is crucial; a learning rate that's too low results in sluggish loss reduction, while one that's too high leads to erratic loss behavior. Advanced techniques, such as cyclical learning rates that alternate between lower and upper bounds, can help navigate complex loss landscapes and escape local minima. Ultimately, the key to finding the ideal learning rate lies in experimentation and careful observation of how different rates impact model training. By combining these methods and maintaining vigilant monitoring of loss curves, practitioners can fine-tune their neural networks for optimal performance across a wide range of tasks and architectures [113–115].

4.14 Cross-Validation in AI/ML Model Training

Cross-validation [116] stands as a cornerstone technique in machine learning, significantly enhancing the reliability and robustness of model evaluations through its multifaceted approach. By evaluating models on multiple subsets of data,

Table 4.1 Types of cross-validation techniques [118]

Cross-validation type	Purpose	Advantages	Disadvantages
K-fold cross validation	Multiple rounds of testing with K segments	Unbiased performance estimation	Computational complexity
Stratified K-fold	Maintains class distribution	Good for imbalanced datasets	Potential data leakage
Leave-one-out (LOO)	Uses one sample for validation	Robust performance estimation	Time-consuming with large datasets
Time series	Maintains sequential order	Preserves temporal relationships	Complex to implement

cross-validation provides a more accurate and comprehensive picture of how a model will perform on unseen data, mitigating the risks associated with overly optimistic or pessimistic performance estimates that can arise from a single train-test split. This method is particularly effective in detecting and mitigating overfitting, as it tests the model's ability to generalize across various held-out sets. Table 4.1 below presents the types of cross-validation techniques and their respective advantages and disadvantages for the model training process.

One of the key strengths of cross-validation lies in its efficient utilization of available data, which is especially valuable when working with limited datasets. Techniques like k-fold cross-validation allow the entire dataset to be used for training and validation, providing a more comprehensive assessment of model performance. This approach also reduces biases that can emerge from specific data partitioning, offering a more stable and reliable performance estimate. Cross-validation is crucial in hyperparameter tuning, as it enables the systematic evaluation of different model configurations without overfitting to a single validation set. It also provides insights into model stability by revealing sensitivity to variations in training data. Stratified cross-validation ensures that each fold maintains the original class distribution for imbalanced datasets, leading to more reliable evaluations. Ultimately, cross-validation simulates real-world performance more accurately, better estimating how the model will perform on new, unseen data in practical applications. By addressing these various aspects, cross-validation significantly enhances the trustworthiness and generalizability of machine learning models, making it an indispensable tool in developing robust and reliable AI systems [117, 118].

4.14.1 Balancing Bias and Variance During Cross-Validation

As shown in Fig. 4.8, balancing bias and variance during cross-validation is a nuanced process that requires a multifaceted approach to ensure robust model performance [119]. The foundation of this balance lies in selecting an appropriate cross-validation technique, such as k-fold cross-validation for general cases, stratified k-fold for imbalanced datasets, or time series cross-validation for temporal data [118]. This selection sets the stage for a comprehensive evaluation of model performance across different data subsets.

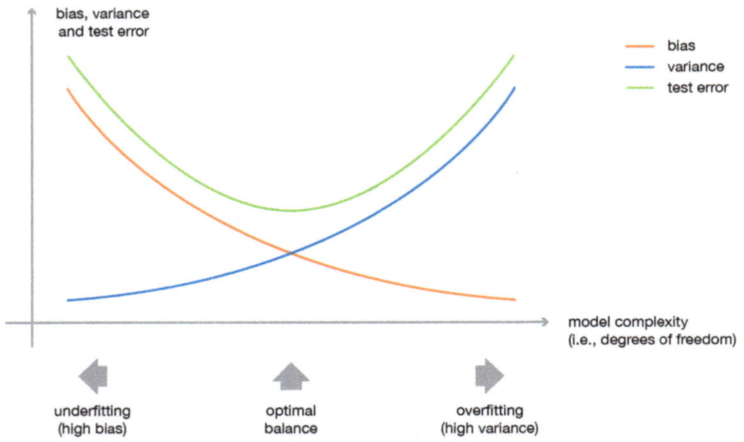

Fig. 4.8 Balancing bias and variance during cross-validation in training process [122]

A key strategy in this balancing act involves carefully adjusting model complexity. Starting with simpler models and gradually increasing complexity while employing regularization techniques, such as L1 or L2 regularization, helps control overfitting and manage the bias-variance trade-off. This approach is complemented by thoughtful feature selection and engineering, which can reduce variance by eliminating irrelevant features and capturing essential data patterns through the creation of new features [119–121].

Ensemble methods play a crucial role in this balance, with techniques like bagging (e.g., Random Forests) effectively reducing variance while boosting methods (e.g., Gradient Boosting) can address both bias and variance simultaneously. Hyperparameter tuning, ideally through nested cross-validation, further refines model performance while minimizing overfitting of the validation data. Continuous monitoring and analysis are essential throughout this process. Observing learning curves helps identify underfitting or overfitting tendencies while analyzing error distributions across folds, which can reveal potential issues of overfitting. Increasing the training data volume can significantly reduce bias and variance [119, 120].

The iterative nature of this process cannot be overstated. Experimenting with different models, techniques, and evaluation metrics while comparing results across multiple runs ensures a comprehensive exploration of the bias-variance landscape. Ultimately, the goal is to achieve a balance where the model generalizes well to unseen data, often requiring careful trade-offs based on the specific characteristics of the problem. This holistic approach to cross-validation enhances model reliability and deepens our understanding of the underlying data patterns and model behaviors [117, 118].

Overfitting is a common challenge in machine learning. It occurs when a model learns the training data too well, including its noise and peculiarities, at the expense of generalizing to new, unseen data. Several key signs indicate a model is overfitting. A significant performance discrepancy between training and test data is a primary indicator, where the model performs notably better on the training set. This is often visible in learning curves, where training and validation curves diverge as training progresses. Overfitting models typically exhibit high variance, showing sensitivity to small fluctuations in the training data and performing inconsistently across different data subsets. Poor generalization is another hallmark, as the model fails to perform well on new data despite achieving high accuracy on the training set. Excessive complexity relative to the data, unstable predictions, and inconsistent feature importance are also tell-tale signs. In cross-validation, significant performance variations across folds can suggest overfitting. In practical applications, rapidly degrading performance over time and identifying implausible relationships in the data are additional red flags. To combat overfitting, it's crucial to employ proper validation techniques, including hold-out validation sets, cross-validation, and learning curve analysis, while regularly monitoring these indicators to ensure models generalize effectively to new data [123, 124].

4.15 Early Stopping and Regularization Techniques

The training process of machine learning models typically exhibits a pattern where the training error continually improves as the model is optimized. If the training error increases, it may indicate an issue with the optimizer. However, the test error usually improves initially but may grow as the network overfits. This pattern is illustrated in the training curves, as shown in Fig. 4.9. This observation leads to a strategy called early Stopping, where training is halted at the point where generalization error starts to increase. To implement this strategy correctly, we use a validation set rather than the test set to determine when to stop, as using the test set would be considered a form of cheating. However, implementing early stopping effectively can be challenging due to fluctuations in training and validation errors caused by stochasticity in the gradients. It can be challenging to distinguish between a genuine increase in error and normal fluctuations. A common approach to address this is to

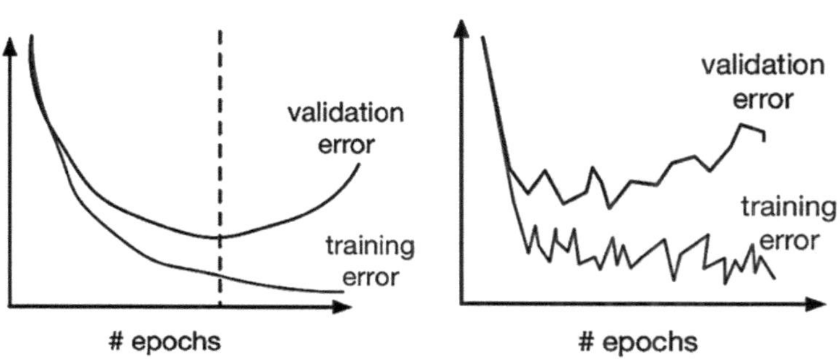

Fig. 4.9 Training curve depicting the training error and validation (tuning) error in ideal condition (left plot) and with the stochasticity in stochastic gradient descent [125]

measure validation error less frequently, such as once per epoch. Training is then stopped if the validation error fails to improve after one or several consecutive epochs. This heuristic helps to smooth out short-term fluctuations and provides a more reliable indicator of when to stop training [125].

Cost functions in machine learning typically consist of the average loss function calculated over the training set. However, incorporating an additional component known as a regularization term or regularizer is often beneficial. This term serves to penalize hypotheses that are considered pathological or unlikely to generalize well to unseen data. By adding a regularizer to the cost function, we can guide the learning process towards accurate solutions on the training data, and we are more likely to perform well on new, unseen examples. This approach helps to mitigate overfitting and improve the model's ability to generalize, striking a balance between fitting the training data and maintaining simplicity or smoothness in the learned function. Regularization techniques are crucial in improving the generalization capabilities of classification machine learning models. These methods aim to prevent overfitting, manage model complexity, and enhance performance on unseen data. Key regularization approaches include L1 (Lasso) and L2 (Ridge) regularization, which add penalties based on the magnitude of coefficients. L1 regularization potentially leads to sparse models, while L2 regularization helps mitigate multicollinearity. The Elastic Net combines these two methods, balancing feature selection with coefficient shrinkage. In neural networks, Dropout is commonly employed (Fig. 4.10), randomly omitting neurons during training to foster more robust feature learning. Early Stopping prevents overfitting by halting training when validation performance begins to decline [125–127].

Data augmentation artificially expands the training set, improving the model's generalization ability. Batch normalization, while primarily used to address internal covariate shifts, also has a slight regularization effect. Weight Decay, similar to L2 regularization, gradually reduces the magnitudes of weights during training. Noise

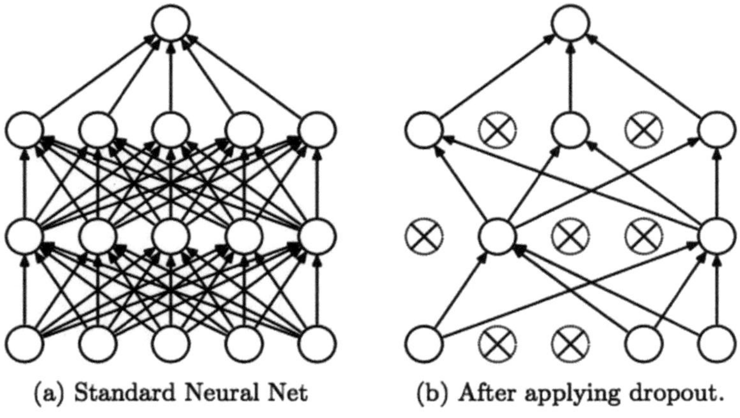

Fig. 4.10 Dropout is a regularization technique used in neural networks to prevent complex co-adaptations from other neurons [127]

injection, which adds random perturbations to inputs or weights, can enhance model robustness. Multi-task learning and transfer learning leverage related tasks or pre-trained models to improve generalization, especially when dealing with limited data. Lastly, cross-validation helps in model selection and hyperparameter tuning by providing a more reliable estimate of generalization performance. The choice of regularization technique depends on the specific problem, dataset characteristics, and model architecture, with each method offering unique benefits in the quest for more generalizable and robust classification models [126, 127].

4.16 Optimize Model Training to Reduce Computational Costs

Optimizing model training and reducing computational costs are crucial for efficient machine learning development. Several key strategies can be employed to achieve these goals [128, 129]. Using older or simpler models for less complex tasks can be a cost-effective approach, as can leveraging batch processing for non-time-sensitive tasks. Hyperparameter optimization, including setting appropriate temperature parameters and utilizing automated tools, is crucial for enhancing efficiency. Implementing caching to store and reuse the results of previous API calls can significantly reduce redundant computations. Model pruning and quantization techniques help decrease model size and memory usage, while early Stopping prevents overfitting and unnecessary computation [129]. Optimizing GPU utilization and reducing hyperparameter search space is essential for maximizing resource efficiency. Data augmentation techniques can artificially increase dataset size without increasing raw data volume, and transfer learning can leverage pre-trained models to reduce training time. Streamlining data processing pipelines, considering knowledge distillation, and using appropriate regularization techniques further contribute to optimization. Monitoring and analyzing training runs with experiment tracking tools helps identify inefficiencies while implementing distributed training, which can parallelize processes across multiple GPUs or machines. By employing these strategies, developers can significantly reduce computational costs associated with model training while maintaining performance and quality, ultimately leading to more efficient and cost-effective machine learning solutions [130–133].

4.16.1 Model Pruning to Reduce Resource Usage During Training

Model pruning is a powerful technique that can significantly reduce resource usage during training in several ways [130]. By removing less important neurons or connections, pruning results in a smaller model with fewer parameters, which reduces the memory footprint and allows training on devices with limited memory capacity. This reduction in model size leads to decreased computational requirements, resulting in faster execution times and lower computational costs for training and inference. Pruned models typically consume less energy, which is particularly important for battery-powered devices and environmentally conscious computing. The reduced number of parameters often leads to faster training times, potentially resulting in cost savings for large-scale neural networks.

Additionally, pruning can act as a form of regularization, helping to prevent overfitting and improving generalization with less extensive training data. The scalability of pruning enables iterative optimization at various stages of model development. Pruned models are often more compatible with hardware accelerators and specialized training engines due to their reduced complexity. While pruning may require some initial computational overhead, the benefits of reduced model size and improved efficiency typically outweigh this cost, especially for large models or when training resources are constrained. Overall, model pruning offers a valuable approach to optimizing resource usage and enhancing the efficiency of machine learning models during training [130, 131, 134].

Neural network pruning encompasses a range of methods designed to reduce model size and computational requirements while maintaining performance. Weight pruning, including magnitude-based and structured approaches, removes individual weights or structural components. Neuron pruning targets individual neurons based on their level of activation or importance to the network's performance. Channel pruning methods, such as L1-norm regularization and dynamic pruning, aim to reduce the number of feature maps. Iterative pruning involves multiple cycles of training and pruning, with the timing of pruning categorized into two approaches: train-time and post-training. Pruning can be applied locally to specific layers or globally across the entire network [135]. Data-driven pruning utilizes activation patterns during training to identify redundant elements. The Lottery Ticket Hypothesis method seeks to identify and retrain sparse "winning ticket" subnetworks [134]. More advanced techniques, such as Optimal Brain Damage/Surgeon, utilize second-order derivative information to determine the importance of weights. In contrast, norm-based pruning relies on the L2 norm of neuron weights [136]. The selection of a pruning method depends on the specific task requirements and network architecture, with each approach offering unique trade-offs between model compression and performance preservation [130, 132, 137].

4.17 Differences in Training and Real-World Populations

The differences between training and real-world populations can significantly impact the performance of AI systems in healthcare settings. One of the primary issues is reduced generalizability, where AI algorithms that perform well on training data may struggle when applied to external real-world data not used in training. This is particularly evident when models trained in limited environments, such as a single hospital, are deployed in diverse clinical settings. A data shift occurs when a mismatch between the training data distribution and the real-world data encountered during deployment can substantially decrease model performance. Biased or inaccurate predictions can result from training data that doesn't adequately represent the real-world population, potentially leading to poor performance in underrepresented groups [138, 139].

This issue is exemplified by cardiovascular risk prediction algorithms that are primarily trained on male patient data, which may be less accurate for female patients. Additionally, performance estimates based on internal test data may overestimate the model's real-world performance on external data, leading to unrealistic expectations. Training-serving skew can occur when a model is deployed on patients with population characteristics that differ from those in the training data, potentially leading to unexpected performance issues in real-world settings. If training data reflects existing healthcare disparities or biases, the AI may learn and perpetuate these biases in its predictions. Moreover, models trained on homogeneous datasets may struggle with the increased heterogeneity in real-world populations. To address these challenges, researchers and developers should ensure diverse and representative training data, conduct thorough external validation on diverse populations, use explainable AI techniques to detect potential biases or data shift issues, implement continuous monitoring and updating of models in real-world settings, and consider the use of transfer learning or domain adaptation techniques to improve generalizability. By addressing these population differences, AI models can be made more robust and reliable for real-world clinical applications, ultimately improving their effectiveness and trustworthiness in healthcare settings [140–142]. This topic is discussed in detail in Chap. 5, Clinical Validation of AI/ML Models.

4.18 Summary

This chapter provided a comprehensive exploration of performance metrics for various machine learning algorithms, including supervised, unsupervised, generative AI, NLP, and reinforcement learning. It delved into the selection of appropriate metrics and addressed critical issues, including overfitting, underfitting, outlier handling, confusion metrics, and monotonicity. The chapter also covered best practices for dataset splitting, cross-validation techniques, and balancing bias and variance. Recent developments in RAG for LLMs and the Hallucination Index Report for foundation models were discussed. An in-depth examination of training methodologies for different machine learning models was presented, along with hyperparameter tuning strategies. The chapter explored techniques such as early stopping and regularization to enhance model generalizability, as well as model pruning and optimization of computational cost.

4.19 Points to Ponder

1. List deep learning libraries for different data modalities for machine learning.
2. List machine learning libraries for dimensionality reduction.
3. List machine learning libraries to check whether the data distribution is the same in the train and test datasets for imaging data.
4. List key performance metrics for image segmentation quality.

4.20 Answers

1. Here is a List of popular deep learning libraries for different data modalities.

Data modality	Top libraries
Tabular	TensorFlow, PyTorch
Image	PyTorch, TensorFlow/Keras, Caffe, MXNet
Text/NLP	PyTorch, TensorFlow/Keras, CNTK
Audio/speech	TensorFlow, PyTorch, MXNet
Time-series/sensor	PyTorch, TensorFlow/Keras, Deeplearning4j
Multimodal	TensorFlow, PyTorch

2. Here is a list of machine learning libraries for dimensionality reduction.

- **Scikit-learn**: This is the most popular Python library for implementing classical dimensionality reduction algorithms, such as Principal Component Analysis (PCA), t-Distributed Stochastic Neighbor Embedding (t-SNE), and Linear Discriminant Analysis (LDA), among others. Its simple API and integration with other scientific libraries make it ideal for prototyping and production workflows.
- **TensorFlow**: While primarily a deep learning framework, TensorFlow supports custom and neural network-based dimensionality reduction methods, such as autoencoders. It is suitable for large-scale and deep learning-based reduction tasks.
- **PyTorch**: Similar to TensorFlow, PyTorch is widely used for implementing neural network-based dimensionality reduction, especially autoencoders and variational autoencoders. Its flexibility and dynamic computation graph are favored for research and custom solutions.
- **Pandas**: Although not explicitly designed for dimensionality reduction, Pandas is often used for preprocessing, feature selection, and data preparation before applying reduction algorithms.
- **FastAI**: built on top of PyTorch, provides high-level abstractions for deep learning, including tools for building and training autoencoders for dimensionality reduction.

- **JAX**: This numerical computing library is used for high-performance machine learning research and can be leveraged for custom dimensionality reduction techniques, especially those requiring automatic differentiation and GPU/TPU acceleration.

3. List machine learning libraries to check whether the data distribution is the same in the train and test datasets for imaging data.

- **genai-evaluation**: This library implements the multivariate Kolmogorov–Smirnov (KS) distance in any dimension, enabling the comparison of the full joint empirical distributions of two datasets, including high-dimensional imaging data. It provides functions to compute the multivariate empirical distribution and the KS statistic, making it suitable for assessing distributional similarity between train and test sets. Install it with pip install genai-evaluation.
- **scipy.stats**: The ks_2samp function from SciPy allows you to perform the two-sample Kolmogorov–Smirnov test, which helps compare the distributions of two datasets. While typically used for one-dimensional data, it can also be applied to summary statistics or specific features extracted from images to assess distributional similarity.
- **Seaborn/matplotlib**: These libraries can help you visually compare the distributions by plotting histograms or density plots of features or pixel intensities from your train and test datasets.

4. Examples of Performance Metrics for Image Segmentation Quality.

- Dice Similarity Coefficient (Dice/F1 Score): Measures the overlap between the predicted and ground truth segmentation masks. It is widely used for both binary and multi-class segmentation tasks and is especially important in medical imaging.

- **Range:** 0 (no overlap) to 1 (perfect overlap)
- Intersection over Union (IoU/Jaccard Index): Quantifies the ratio of the intersection to the union of the predicted and ground truth masks. This is a standard metric for evaluating the accuracy of segmentation.
- **Range:** 0 (no overlap) to 1 (perfect overlap)
- Sensitivity (Recall): Assesses the proportion of actual positives correctly identified by the segmentation.
- **Range:** 0 (no true positives detected) to 1 (all true positives detected)
- Specificity: Measures the proportion of actual negatives correctly identified.
- **Range:** 0 (no true negatives detected) to 1 (all true negatives detected)
- Precision: Evaluates the proportion of predicted positives that are truly positive.
- **Range:** 0 (no correct positive predictions) to 1 (all positive predictions are correct)
- Cohen's Kappa: Assesses agreement between predicted and reference segmentations, accounting for chance agreement.

- **Range:** −1 (complete disagreement) to 1 (perfect agreement), with 0 indicating chance-level agreement.
- Rand Index/Adjusted Rand Index: Measures the similarity between two clusterings, helpful in comparing segmentation masks.
- **Range:** 0 (no similarity) to 1 (identical clusterings).
- Hausdorff Distance: Captures the largest distance between the boundaries of the predicted and ground truth masks, highlighting spatial discrepancies.
- **Range:** 0 (perfect boundary alignment) to infinity (greater values indicate larger spatial discrepancies).
- Structural Similarity Index Measure (SSIM): Evaluates the perceptual similarity between segmentation masks, taking into account structural information.
- **Range:** −1 (no similarity) to 1 (identical structure).
- Global/Per-Class Accuracy: Indicates the proportion of correctly classified pixels overall or within each class.
- **Range:** 0 (no correct classifications) to 1 (all correct).

References

1. Machine learning foundational courses. https://developers.google.com/machine-learning/intro-to-ml/supervised
2. Supervised vs. unsupervised learning. https://cloud.google.com/discover/supervised-vs-unsupervised-learning
3. Clustering. https://developers.google.com/machine-learning/clustering
4. https://www.ibm.com/think/topics/dimensionality-reduction
5. https://www.datacamp.com/tutorial/association-rule-mining-python
6. What is reinforcement learning? https://aws.amazon.com/what-is/reinforcement-learning/
7. What is semi-supervised learning? https://cloud.google.com/discover/supervised-vs-unsupervised-learning#what-is-semi-supervised-learning
8. What is deep learning? https://azure.microsoft.com/en-us/resources/cloud-computing-dictionary/what-is-deep-learning
9. Ensemble machine learning. https://www.machinelearningmastery.com/tour-of-ensemble-learning-algorithms/
10. https://www.fda.gov/medical-devices/software-medical-device-samd/transparency-machine-learning-enabled-medical-devices-guiding-principles
11. https://www.fda.gov/medical-devices/medical-device-regulatory-science-research-programs-conducted-osel/evaluation-methods-artificial-intelligence-ai-enabled-medical-devices-performance-assessment-and
12. https://www.fda.gov/medical-devices/medical-device-regulatory-science-research-programs-conducted-osel/performance-evaluation-methods-evolving-artificial-intelligence-ai-enabled-medical-devices
13. https://www.fda.gov/medical-devices/medical-devices-news-and-events/cdrh-issues-guiding-principles-predetermined-change-control-plans-machine-learning-enabled-medical
14. https://www.news-medical.net/news/20241016/FDA-strengthens-AI-regulation-to-ensure-patient-safety-and-innovation-in-healthcare.aspx
15. https://www.fda.gov/files/medical%20devices/published/US-FDA-Artificial-Intelligence-and-Machine-Learning-Discussion-Paper.pdf

16. https://namsa.com/resources/blog/fdas-regulation-of-ai-ml-samd/
17. Cabitza F et al (2020) As if sand were stone. New concepts and metrics to probe the ground on which to build trustable AI
18. https://svitla.com/blog/ai-ml-performance-metrics/
19. https://bmcmedinformdecismak.biomedcentral.com/articles/10.1186/s12911-021-01634-3
20. Topaloglu M et al (2021) In the pursuit of privacy: the promises and predicaments of federated learning in healthcare
21. Lal A et al (2022) Regulatory oversight and ethical concerns surrounding software as medical device (SaMD) and digital twin technology in healthcare
22. https://neptune.ai/blog/how-to-monitor-your-models-in-production-guide
23. https://stats.stackexchange.com/questions/595983/precision-vs-specificity
24. https://developers.google.com/machine-learning/crash-course/classification/accuracy-precision-recall
25. https://stats.stackexchange.com/questions/362332/is-there-any-difference-between-sensitivity-and-recall
26. https://www.numpyninja.com/post/recall-specificity-precison-f1-scores-and-accuracy
27. https://www.geeksforgeeks.org/visualize-confusion-matrix-using-caret-package-in-r/
28. https://www.kaggle.com/code/dansbecker/what-is-log-loss
29. https://developers.google.com/machine-learning/crash-course/classification/roc-and-auc/
30. https://developer.nvidia.com/blog/a-comprehensive-overview-of-regression-evaluation-metrics/
31. https://towardsdatascience.com/ways-to-evaluate-regression-models-77a3ff45ba70
32. https://www.ejable.com/tech-corner/ai-machine-learning-and-deep-learning/performance-metrics-for-regression/
33. https://c3.ai/introduction-what-is-machine-learning/regression-performance/
34. https://docs.aws.amazon.com/machine-learning/latest/dg/model-fit-underfitting-vs-overfitting.html
35. https://docs.aws.amazon.com/machine-learning/latest/dg/splitting-the-data-into-training-and-evaluation-data.html
36. https://www.linkedin.com/advice/3/how-do-outliers-impact-linear-regression-dz0ff
37. https://whanchoi.people.uic.edu/TheEffectofOutliersonRegressionAnalysis.pdf
38. https://developer.nvidia.com/blog/dealing-with-outliers-using-three-robust-linear-regression-models/
39. https://www.institutedata.com/us/blog/evaluating-regression-models/
40. https://www.ibm.com/think/topics/confusion-matrix
41. https://www.v7labs.com/blog/confusion-matrix-guide
42. https://lightning.ai/docs/torchmetrics/stable/classification/confusion_matrix.html
43. https://viso.ai/deep-learning/confusion-matrix/
44. https://learn.microsoft.com/en-us/azure/machine-learning/component-reference/one-vs-all-multiclass?view=azureml-api-2
45. https://machinelearningmastery.com/one-vs-rest-and-one-vs-one-for-multi-class-classification/
46. https://www.kdnuggets.com/2023/04/exploring-unsupervised-learning-metrics.html
47. https://scikit-learn.org/1.6/auto_examples/cluster/plot_adjusted_for_chance_measures.html
48. https://www.linkedin.com/pulse/uncertainty-unsupervised-deep-learning-challenges-gplkc
49. https://eitca.org/artificial-intelligence/eitc-ai-adl-advanced-deep-learning/unsupervised-learning/unsupervised-representation-learning/examination-review-unsupervised-representation-learning/what-are-the-challenges-associated-with-evaluating-the-effectiveness-of-unsupervised-learning-algorithms-and-what-are-some-potential-methods-for-this-evaluation/
50. https://cloud.google.com/transform/kpis-for-gen-ai-why-measuring-your-new-ai-is-essential-to-its-success
51. https://learn.microsoft.com/en-us/azure/ai-studio/concepts/evaluation-metrics-built-in

52. https://research.aimultiple.com/large-language-model-evaluation/
53. https://www.cs.cmu.edu/~roni/papers/eval-metrics-bntuw-9802.pdf
54. https://cloud.google.com/vertex-ai/generative-ai/docs/models/determine-eval
55. https://developer.nvidia.com/blog/measuring-generative-ai-model-performance-using-nvidia-genai-perf-and-an-openai-compatible-api/
56. https://www.galileo.ai/hallucinationindex-2023/methodology
57. Tivnan M et al (2024) Hallucination index: an image quality metric for generative reconstruction models
58. Zero-shot Learning (2024) https://www.ibm.com/think/topics/zero-shot-learning
59. Oliver M et al (2024) Crafting efficient fine-tuning strategies for large language models. https://arxiv.org/abs/2407.13906
60. What is RAG? https://aws.amazon.com/what-is/retrieval-augmented-generation/?nc1=h_ls
61. Yu T et al (2024) In defense of RAG in the era of long-context language models. https://arxiv.org/abs/2409.01666
62. https://cloud.google.com/use-cases/retrieval-augmented-generation?hl=en
63. https://www.smashingmagazine.com/2024/01/guide-retrieval-augmented-generation-language-models/
64. https://snorkel.ai/large-language-models/rag-retrieval-augmented-generation/
65. https://www.fingoweb.com/blog/rag-vs-fine-tuning-alternative-solutions-for-ai-model-customization/
66. Barnett S et al (2024) Seven failure points when engineering a retrieval augmented generation system. https://arxiv.org/html/2401.05856v1
67. https://irisagent.com/blog/enhancing-large-language-models-a-deep-dive-into-rag-llm-technology/
68. Hallucination Index Report (2024) https://www.galileo.ai/hallucinationindex
69. Blagec K (2022) A global analysis of metrics used for measuring performance in natural language processing. https://arxiv.org/abs/2204.11574
70. https://deconvoluteai.com/blog/evaluating-nlp-models
71. Jordan S et al (2020) Evaluating the performance of reinforcement learning algorithms. https://proceedings.mlr.press/v119/jordan20a/jordan20a.pdf
72. https://www.linkedin.com/advice/0/how-can-you-measure-success-reinforcement-learning
73. Dayal A et al (2021) Reward criteria impact on the performance of reinforcement learning agent for autonomous navigation. https://www.sciencedirect.com/science/article/pii/S1568494622004586
74. Abouelazm A et al (2024) A review of reward functions for reinforcement learning in the context of autonomous driving. https://arxiv.org/pdf/2405.01440v1
75. https://www.geeksforgeeks.org/how-to-make-a-reward-function-in-reinforcement-learning/
76. https://www.alphanome.ai/post/the-critical-importance-of-the-reward-function-in-reinforcement-learning
77. https://towardsdatascience.com/reinforcement-learning-explained-visually-part-4-q-learning-step-by-step-b65efb731d3e
78. https://blog.datadive.net/monotonicity-constraints-in-machine-learning/
79. https://www.geeksforgeeks.org/why-do-activation-functions-have-to-be-monotonic/
80. https://www.toptal.com/machine-learning/monotonic-ai-models
81. Kotlowski W et al (2009) Rule learning with monotonicity constraints
82. https://community.alteryx.com/t5/Data-Science/New-ML-Model-Macro-With-Monotonic-Constraint-Capabilities/ba-p/1095691
83. https://scikit-learn.org/stable/auto_examples/ensemble/plot_monotonic_constraints.html
84. https://towardsdatascience.com/training-regression-models-bfd779a73a60
85. Supervised machine learning: classification. https://www.coursera.org/learn/supervised-machine-learning-classification/
86. https://dagshub.com/blog/how-to-improve-machine-learning-model-accuracy/
87. https://aws.amazon.com/what-is/hyperparameter-tuning/
88. https://serokell.io/blog/hyperparameter-tuning-in-ml

89. Splitting into train, dev and test sets. https://cs230.stanford.edu/blog/split/
90. Datasets: dividing the original dataset. https://developers.google.com/machine-learning/crash-course/overfitting/dividing-datasets
91. Real life: Should your training/valid/test set come from the same distribution?. https://forums.fast.ai/t/real-life-should-your-training-valid-test-set-come-from-the-same-distribution/77709
92. What to do when your training and testing data come from different distributions. https://www.freecodecamp.org/news/what-to-do-when-your-training-and-testing-data-come-from-different-distributions-d89674c6ecd8/
93. https://towardsdatascience.com/two-rookie-mistakes-i-made-in-machine-learning-improper-data-splitting-and-data-leakage-3e33a99560ea
94. https://encord.com/blog/train-val-test-split/
95. Naderalvojoud B et al. Improving machine learning with ensemble learning on observational healthcare data. https://pmc.ncbi.nlm.nih.gov/articles/PMC10785929/pdf/146.pdf
96. A comprehensive guide to ensemble learning: what exactly do you need to know. https://neptune.ai/blog/ensemble-learning-guide
97. A gentle introduction to ensemble learning algorithms. https://machinelearningmastery.com/tour-of-ensemble-learning-algorithms/
98. https://developers.google.com/machine-learning/resources/intro-llms
99. https://www.elastic.co/what-is/large-language-models
100. https://snorkel.ai/blog/large-language-model-training-three-phases-shape-llm-training/
101. https://aws.amazon.com/what-is/large-language-model/?nc1=h_ls
102. https://en.wikipedia.org/wiki/Large_language_model
103. https://www.labellerr.com/blog/challenges-in-development-of-llms/
104. https://www.oracle.com/artificial-intelligence/ai-model-training-challenges/
105. https://www.shiksha.com/online-courses/articles/challenges-of-training-large-language-models-an-in-depth-look/
106. https://wandb.ai/mostafaibrahim17/ml-articles/reports/An-Introduction-to-Unsupervised-Learning%2D%2DVmlldzozMjYzNzcw
107. https://www.ibm.com/think/topics/unsupervised-learning
108. https://en.wikipedia.org/?title=Unsupervised_learning
109. https://www.codecademy.com/article/deep-learning-workflow
110. https://developers.google.com/machine-learning/crash-course/neural-networks/back-propagation
111. https://towardsdatascience.com/how-do-we-train-neural-networks-edd985562b73
112. https://www.simplilearn.com/tutorials/deep-learning-tutorial/neural-network
113. https://machinelearningmastery.com/learning-rate-for-deep-learning-neural-networks/
114. https://www.jeremyjordan.me/nn-learning-rate/
115. https://towardsdatascience.com/estimating-optimal-learning-rate-for-a-deep-neural-network-ce32f2556ce0
116. https://docs.aws.amazon.com/machine-learning/latest/dg/cross-validation.html
117. https://graphite-note.com/the-importance-of-cross-validation-in-machine-learning/
118. https://editverse.com/cross-validation-techniques-ensuring-model-reliability-in-2024/
119. Krstajic D et al (2014) Cross-validation pitfalls when selecting and assessing regression and classification models. https://jcheminf.biomedcentral.com/articles/10.1186/1758-2946-6-10
120. https://www.botcampus.ai/understanding-the-bias-variance-tradeoff-in-machine-learning
121. https://harksys.com/blog/understanding-the-bias-variance-trade-off-2/
122. https://duchesnay.github.io/pystatsml/_images/bias_variance.png?sa=X&ved=2ahUKEwiZ9ZrFw9CKAxU7lYkEHbT0C54Q_B16BAgCEAI
123. https://www.pecan.ai/blog/machine-learning-model-underfitting-and-overfitting/
124. Reps J et al (2021) Investigating the impact of development and internal validation design when training prognostic models using a retrospective cohort in big US observational healthcare data
125. Roger Grosse, Generalization. https://www.cs.toronto.edu/~lczhang/321/notes/notes09.pdf
126. https://www.ibm.com/think/topics/regularization

127. Regularization: make your machine learning algorithms “Learn”, not “Memorize”. https://www.einfochips.com/blog/regularization-make-your-machine-learning-algorithms-learn-not-memorize/
128. https://www.cloudzero.com/blog/openai-cost-optimization/
129. https://www.comet.com/site/blog/reduce-model-training-spending/
130. https://neptune.ai/blog/deep-learning-model-optimization-methods
131. https://www.appliedaicourse.com/blog/issues-in-machine-learning/
132. https://www.index.dev/blog/ai-model-optimization-techniques
133. https://keymakr.com/blog/maximizing-performance-ai-model-optimization-techniques/
134. Diving into model pruning in deep learning. https://wandb.ai/authors/pruning/reports/Diving-Into-Model-Pruning-in-Deep-Learning%2D%2DVmlldzoxMzcyMDg
135. Decoding efficiency in deep learning, a guide to neural network pruning in big data mining. https://www.red-gate.com/simple-talk/development/python/decoding-efficiency-in-deep-learning-a-guide-to-neural-network-pruning-in-big-data-mining/
136. Pruning: the history and overview. https://polukhin.tech/2022/10/27/pruning-the-history-and-overview
137. https://deepgram.com/learn/model-pruning-distillation-and-quantization-part-1
138. Choi Y et al. Translating AI to clinical practice: overcoming data shift with explainability. https://pmc.ncbi.nlm.nih.gov/articles/PMC10190133/
139. Ueda D et al. Fairness of artificial intelligence in healthcare: review and recommendations. https://pmc.ncbi.nlm.nih.gov/articles/PMC10764412/
140. Akosa J. Predictive accuracy: a misleading performance measure for highly imbalanced data
141. Chen D et al. Deep learning and alternative learning strategies for retrospective real-world clinical data. https://www.nature.com/articles/s41746-019-0122-0
142. Park SH. Methods for clinical evaluation of artificial intelligence algorithms for medical diagnosis. https://pubs.rsna.org/doi/full/10.1148/radiol.220182

Chapter 5
Transfer Learning and Multimodal Artificial Intelligence

Abstract This chapter presents a thorough exploration of Transfer Learning, Foundation models, and Multimodal AI, with a specific focus on their use and the hurdles they present in Medical Device AI models. It begins by discussing the methods of Transfer Learning and their shortcomings. It then closely examines how to select and fine-tune foundation models, also addressing the difficulties associated with them, such as fine-tuning issues and the problem of catastrophic forgetting. A considerable part of the chapter is devoted to multimodal machine learning models, which are highly applicable to medical devices. It emphasizes recent progress and existing challenges in this area. The chapter wraps up by providing an overview of the latest developments in Multimodal Generative AI and Massively Multimodal Masked Modeling (4 M), accompanied by a concise illustration of Multimodal AI.

Keywords Supervised · Unsupervised · Generative AI · Large language models (LLMs) · Natural language processing (NLP) · Reinforcement learning · Classification · Regression · Deep learning · Transfer learning · Foundation models · Fine-tuning · Catastrophic forgetting · Multimodal machine learning model · Multimodel AI · Any-to-any multimodal models · AI/ML-enabled medical devices · AI/ML-enabled medical device performance · Training · Evaluation · Performance metrics · Evaluation metrics · Massively multimodal masked modeling · 4M Multimodal AI · Bidirectional Encoder Representations from Transformers (BERT) · Generative Pre-trained Transformer (GPT) · Amazon Titan · AI21 Jurassic · Claude · Claude 3.5 Sonnet · Claude 3 Opus · Claude 3 Haiku · Cohere · Stable diffusion · Bloom and hugging face · ChatGPT

5.1 Transfer Learning in Medical Device AI/ML Models

By leveraging transfer learning techniques, medical device AI models can achieve high performance with limited data, accelerate development timelines, and enhance accuracy in tasks such as medical image classification and analysis. Transfer learning can be effectively utilized in medical device AI models in several ways [1–4]:

A. Pandey et al., *Artificial Intelligence in Medical Software*,
https://doi.org/10.1007/978-3-032-10808-1_5

5.1.1 Data Efficiency

Transfer learning provides a powerful solution to the challenge of limited medical datasets, enabling the development of accurate AI models for healthcare applications. This approach leverages pre-trained models created using large, general image datasets, such as ImageNet. By utilizing these pre-trained models as a starting point, transfer learning significantly reduces the number of labeled medical images required to achieve good performance. This is particularly valuable in the medical field, where obtaining large, high-quality labeled datasets can be challenging due to privacy concerns, the rarity of certain conditions, and the expertise required for accurate labeling. As a result, transfer learning enables researchers and developers to create precise and reliable models even when working with relatively small domain-specific datasets, making it an essential technique in advancing AI applications in medicine.

5.1.2 Model Architecture

When applying transfer learning to medical device AI, selecting the model architecture is crucial for leveraging pre-existing knowledge for specific medical tasks. Deep convolutional neural network (CNN) architectures, such as ResNet or Inception, are commonly used as base models due to their proven effectiveness in image recognition tasks. The key strategy involves retaining the pre-trained convolutional layers of these networks as feature extractors, preserving the general image understanding capabilities developed on large datasets. The final classification layers are then replaced and explicitly retrained for the medical task. This approach enables the model to leverage the rich feature representations learned from diverse images while adapting to the nuances of medical imaging. By fine-tuning only the task-specific layers, this method efficiently balances the use of generalized knowledge with the specialization required for medical applications, enabling the development of accurate models with limited domain-specific data.

5.1.3 Fine-Tuning Approaches

Fine-tuning approaches are crucial in adapting pre-trained models to specific medical tasks. Three main strategies are commonly employed: feature extraction, partial fine-tuning, and full fine-tuning. The pre-trained model is a fixed feature extractor in feature extraction, with only the new classification layers being trained on the medical data. Partial fine-tuning involves gradually unfreezing and fine-tuning deeper layers of the pre-trained model, allowing for a more tailored adaptation to the medical domain. Full fine-tuning, the most comprehensive approach, retrains the entire network on the medical data, using pre-trained weights as initialization. This method offers incredible flexibility but requires more computational resources and a larger dataset. The choice among these strategies depends on the dataset size, task complexity, and available computational resources, allowing researchers to optimize the balance between leveraging pre-existing knowledge and adapting to specific medical imaging tasks.

5.1.4 Task-Specific Optimization

Task-specific optimization is crucial when adapting foundation models for medical applications. This process involves several key strategies to ensure the model's effectiveness and reliability in healthcare settings. First, it's essential to use appropriate evaluation metrics such as sensitivity and specificity, which are particularly relevant in medical diagnostics. These metrics provide a more nuanced understanding of the model's performance in detecting conditions and avoiding false positives or negatives. Second, implementing data augmentation techniques tailored to medical imaging can significantly enhance the model's robustness and generalization capabilities. These techniques might include realistic transformations that mimic variations in medical imaging procedures or patient anatomies. Finally, collaboration with medical experts is invaluable in this optimization process. Their domain knowledge can be incorporated into the model design and training process, ensuring that the AI system aligns with clinical practices and addresses real-world medical challenges. This interdisciplinary approach helps create technically proficient, clinically relevant, and trustworthy models.

5.1.5 Regulatory Considerations

Regulatory considerations are paramount when developing medical AI models using transfer learning to ensure compliance, safety, and efficacy. A comprehensive approach is necessary, starting with thorough documentation of the transfer learning process. This documentation is crucial for regulatory compliance and transparency, allowing authorities to review and validate the model's development. Additionally, it's essential to validate the model's performance on diverse datasets encompassing various patient demographics, imaging equipment, and clinical settings. This step ensures the model's generalizability and robustness across different scenarios.

Furthermore, conducting clinical validation studies is critical to demonstrate the model's real-world effectiveness. These studies provide empirical evidence of the AI model's performance in clinical settings, addressing potential biases and limitations. By adhering to these regulatory considerations, developers can create medical AI models that meet technical standards and comply with healthcare regulations, ultimately leading to safer and more reliable AI-assisted medical devices.

5.1.6 Transfer Learning Methods in AI/ML-Enabled Medical Devices

Figure 5.1 illustrates the most common transfer learning methods employed in medical AI, encompassing several key approaches [2]. Feature extraction is a widely used technique in which pre-trained convolutional neural networks (CNNs) serve as fixed feature extractors, utilizing their convolutional layers to extract relevant features from medical images. These extracted features are then used to train new classifiers for specific medical tasks. Fine-tuning is another widely adopted method that involves adapting pre-trained models to medical datasets. This can be done through partial fine-tuning, where only some layers are retrained, or full fine-tuning, where the entire network is retrained on medical data [5].

Domain adaptation is employed when transferring knowledge between different but related domains, such as adapting models trained on one medical imaging modality to another. Multi-task learning, which involves training models to perform multiple related diagnostic tasks simultaneously, is also utilized in medical AI. Some researchers apply image-based models to non-image medical data by first transforming the data into image-like formats, such as converting time series data into spectrograms. Regarding model selection, frequently used pre-trained models for transfer learning in medical image analysis include Inception, AlexNet, VGG, and ResNet, which are often initially trained on large datasets like ImageNet before being adapted to medical tasks [6, 7].

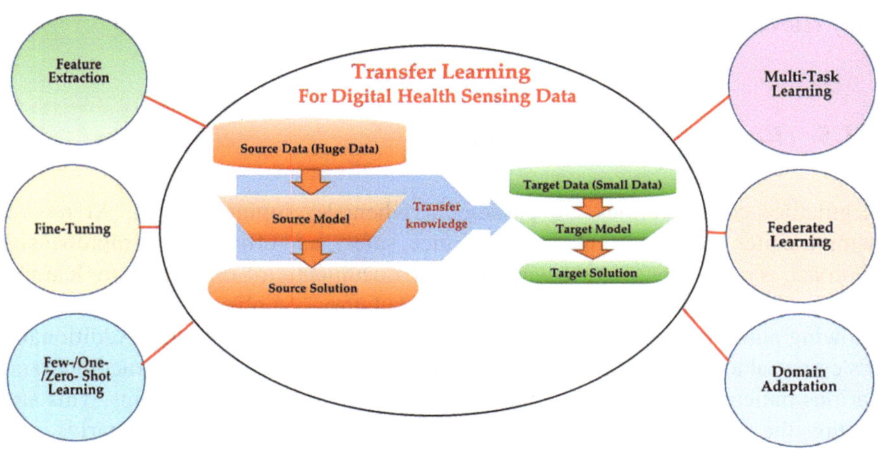

Fig. 5.1 Transfer learning methods for digital health sensing data [3]

5.1.7 Limitations of Transfer Learning for Medical Data

Transfer learning, while offering significant benefits, faces several important limitations. A primary challenge is the domain mismatch between natural images, often used in pre-trained models and medical images. Medical images possess unique characteristics in terms of shape, color, resolution, and dimensionality, which can lead to suboptimal performance or negative transfer when using models pre-trained on natural images. The transferability of features also decreases as the distance between the base task and target task increases, with lower layers capturing more generic, transferable features compared to higher, task-specific layers [5].

The risk of overfitting is another concern, particularly when working with small medical datasets and complex pre-trained models, which can lead to poor generalization of unseen data. Effective transfer learning often requires extensive domain-specific fine-tuning, which can be challenging if insufficient labeled medical data is available. Additionally, different medical imaging modalities (such as X-rays, MRI, and CT scans) present unique challenges, potentially limiting the transferability of models across modalities [7, 8].

Performance limitations have been observed in some cases, where lightweight models trained from scratch on medical images perform similarly to large pre-trained models. This suggests that the benefits of transferring learning from natural images to medical domains may be limited in specific scenarios. Data quality and annotation issues in medical datasets can further hinder the effectiveness of transfer learning. Lastly, ethical concerns and potential biases in pre-trained models or fine-tuning datasets can lead to performance disparities across different patient demographics. Understanding and addressing these limitations is crucial for the practical application of transfer learning in medical imaging [5–7, 9].

5.2 Selection of Foundation AI/ML Models

Foundation models (FMs) have revolutionized the field of machine learning and artificial intelligence. These large, deep learning neural networks, trained on extensive datasets, serve as a powerful starting point for data scientists developing new ML applications. The term "foundation model," introduced by researchers, describes ML models trained on a diverse range of generalized and unlabeled data. These models demonstrate remarkable versatility, performing various general tasks, including language comprehension, text and image generation, and natural language conversation. By leveraging foundation models, data scientists can now create sophisticated ML models for new applications more efficiently and cost-effectively, significantly streamlining the AI development process compared to building systems from scratch [10]. Here are some popular examples of foundation models: Bidirectional Encoder Representations from Transformers (BERT), Generative Pre-trained Transformer (GPT), Amazon Titan, AI21 Jurassic, Claude,

Claude 3.5 Sonnet, Claude 3 Opus, Claude 3 Haiku, Cohere, Stable Diffusion, Bloom, and Hugging Face [10].

When choosing a foundation model for a specific task, several key factors need to be carefully considered to ensure optimal performance and alignment with project requirements and by thoroughly evaluating these factors, discussed in Sects. 5.2.1–5.2.6 [10–12], an informed decision when selecting a foundation model that best fits specific task requirements, resource constraints, and ethical considerations.

5.2.1 *Task Requirements*

When evaluating task requirements for selecting a foundation model, it's crucial to consider three main aspects: the nature of the task, data type, and level of complexity. The nature of the task involves clearly defining the specific functions your application needs to perform (Fig. 5.2), such as text generation, language translation, image captioning, code completion, sentiment analysis, or named entity recognition. Data type considerations focus on determining the input and output modalities your application will process and generate, which could be text-only, images, audio, video, or a combination of these in multimodal applications. The level of complexity assessment involves gauging the difficulty of your tasks and considering whether larger, more sophisticated models are necessary for tasks that are more complex. Additionally, it's essential to evaluate if your main task comprises multiple subtasks

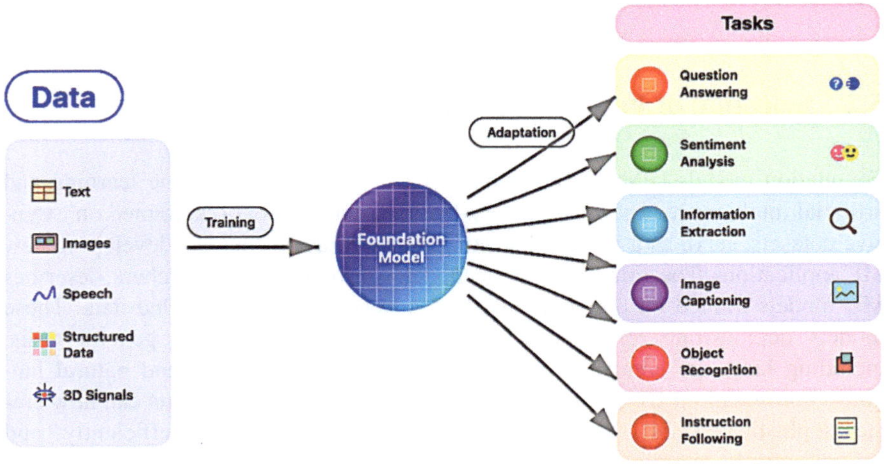

Fig. 5.2 A foundation model can consolidate information from diverse data modalities, allowing it to be adapted for various downstream tasks [11]

that the model needs to handle simultaneously. By thoroughly analyzing these task requirements, you can better align your choice of foundation model with your application's specific needs and ensure optimal performance.

5.2.2 *Model Characteristics*

When selecting a foundation model, it's essential to consider various model characteristics. Model size and architecture play a crucial role, with larger models, such as GPT-3, often performing better but requiring more resources. The choice between transformer-based models, CNNs, or hybrid architectures depends on your task. Performance metrics, including accuracy on relevant benchmark datasets, task-specific metrics like F1 Score or BLEU Score, and robustness across diverse inputs and edge cases, are equally important. Inference options should be evaluated based on deployment methods, such as self-hosted solutions that offer more control, cloud-based APIs for a more straightforward implementation, or edge deployment for low-latency applications. The context window is another critical factor, determining how much input the model can process and the maximum length of outputs it can generate. Lastly, latency considerations involve assessing the model's response time and whether your application requires near-instantaneous responses. By carefully weighing these characteristics, you can select a foundation model that best aligns with your project's requirements and constraints.

5.2.3 *Resources and Constraints*

Several key factors must be considered when evaluating resources and constraints for implementing a foundation model. Computational resources play a crucial role, including assessing the GPU/TPU requirements for efficient model operation, considering RAM needs (especially for larger models), and evaluating the capacity of your current infrastructure to handle the chosen model. Inference costs are another critical aspect, involving the calculation of per-query costs to understand the financial implications of running the model at scale and comparing different pricing tiers offered by model providers. Scalability is equally important, ensuring the model can handle your expected query volume and perform well under varying loads with concurrent users. By carefully analyzing these resources and constraints, you can make an informed decision that balances the model's capabilities with your organization's available resources and operational requirements, ensuring a sustainable and efficient implementation of the foundation model in your project or application.

5.2.4 Data Considerations

When selecting a foundation model, data considerations are paramount. The quality and quantity of data play a crucial role in model performance. It's essential to assess the relevance and quality of the data the model was trained on, as this directly impacts its capabilities and potential biases. Additionally, if you plan to fine-tune the model, evaluate whether you have sufficient high-quality data. Domain specificity is another critical factor. You'll need to decide between general-purpose models, such as GPT-3, which offer broad applicability, and domain-specific models, like BioBERT, which provide specialized knowledge in particular fields, like biomedical tasks. Consider the transfer learning potential of the model—how well its pre-trained knowledge can be applied to your specific domain. This assessment helps determine whether a general model can be effectively adapted to your needs or if a specialized model is necessary for optimal performance in your particular use case.

5.2.5 Customization and Integration

Customization and integration are crucial aspects to consider when selecting a foundation model. The level of customization available can significantly impact the model's effectiveness for your specific use case. Assess the ease of fine-tuning the model on your data, consider its responsiveness to well-crafted prompts, and evaluate the customization options provided through the model's API. Integration complexity is equally essential. Review the quality and comprehensiveness of the API documentation to ensure smooth implementation. Check for official SDKs in your preferred programming language, as these can streamline the integration process. Additionally, assessing the active community and available resources for implementation support, as a robust ecosystem, can be invaluable when facing challenges or seeking optimizations. By carefully considering these customization and integration factors, you can ensure that the chosen foundation model meets your current needs and offers the flexibility to adapt to future requirements, seamlessly integrating into your existing workflows.

5.2.6 Legal and Ethical Considerations

Legal and ethical considerations are paramount when selecting and implementing a foundation model. Licensing agreements must be carefully reviewed, with close attention paid to any usage restrictions on commercial applications and understanding the implications for data ownership and privacy. Data privacy and security are critical, ensuring the model and its deployment method comply with relevant regulations, such as GDPR or HIPAA. Implementing secure processing measures, such as end-to-end encryption, is essential, especially when handling sensitive data.

Ethical AI practices should also assess the model's known biases and available mitigation strategies. Transparency is crucial, considering the level of information provided about the model's training data and decision-making processes. By thoroughly addressing these legal and ethical considerations, organizations can ensure responsible and compliant use of foundation models, mitigating risks and building trust with users and stakeholders.

5.2.7 Challenges with Fine-Tuning a Foundation Model

When fine-tuning a foundation model, several common challenges arise [11, 12]:

- Overfitting: Fine-tuning on small, specialized datasets can lead the model to learn noise rather than meaningful patterns, thereby reducing its ability to generalize to new data [11, 14].
- Catastrophic forgetting: The model may lose some of its pre-trained knowledge, resulting in decreased performance on the original tasks [13].
- Infrastructure requirements: Fine-tuning large models requires substantial computational resources and can be both expensive and time-consuming [10].
- Hyperparameter tuning complexity: Identifying optimal settings for hyperparameters, such as learning rate and batch size, is time-consuming and computationally expensive [11].
- Maintaining model reliability: Fine-tuned models may produce unreliable or inappropriate outputs, particularly when encountering unfamiliar contexts [11, 14, 15].
- Bias preservation or introduction: Models can retain or introduce biases in the training data, requiring careful data filtering and bias mitigation strategies [10].
- Integration challenges: Developers must effectively integrate fine-tuned models into existing software stacks, including tools for prompt engineering and pipeline management [11, 14, 15].
- Balancing task-specific performance with general knowledge: Ensuring the model excels at the target task while retaining its broad capabilities can be challenging [11].

To address these challenges, best practices include careful dataset curation, regular evaluation during training, and implementing techniques like LDIFS (L2 distance in feature space) to preserve pre-trained knowledge while learning new concepts.

5.2.8 Catastrophic Forgetting in Fine-Tuning

Several strategies can be employed to mitigate catastrophic forgetting when fine-tuning large language models [16]:

- **Regularization Techniques:** Dropout and weight decay help preserve essential characteristics from prior tasks by penalizing significant weight changes during training [11, 16].
- **Rehearsal Strategies:** Periodically retraining the network using a mix of old and new data helps reinforce previously learned representations while integrating new information [11, 16].
- **Dynamic Structures:** Approaches such as synaptic intelligence (SI) and elastic weight consolidation (EWC) protect critical weights by dynamically adjusting their values based on their importance to previous tasks [11, 16].
- **Parameter Sharing:** Utilizing shared representations across tasks promotes transfer learning and reduces interference between task-specific knowledge [11].
- **Knowledge Distillation:** Transferring knowledge from a more extensive, pre-trained network to a smaller one can reduce forgetting and enhance overall performance [16].
- **Gradient Clipping:** Limiting the magnitude of gradients during fine-tuning prevents drastic changes to the model's weights, thereby helping to preserve previously learned patterns [11].
- **Selective Fine-Tuning:** Updating only particular layers, typically those higher up, allows the model to adapt to new tasks while maintaining foundational knowledge in lower layers [16].
- **Preserving Principal Subspaces:** Using Singular Value Decomposition (SVD) to identify and preserve key components within weight matrices, focusing updates on lower-impact directions [16].
- **Multi-Task Fine-Tuning:** Training the model on multiple tasks simultaneously can help maintain performance across a broader range of capabilities [16].

These strategies strike a balance between acquiring new knowledge and preserving previously learned information, thereby addressing the challenge of catastrophic forgetting in large language models.

5.3 Multimodal AI/ML in Healthcare

Current AI in medicine often focuses on single-modality data for specific tasks, unlike clinicians who integrate multiple data sources for comprehensive patient care. AI assessments typically provide static snapshots, whereas health is a continuous state. Ideally, AI models should utilize all available clinical data sources and even specialized knowledge beyond typical clinical expertise, mirroring and potentially enhancing the holistic nature of medical decision-making [17].

The biomedical field has seen a surge in the application of machine learning algorithms, driven by the proliferation of wearable sensors, enhanced data capture and analysis capabilities, and reduced costs of genomic sequencing and related 'omics' technologies. This convergence of factors has created both an opportunity and a necessity for innovative tools that can process vast amounts of data from diverse sources. Such tools are poised to deliver significant value across the entire spectrum of biomedicine, from fundamental research and discovery to clinical applications in diagnosis, prognosis, treatment, and preventive care. By harnessing the power of multimodal data integration, these advanced algorithms have the potential to revolutionize healthcare and accelerate biomedical progress [17, 18].

Multimodal AI models integrate diverse data sources (Fig. 5.3), including biosensors, genetic, imaging, clinical, and environmental data. This approach bridges the gap between AI and clinical practice, enabling applications like personalized medicine and real-time pandemic monitoring. Multimodal machine learning aims to process and relate information from multiple modalities, mirroring human perception of the world. The field faces challenges in representation, translation, alignment, fusion, and co-learning, extending beyond traditional early and late fusion categorizations. This comprehensive approach has significant potential to advance AI's understanding of complex, real-world scenarios [18]. Multimodal machine learning is a subfield of artificial intelligence that focuses on developing models capable of processing and integrating information from multiple modalities, such as text, images, audio, and video. Figures 5.4 and 5.5 illustrate that this approach aims to create a more comprehensive understanding of data by combining inputs from various sources, mimicking how humans perceive and interpret the world around them [19].

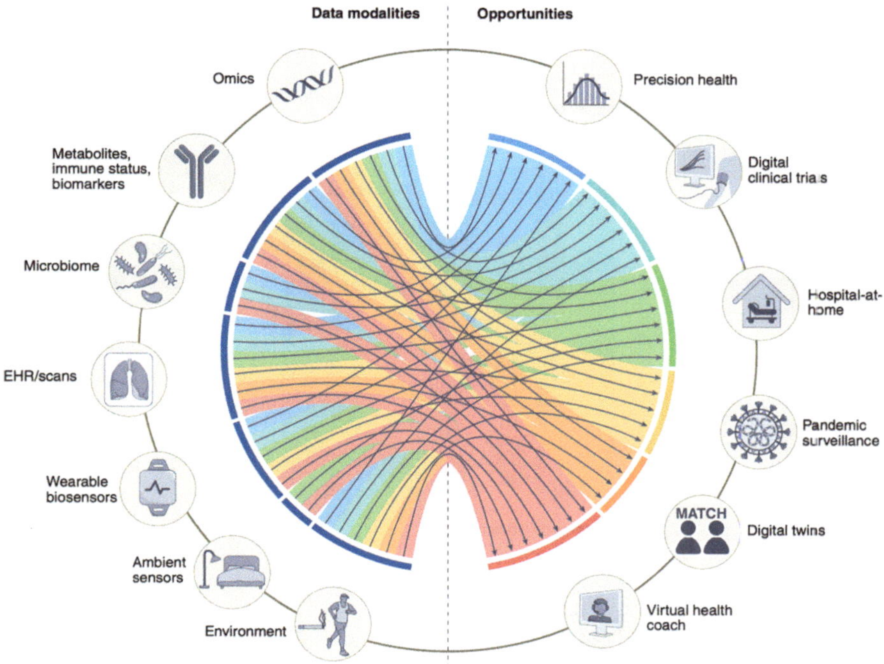

Fig. 5.3 Data modalities and opportunities for multimodal biomedical AI [17]

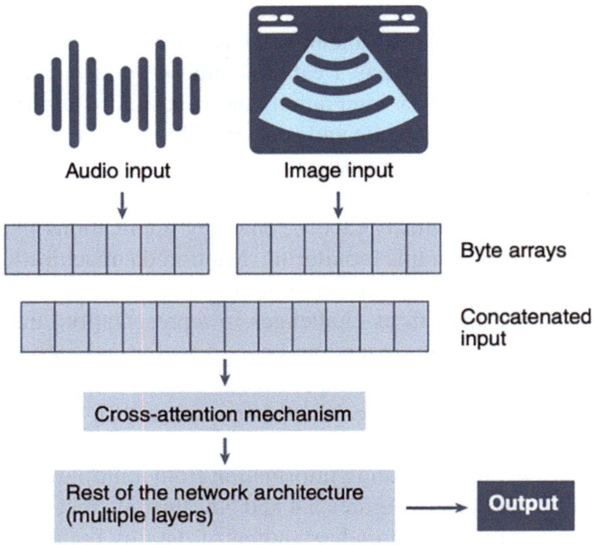

Fig. 5.4 Simplified depiction of multimodal AI/ML framework [17]

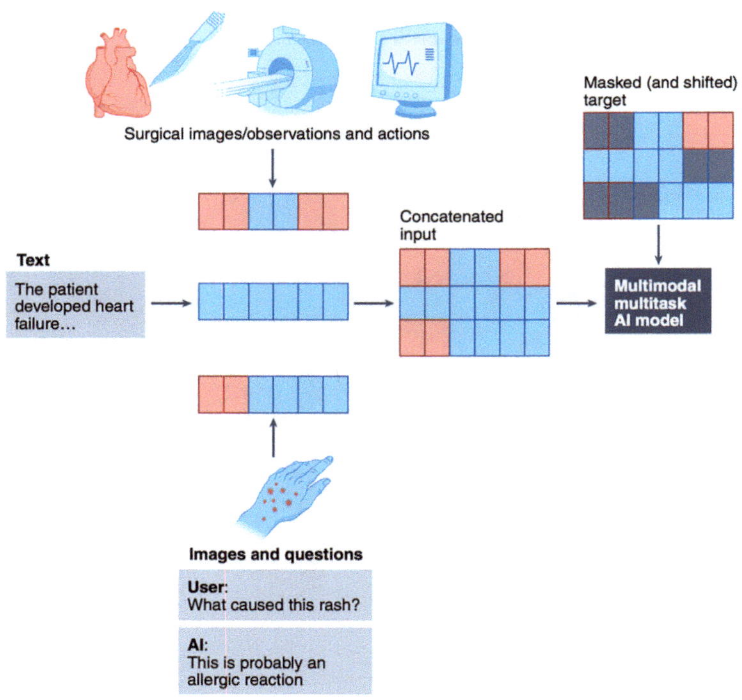

Fig. 5.5 Simplified depiction of multimodal and multi-task AI/ML framework [17]

The key components of multimodal machine learning include:

- Unimodal encoders: Separate neural networks that process each input modality individually.
- Fusion Network: Combines features extracted from different modalities using techniques such as concatenation or attention mechanisms.
- Classifier: Accepts the fused data and makes predictions based on the integrated information.

Multimodal learning addresses several core challenges in the field of artificial intelligence. These challenges include representation, which involves learning computer-interpretable descriptions of heterogeneous data from multiple modalities; translation, which focuses on converting data from one modality to another, such as image captioning; alignment, which aims to identify relationships between elements from different modalities; fusion, which involves joining information from multiple modalities to perform prediction tasks; and co-learning, which deals with transferring knowledge between modalities and their representations. By integrating data from various sources, multimodal machine learning significantly enhances the depth and accuracy of AI systems, leading to improved decision-making and predictive abilities in complex tasks. This innovative approach has applications in diverse fields, including self-driving cars, robotics, healthcare, and emotion recognition, demonstrating its versatility and potential to revolutionize numerous industries [19].

The clinical diagnostic process integrates diverse data types:

1. Image: CT scans, X-rays, MRIs, and digital pathology visually represent internal structures.
2. Text: Electronic health records and clinical notes offer patient history and context.
3. Speech: Recordings capture nuances in patients' experiences and symptoms.
4. Genetic data reveals predispositions and genetic markers that influence diseases.
5. Physiological signals: Real-time cardiac and neural activity data capturing temporal variations.

Figures 5.6 and 5.7 illustrate how each modality contributes unique insights, collectively enabling a comprehensive understanding of patients' physiological states for accurate diagnosis using multimodal machine learning and deep learning [20]. Multimodal AI/ML training follows similar steps, involving data collection, preprocessing, feature engineering, and algorithm selection.

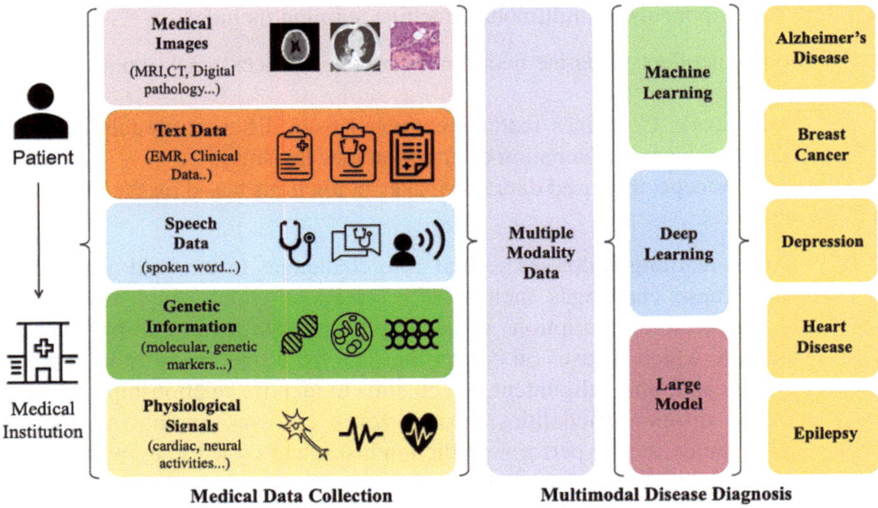

Fig. 5.6 Diverse medical data types in the clinical diagnostic process [20]

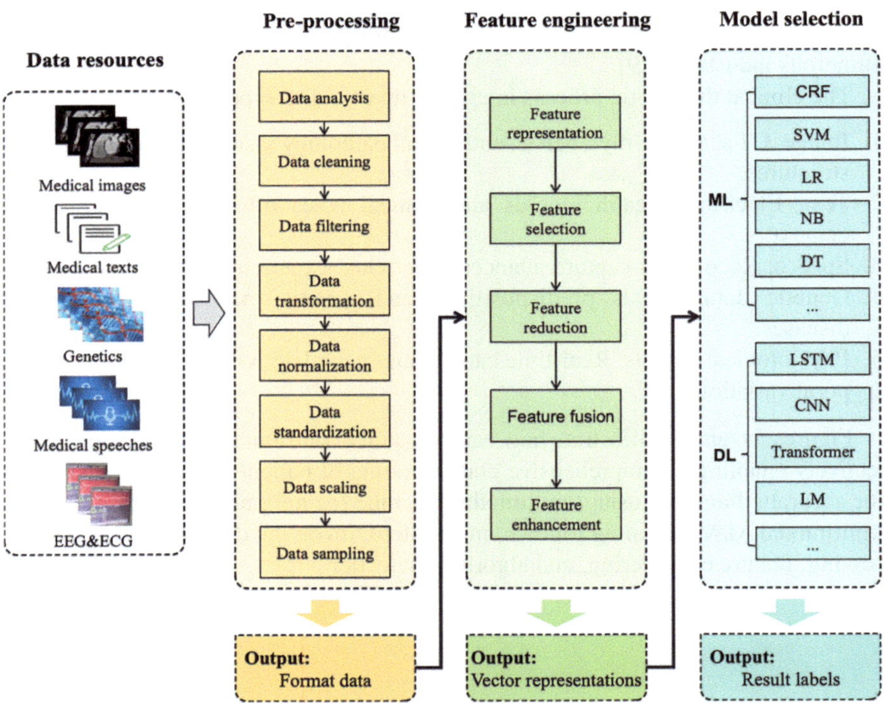

Fig. 5.7 The framework for AI in disease diagnosis modeling [20]

5.4 Challenges in Multimodal AI/ML Algorithm Development

In multimodal machine learning, several interconnected challenges emerge [17, 21]: representation, fusion, translation, alignment, and co-learning. These challenges are shown graphically in Figs. 5.8 and 5.9. Representation involves determining the geometric depiction of multiple modalities and their intrinsic relationships, which can be either joint (combined into a single vector) or coordinated (separately constructed but mutually influential). Fusion tackles the challenge of integrating multiple modalities into a predictive model, employing either model-based or model-agnostic approaches. Translation focuses on mapping one modality to another, while alignment attempts to synchronize separate modalities spatially or temporally. Co-learning involves using one modality to assist in the learning of another. This framework [21] provides a comprehensive overview of multimodal machine learning challenges, which can be applied to various fields, including

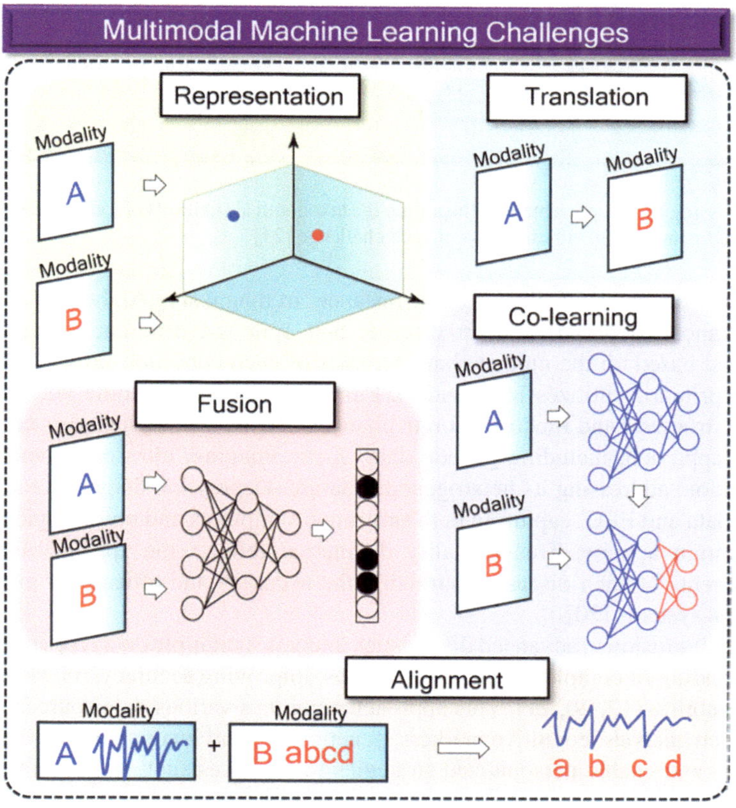

Fig. 5.8 Challenges in multimodal machine learning comprised of representation, fusion, translation, co-learning, and alignment for output [21]

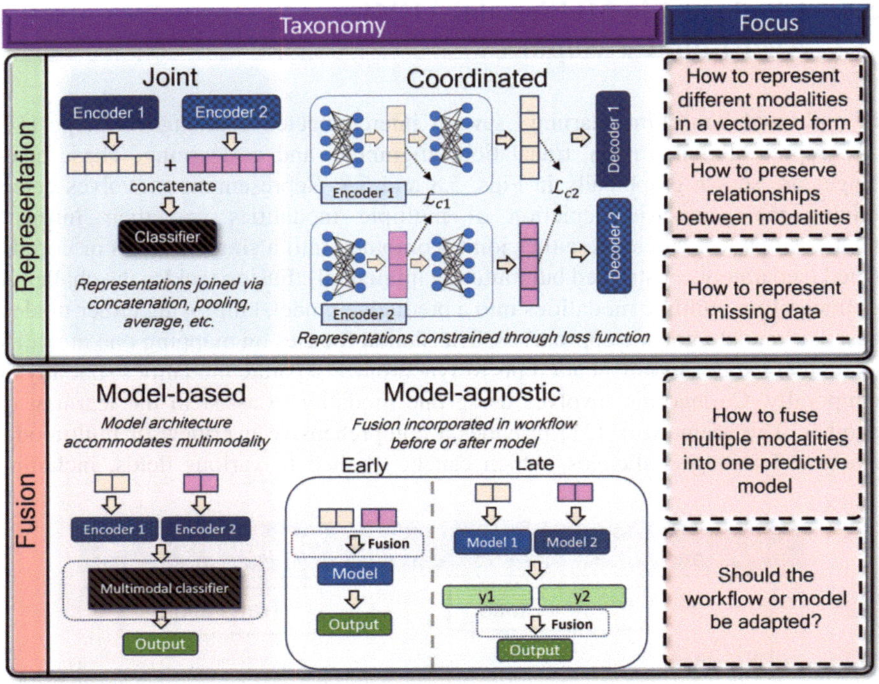

Fig. 5.9 A graphical representation illustrating the taxonomical sublevels of multimodal representation and fusion, along with the focus of each challenge [21]

automated diagnosis of diseases. For instance, in diagnosing Alzheimer's disease, breast cancer, depression, heart disease, and epilepsy, different modalities are employed based on the unique characteristics of each condition [20]. Alzheimer's disease primarily utilizes MRI and PET images, along with some voice data, to capture structural and functional brain changes. Breast cancer diagnosis employs a diverse approach, including genetic data, X-ray imaging, ultrasound, and textual information, addressing its heterogeneous nature. Depression diagnosis focuses on textual data and EEG, capturing both subjective symptoms and physiological markers of brain activity. This modality distinction reflects the specific diagnostic requirements of each disease, optimizing the accuracy and efficacy of automated diagnosis systems [20].

Modality fusion in advanced diagnostics integrates multiple data types to enhance understanding of complex medical conditions, improving accuracy, robustness, and interpretability [17, 20, 21]. This approach combines various data sources, including speech analysis, cognitive markers, genetic data, and imaging, to provide holistic views and enable personalized strategies for diseases such as Alzheimer's and cancer. Multimodal fusion in machine learning encompasses two primary methods: model-agnostic (including early, late, and hybrid fusion) and model-based approaches, which modify predictive models to handle each modality uniquely.

Recent trends favor model-based approaches, particularly in medical imaging applications, aiming to harness complementary information from different modalities for improved performance in complex clinical scenarios [21]. While implementation challenges persist, recent AI advancements, such as DALL-E and Gato, demonstrate the potential of concatenating diverse data types, offering promising solutions for biomedical applications and advancing medical technology towards more comprehensive assessments [17].

5.5 Multimodal Generative AI

Multimodal generative AI systems [22] integrate diverse inputs such as images, videos, audio, and text prompts to produce varied outputs. These models are trained on vast datasets, learning patterns, and associations between different modalities. The process begins with the provision of input and safety checks to detect inappropriate content. The core model processing stage involves interpreting inputs using pre-trained knowledge, combining information from all input types, and generating output through iterative prediction. A language model predicts each subsequent word based on the combined input and previously generated words, repeating until the response is complete. Finally, the output undergoes refinement and additional safety measures to ensure quality and appropriateness. This sophisticated process enables these systems to create coherent, relevant, diverse outputs across multiple modalities.

5.6 4M: Massively Multimodal Masked Modeling

Recent advances in multimodal AI [23] introduce a versatile "any-to-any"4M model, depicted in Fig. 5.10, which is trained on diverse modalities, including images, text, semantic and geometric data, feature maps, and novel steering modalities such as metadata and color palettes. 4M facilitates the development of versatile multimodal and multi-task models that can immediately perform various vision tasks, including multimodal conditional generation. This capability, combined with in-painting functionality, enables advanced image editing. These generalist models demonstrate excellent transfer learning capabilities for various downstream tasks and new modalities, and can be easily fine-tuned for more specialized applications. It has been shown that training a single model on tens of highly diversified modalities without a loss in performance compared to specialized single or task models. Key innovations include:

1. Co-training on large-scale multimodal datasets and text corpora
2. Discrete tokenization of various data types
3. Integration of state-of-the-art models and pseudo-labels

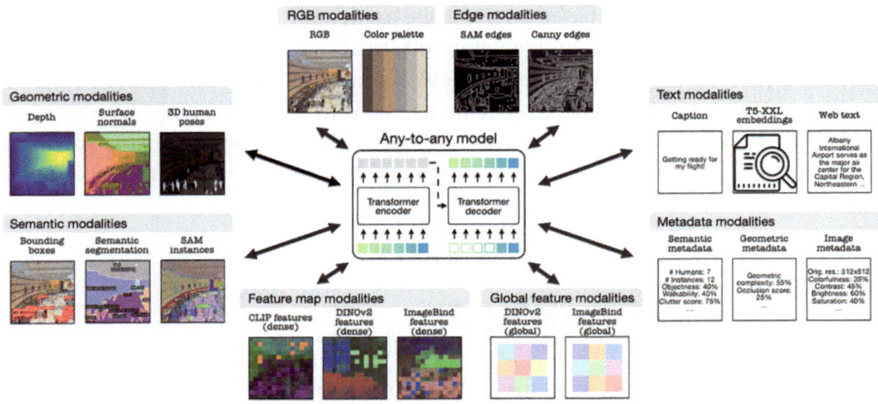

Fig. 5.10 A single, versatile any-to-any model capable of training on and performing dozens of diverse modalities and tasks, maintaining performance levels comparable to specialized models designed for single or limited task sets [23]

The resulting model handles three times more tasks and modalities than existing models without performance loss, enabling fine-grained multimodal generation and unified model distillation. Trained at 3 billion parameters, the model and code are open-sourced [24].

5.7 Summary

This chapter presented a thorough exploration of transfer learning, foundation models, and multimodal AI, with a specific focus on their use and the hurdles they presented in medical device AI models. It began by discussing the methods of transfer learning and their shortcomings. It then closely examined how to select and fine-tune foundation models, also addressing the difficulties associated with them, such as fine-tuning issues and the problem of catastrophic forgetting. A considerable part of the chapter was devoted to multimodal machine learning models, which are highly applicable to medical devices. It emphasized recent progress and existing challenges in this area. The chapter wrapped up by providing an overview of the latest developments in multimodal generative AI and massively multimodal masked modeling (4M), accompanied by a concise illustration of multimodal AI.

5.8 Points to Ponder

1. Briefly discuss data fusion techniques for multimodal model training.

5.9 Answers

1. Data fusion techniques for multimodal model training [25–28].

   Multimodal data fusion integrates information from multiple data types, including sound, text, images, and numeric data, to create richer and more robust machine learning models. The main fusion strategies are:

   - Early Fusion (Feature-level Fusion):
   - Combines raw or pre-processed features from all modalities at the input stage. The model receives a single, concatenated feature vector representing all modalities. This approach requires careful alignment and normalization of data but allows the model to learn joint representations from the outset.
   - Intermediate Fusion (Representation-level Fusion):
   - Processes each modality separately to extract features, then fuses these intermediate representations. This method balances modality-specific processing with joint learning, often implemented using neural network architectures that merge embeddings or feature maps from different modalities.
   - Late Fusion (Decision-level Fusion):
   - Trains separate models for each modality and combines their outputs (e.g., via averaging, voting, or a meta-model). This approach is robust to missing modalities and allows for asynchronous data, but may not capture deep cross-modal relationships.
   - Hybrid Fusion:
   - Integrates aspects of early, intermediate, and late fusion, sometimes in a hierarchical or staged manner, to leverage the strengths of each technique.
   - Advanced Techniques:
     - Multimodal Embeddings: Learn shared latent spaces for different modalities.
     - Attention Mechanisms and Transformers: Model complex cross-modal interactions, allowing the model to focus on relevant information across modalities.
     - Specialized Neural Architectures: Use CNN-RNN hybrids or Graph Neural Networks (GNNs) for structured multimodal data.

Python Libraries for Multimodal Data Fusion

- Fusilli:
- A dedicated Python package offering 23 deep-learning-based multimodal data fusion methods. It supports tabular-tabular and tabular-image fusion (2D/3D), and is suitable for combining image, numeric, and other data types in regression and classification tasks.
- PyTorch & TensorFlow:
- Both frameworks are highly flexible for building custom multimodal models. They support early, intermediate, and late fusion strategies, and provide modules for combining data types such as images, text, and audio.
- Hugging Face Transformers:

- Offers models like CLIP and BLIP that natively handle text-image fusion and can be extended for other modalities.
- AutoGluon:
- Provides automated machine learning with native support for multimodal data, including tabular, text, and image modalities.
- Deep Multimodal Fusion Libraries:
- Many research-oriented libraries and toolkits (often built on PyTorch/TensorFlow) implement advanced fusion strategies, including attention mechanisms and multimodal embeddings.
- Scikit-Learn:
- While not specialized for deep multimodal fusion, it supports late fusion via ensemble methods and can be used for simpler multimodal tasks.
- FuseLib:

- An older but still relevant library, FuseLib, provides probabilistic and feature-level fusion models in Python and MATLAB. It supports both early (feature concatenation) and late (decision-level) fusion, making it helpful for prototyping and benchmarking multimodal fusion algorithms.

References

1. Transfer learning. https://aws.amazon.com/what-is/transfer-learning/
2. https://www.spiceworks.com/tech/artificial-intelligence/articles/articles-what-is-transfer-learning/
3. Chato L et al (2023) Survey of transfer learning approaches in the machine learning of digital health sensing data. J Pers Med 13(12):1703. https://pmc.ncbi.nlm.nih.gov/articles/PMC10744730/
4. https://www.matellio.com/blog/transfer-learning-use-cases/
5. Kim H (2022) Transfer learning for medical image classification: a literature review. BMC Med Imaging 22:69. https://bmcmedimaging.biomedcentral.com/articles/10.1186/s12880-022-00793-7
6. https://www.scribbledata.io/blog/transfer-learning-in-ai-a-complete-guide/
7. Elyan E et al (2022) Computer vision and machine learning for medical image analysis: recent advances, challenges, and way forward. Art Int Surg 2:24–45. https://www.oaepublish.com/articles/ais.2021.15
8. Davila A et al (2024) Comparison of fine-tuning strategies for transfer learning in medical image classification. https://arxiv.org/pdf/2406.10050v1
9. Davila A et al (2024) Comparison of fine-tuning strategies for transfer learning in medical image. https://arxiv.org/pdf/2406.10050v1
10. https://aws.amazon.com/blogs/startups/selecting-the-right-foundation-model-for-your-startup/
11. Bommasani R et al (2022) On the opportunities and risks of foundation models. https://arxiv.org/pdf/2108.07258
12. https://research.aimultiple.com/foundation-models/
13. Understanding fine-tuning in AI models (2024) https://telnyx.com/resources/what-is-fine-tuning-ai
14. https://hackernoon.com/the-challenges-costs-and-considerations-of-building-or-fine-tuning-an-llm

15. https://www.acorn.io/resources/learning-center/fine-tuning-llm/
16. Catastrophic forgetting: understanding AI memory loss. https://www.projectpro.io/article/catastrophic-forgetting/1034
17. Acosta JN et al (2022) Multimodal biomedical AI. Nat Med 28(9):1773–1784
18. Baltrusaitis T et al (2019) Multimodal machine learning: a survey and taxonomy. IEEE Trans Pattern Anal Mach Intell 41(2):423–443
19. http://multicomp.cs.cmu.edu/multimodal-machine-learning/
20. Xu X et al (2024) A comprehensive review on synergy of multi-modal data and AI technologies in medical diagnosis. Bioengineering (Basel) 11(3):219
21. Warner E et al (2024) Multimodal machine learning in image-based and clinical biomedicine: survey and prospects. Int J Comput Vis 132:3753–3769
22. https://ai.meta.com/tools/system-cards/multimodal-generative-ai-systems/
23. Bachmann R et al. 4M-21: an any-to-any vision model for tens of tasks and modalities
24. https://4m.epfl.ch
25. Pawłowski M, Wróblewska A, Sysko-Romańczuk S (2023) Effective techniques for multimodal data fusion: a comparative analysis. Sensors (Basel) 23(5):2381. https://doi.org/10.3390/s23052381. PMID: 36904585; PMCID: PMC10007548
26. Zhao F et al (2024) Deep multimodal data fusion. ACM Comput Surv 56:1–36. https://doi.org/10.1145/3649447
27. Nikolaou N, Salazar D, RaviPrakash H et al (2025) A machine learning approach for multimodal data fusion for survival prediction in cancer patients. NPJ Precis Oncol 9:128. https://doi.org/10.1038/s41698-025-00917-6
28. Mastering multimodal data fusion: how to combine text, images, Audio & More. https://www.sapien.io/blog/mastering-multimodal-data-fusion

Chapter 6
Clinical AI/ML Model Testing and Validation

Abstract This chapter comprehensively explores AI-enabled medical device testing and clinical validation. It begins by outlining best practices for AI model testing frameworks, followed by an in-depth discussion on AI Model Robustness testing. The chapter then delves into the crucial concepts of AI Model Transparency, explainability, and interpretability, addressing how these can be implemented and how developers can strike a balance between interpretability and performance. A thorough examination of various types of bias and methods for testing bias is presented. The chapter introduces and clearly defines the concepts of human-in-loop and human-in-command, explaining how developers can leverage these approaches to enhance ethical considerations, mitigate bias, and reduce product liability while improving regulatory compliance. The discussion then transitions to clinical validation, offering strategies for successful clinical studies and improving AI model generalization. A review of real-world evidence testing is provided, and the chapter concludes with an analysis of two crucial regulatory requirements: the US FDA and EU MDR for Clinical evaluation and related documentation. This comprehensive overview equips readers with the essential knowledge needed to navigate the complex landscape of AI-enabled medical device development and validation.

Keywords AI model testing · Model evaluation · Data management · Human–AI interaction · AI Act · FDA · MDR · Clinical evaluation · Robustness · Transparency · Explainability · Interpretability · Bias · Generalizability · Clinical validation · Retrospective clinical validation · Prospective clinical validation · Random clinical trial · Real-world evidence · Human-in-loop · Human-in-command · MDCG guidance · MDSW · SaMD · Performance characteristics · Design for interpretability · Design for generalization

A. Pandey et al., *Artificial Intelligence in Medical Software*,
https://doi.org/10.1007/978-3-032-10808-1_6

6.1 Testing of AI/ML Model for Medical Device Application

Testing machine learning and AI models for medical devices requires a comprehensive approach to ensure safety, effectiveness, and regulatory compliance. It's important to note that this field is rapidly evolving, and staying current with the latest regulatory guidance and best practices is crucial. Here are key considerations for testing AI/ML models in medical device applications:

6.1.1 Data Management

Data management and validation are crucial to developing and testing ML models for medical devices. A fundamental principle is ensuring a complete separation between training and test datasets, which is essential for unbiased performance assessment and avoiding overfitting. When separating datasets, it's important to consider patient overlap, data acquisition methods, and the data collection site. This independence allows for a more accurate evaluation of the model's performance in real-world scenarios. Equally important are well-curated, labeled, and representative datasets that accurately reflect the intended use and patient population. While this process can be challenging and resource-intensive, it is vital for robust model evaluation. High-quality, representative data ensures that the ML model can generalize well to diverse patient populations and clinical settings, which is particularly critical in the medical field, where patient safety is paramount. By adhering to these data management and validation principles, developers can create ML models for medical devices that are accurate but also reliable and safe for clinical use [1–3].

6.1.2 Model Evaluation

Model evaluation is critical in developing and testing ML models for medical devices. This process involves assessing clinically relevant performance metrics, such as accuracy, sensitivity, specificity, and the area under the ROC curve, with a focus on metrics directly related to the device's intended use and clinical benefits. Developing and executing test plans based on sound statistical principles ensure adequate sample sizes and appropriate statistical methods for performance evaluation [2, 4].

6.1.3 Clinical Validation

Furthermore, the model should be tested under conditions that simulate real-world clinical use. This includes considering the intended patient populations and important subgroups, as well as clinical environments, workflows, variability in measurement inputs, and potential confounding factors. By rigorously evaluating the model under these clinically relevant conditions, developers can ensure that the ML-based medical device will perform reliably and effectively in healthcare settings. This comprehensive approach to model evaluation enhances the device's performance and builds confidence in its ability to provide accurate and meaningful results in diverse clinical scenarios [5].

6.1.4 Human–AI Interaction

Human–AI interaction is crucial for evaluating machine learning (ML) models used in medical devices. Instead of assessing the AI model in isolation, it's essential to evaluate the performance of the human AI team as a whole. This approach considers the user's skills and expertise, their understanding of model outputs and limitations, and the potential for user error or misuse. Additionally, assessing how the AI model integrates into existing clinical workflows and decision-making processes is vital. This comprehensive evaluation ensures that the ML-based medical device performs well technically and enhances clinical practice when used by healthcare professionals. By considering the human element alongside the AI model, developers can create medical devices more likely to be adopted successfully and used effectively in real-world healthcare settings. This holistic approach to human-AI interaction maximizes the benefits of AI in medicine while minimizing the potential risks associated with its implementation [4, 6–8].

6.1.5 Bias and Fairness

Addressing bias and ensuring fairness are critical considerations in developing and testing ML models for medical devices. Designing trials that can effectively identify and mitigate algorithmic biases is essential. This involves testing the model across diverse datasets to ensure equitable performance across different patient groups, which is crucial for maintaining fairness in healthcare applications. Additionally, evaluating the model's ability to generalize across relevant patient populations, clinical environments, and use cases is paramount. This assessment of generalizability helps ensure that the ML-based medical device can perform consistently and accurately in various real-world scenarios, not just in the specific conditions under which it was trained. By rigorously examining bias and fairness, developers can create more robust and reliable ML models for medical devices that provide equitable and effective care across diverse patient populations and healthcare settings. This approach enhances the model's performance and builds trust in AI-driven medical technologies among healthcare professionals and patients [2].

6.1.6 Continuous Monitoring and Improvement

Continuous monitoring and improvement are essential to managing ML models for medical devices throughout their lifecycle. Implementing methods for ongoing monitoring of the model's performance in real-world clinical settings after deployment is crucial to ensure sustained effectiveness and safety. This real-world performance tracking allows developers to identify discrepancies between the model's behavior in controlled testing environments and its performance in diverse clinical scenarios. For models that may be updated or retrained after deployment, it's vital to establish robust controls to manage risks such as overfitting, unintended bias, or performance degradation. These controls help maintain the model's integrity and reliability over time, even as it adapts to new data or changing clinical needs. By adopting a proactive approach to continuous monitoring and improvement, developers can ensure that ML-based medical devices remain accurate, fair, and effective throughout their operational life, thereby maintaining patient safety and clinical efficacy in the long term [2, 4].

6.1.7 Regulatory Considerations

Regulatory considerations are pivotal in developing and testing ML models for medical devices. Adhering to Good Machine Learning Practice (GMLP) principles, as outlined by regulatory bodies such as the FDA, Health Canada, and MHRA, is essential for ensuring the quality and reliability of ML-based medical devices. These principles provide a framework for the responsible development and deployment of AI in healthcare. Equally important is maintaining comprehensive documentation throughout the model development process, including testing procedures and results. This thorough documentation is crucial for supporting regulatory submissions and demonstrating compliance with relevant standards. It provides transparency into the model's development, validation, and performance, which is necessary for regulatory approval and ongoing oversight. By prioritizing these regulatory considerations, developers can navigate the complex landscape of healthcare regulations and create ML-based medical devices that meet stringent safety and efficacy standards, ultimately facilitating their adoption and use in clinical practice [2, 4, 6, 9].

6.2 Testing the Robustness of AI Models

Testing the robustness of machine learning and AI models involves a multifaceted approach that combines various techniques. Cross-validation is a fundamental method that splits data into multiple subsets to estimate generalization error. Separate validation and test sets are crucial for evaluating performance on unseen data. Learning curves help visualize model convergence and generalization by plotting training and validation errors against training set size. Performance metrics such as accuracy, precision, recall, and F1-score provide quantitative measures of model effectiveness across different datasets [10, 11].

Robustness curves plot model performance against data quality parameters, helping to identify strengths and weaknesses. Adversarial testing assesses a model's resilience to malicious or perturbed inputs, while data shift testing evaluates performance across different distributions or domains. Specific robustness metrics, such as robustness score and mean corruption error, provide stability measures under various perturbations. Stress testing subjects' models to extreme conditions reveals failure modes and limitations [12].

Explainability analysis guarantees consistent and logical reasoning across inputs, while continuous monitoring assesses performance over time in real-world deployments. By combining these methods, developers can thoroughly evaluate and enhance the robustness of AI/ML models across various scenarios and conditions. The testing strategy should be tailored to the AI application's specific use case and requirements, ensuring a thorough evaluation of the model's strengths and weaknesses.

Utilizing a robustness score to evaluate machine learning models provides numerous advantages, enhancing the creation of more dependable and effective AI systems. This quantitative assessment objectively measures a model's capacity to maintain performance in the face of various perturbations or alterations in input data, facilitating meaningful comparisons between different models. A high robustness score strongly indicates a model's potential to generalize well to unseen data and real-world situations, reaching beyond the limits of training and validation sets. It also showcases the model's resilience against noise, outliers, and potential adversarial attacks, which are essential for practical applications where data imperfections and security concerns are prevalent. However, a trade-off between robustness and accuracy must be considered, as illustrated in Fig. 6.1.

Robustness scores measure consistency, evaluating a model's ability to produce stable results across different data distributions and slight variations in input. This stability is essential for reliable deployment and can indicate less bias across various subgroups or demographics, potentially serving as a measure of fairness. Models with high robustness scores are generally more trustworthy and reliable, a critical factor in applications where decisions have significant consequences. In certain industries, demonstrating model robustness may be necessary for regulatory compliance purposes.

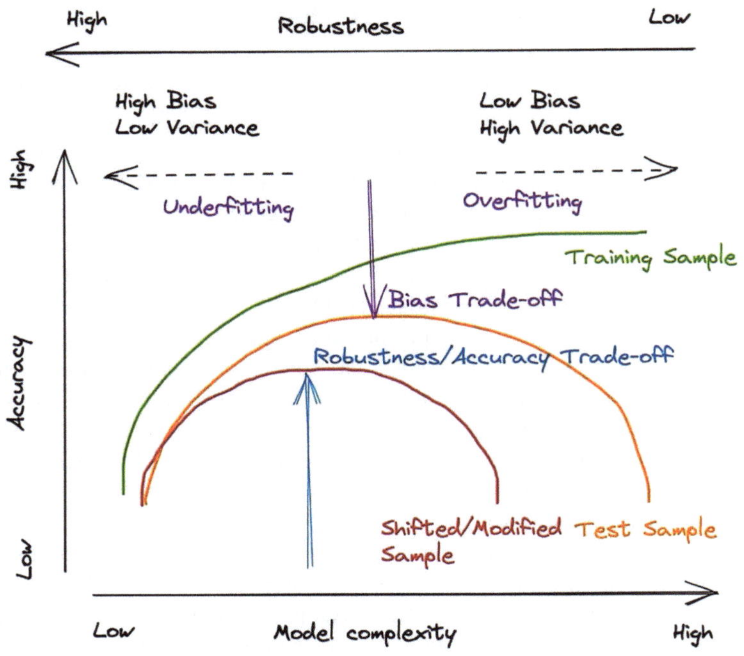

Fig. 6.1 The trade-off between robustness vs accuracy [12]

Additionally, robust models necessitate less frequent retraining or updates, which could lower long-term maintenance costs. By incorporating robustness scores into their evaluation process, data scientists and machine learning engineers can create AI systems that are not only more dependable and stable but also more trustworthy and capable of excelling in various real-world conditions. This comprehensive approach to model assessment facilitates the development of AI solutions that are better equipped to navigate the complexities and uncertainties of practical applications [11].

6.3 Transparency, Explainability, and Interpretability

The primary distinctions among transparency, explainability, and interpretability in AI represent unique yet interconnected approaches to enhancing the understanding and trustworthiness of AI systems. Transparency refers to the overall openness in designing, developing, and deploying AI systems, concentrating on making mechanisms, data sources, and decision-making processes readily available and comprehensible. It aims to foster trust in the entire system, exemplified by open-source AI projects that feature publicly accessible code, data, and documentation.

Conversely, explainability articulates how an AI system arrives at a particular decision or output. It emphasizes the logic or reasoning underlying individual AI decisions, making AI processes more comprehensible and relatable to end users. Explainability often employs post-hoc techniques to illuminate the inner workings of "black box" models, such as a credit-scoring AI that provides explanations for rejecting a loan application.

Interpretability goes beyond explainability by concentrating on understanding the algorithm's inner workings. It comprehensively analyzes the model's internal processes, including parameters, functions, and connections, to view the algorithm's logic. Typically, interpretability is integrated into the model, facilitating easier human comprehension, as demonstrated by decision trees in medical diagnosis tools.

The key distinctions are in their scope and approach: transparency addresses overall system openness, explainability focuses on individual decisions, and interpretability explores the model's internal logic. While explainability often employs post hoc techniques on complex models, interpretability is generally built into more straightforward, transparent models. These concepts represent different levels of making AI systems understandable, with transparency being the broadest, explainability concentrating on specific outputs, and interpretability offering the most profound insight into the model's inner workings [13, 14].

6.3.1 Transparency for ML-Enabled Medical Devices

In 2021, Health Canada, the US FDA, and the UK's MHRA collaborated to establish 10 guiding principles for good machine learning practice (GMLP), supporting the development of safe, effective, and high-quality AI/ML technologies in healthcare. Building upon these GMLP principles, particularly those focusing on human–AI team performance and clear user information, the agencies have further developed guiding principles for transparency in machine learning-enabled medical devices (MLMDs). Transparency in this context refers to the clear communication of essential information about an MLMD, including its intended use, development, performance, and, when possible, its logic. The concept of "logic" encompasses information about how a device reaches its outputs or decisions, with "explainability"referring to the degree to which this logic can be understood by a person [15].

Adequate transparency in MLMDs ensures that information crucial to risks and patient outcomes is appropriately communicated, considering the intended audience's needs and the context of use. It employs optimal media, timing, and strategies for successful communication, relying on a comprehensive understanding of users, environments, and workflows. Additionally, "human-centered design" is vital in achieving transparency, involving an iterative process that addresses the entire user experience and incorporates input from relevant parties throughout the design and development stages. This approach to transparency and design in MLMDs aims to enhance the safety, efficacy, and overall quality of AI/ML technologies in healthcare while promoting clear communication and understanding among users and stakeholders.

6.3.1.1 The Guiding Principles [15]

The guiding principles for transparency of MLMDs consider the following:

- Who (relevant audiences)
- Why (motivation)
- What (relevant information)
- Where (placement of information)
- When (timing)
- How (methods used to support transparency)

Table 6.1 contains a detailed explanation of the guiding principles.

6.3.2 AI Model Interpretability

Achieving model interpretability in machine learning necessitates a multifaceted approach that strikes a balance between complexity, performance, and understandability. Starting with simpler models, such as linear regression or decision trees, provides an inherently interpretable foundation when possible. Careful feature selection and engineering, guided by domain knowledge, can create meaningful and interpretable inputs. For more complex models, techniques such as SHAP values or LIME can be employed to interpret individual predictions [14, 16].

Visualizations are crucial in making models more understandable, with tools such as feature importance plots and partial dependence plots providing valuable insights into model behavior. Explaining individual predictions, especially for high-stakes decisions, enhances transparency and trust. Evaluating interpretability through metrics and user studies with domain experts ensures that the model's logic is comprehensible [17, 18].

Balancing performance and interpretability is key, often requiring a trade-off between model complexity and ease of understanding. Documenting model assumptions, limitations, and intended use cases provides essential context. Continuously monitoring and validating the model's interpretability as new data arrives helps maintain its relevance and reliability over time.

Regulatory considerations are also important, as many industries have specific requirements for model interpretability. By adhering to these practices, developers can create machine learning models that are accurate, interpretable, and explainable to stakeholders and end-users. This fosters trust and enables more informed decision-making based on AI outputs [18].

Table 6.1 Summary of transparency guiding principles [15]

Who: Relevant audiences for transparency	Transparency is relevant to all parties involved in a patient's health care, including those intended to: • Use or receive health care with the device • Make decisions about the device to support patient outcomes
Why: Motivation for transparency	Transparency supports: • Safe and effective use • Patient-centered care • Identification and evaluation of risks and benefits of a device • Informed decision-making and risk management • Device maintenance and detection of errors or performance degradation • Health equity through identifying bias • Increased fluency and confidence in MLMD use and increased technology adoption
What: Relevant information	Enabling an understanding of the MLMD includes sharing relevant information on: • Device characterization and intended use • How the device fits into health care workflow, including its intended impact on a health care professional's judgment • Device performance • Device benefits and risks • Product development and risk management activities across the lifecycle • Logic of the model, when available • Device limitations, including biases, confidence intervals, and data characterization gaps • How safety and effectiveness are maintained across the lifecycle
Where: Placement of information	Maximizing the utility of the software user interface can: • Make information more responsive • Allow information to be personalized, adaptive, and reciprocal • Address user needs through a variety of modalities
When: Timing of communication	Timely communication can support successful transparency, such as: • Considering information needs at different stages of the total product lifecycle • Providing notifications of device updates • Providing targeted information when it's required for the workflow
How: Methods to support transparency	Human-centered design principles can support transparency

6.3.2.1 SHAP (Shapley Additive exPlanations)

SHAP (SHapley Additive exPlanations) is a powerful technique that significantly enhances model interpretability in machine learning through various key aspects. It provides a feature importance measure, allowing users to understand which variables have the most significant impact on model outputs. SHAP's ability to explain individual predictions offers crucial local interpretability, enabling a case-by-case understanding of model behavior. Based on Shapley values from game theory,

SHAP provides a consistent and theoretically sound approach to feature attribution. Its additive property makes comprehending how features collectively influence predictions easier, as the sum of feature attributions equals the difference between the model output and the average prediction [16].

SHAP's versatility is evident in its model-agnostic nature, making it applicable to any machine-learning model. It offers visualization techniques, such as summary and force plots, which intuitively illustrate the impacts and interactions of features. SHAP can reveal complex feature interactions that might not be apparent from more straightforward interpretability methods. It also aids in bias detection by examining values across different subgroups. For model debugging, SHAP values can pinpoint reasons for errors, facilitating targeted improvements. By providing global and local insights, SHAP offers a comprehensive view of model behavior, encompassing overall feature importance and per-prediction explanations. These detailed and theoretically grounded explanations significantly enhance model interpretability, making it easier to understand, validate, and improve machine learning models, ultimately fostering trust and enabling more informed decision-making based on AI outputs.

6.3.2.2 Saliency Maps and Other Techniques

AI transparency and explainability for healthcare encompass a range of strategies aimed at ensuring AI algorithms are understandable and trustworthy. A key focus is on developing interpretable algorithms, emphasizing the creation of inherently more interpretable models whenever possible and utilizing techniques such as SHAP values for complex models. Providing clear, tailored explanations for AI decisions and recommendations, along with visual aids, is crucial for different stakeholders. Comprehensive documentation of the AI development process, including data sources, model architecture, and validation methods, is essential for transparency [19].

Saliency maps [20] play a crucial role in enhancing the interpretability of neural networks by offering visual explanations of the model's decision-making process. These maps effectively highlight the most significant regions or features in an input image that influences the model's predictions, facilitating human understanding and trust in the model's reasoning. The key benefits of saliency maps include visualizing important features, reducing noise, aligning with human perception, and improving model transparency. By highlighting the areas of an input image that are most influential in the model's decision, saliency maps enable humans to see what the model focuses on. Figure 6.2 contains a diagram of an interpretable lung disease classification pipeline [20]. After classifying a chest radiograph (CXR) using a vision transformer network, an attention-based saliency map, transformer multimodal explainability (TMME), is generated to aid the radiologist in interpreting the model's prediction. The saliency map based on TMME was compared with the conventional gradient-weighted class activation map (GradCAM).

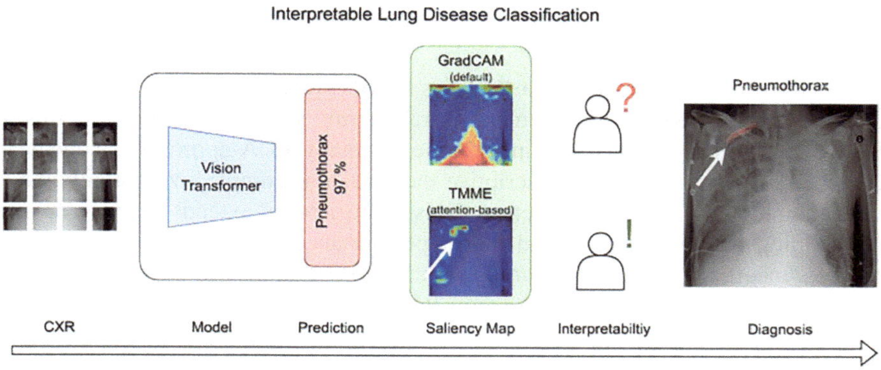

Fig. 6.2 Diagram of an interpretable lung disease classification pipeline [20]

Advanced techniques, such as transformer multimodal explainability (TMME) [20], can eliminate noise and produce sparser, more faithful representations of the model's attention. Furthermore, incorporating human-defined saliency into the training process guides models to focus on visually salient elements that align with human intuition. Saliency maps have proven valuable in critical domains, including medical imaging and pathological diagnosis, autonomous driving, financial analysis, and quantitative trading. By improving the interpretability of neural networks through saliency maps, practitioners can gain deeper insights into model behavior, enhance model performance, and increase the transparency and explainability of AI systems, ultimately bridging the gap between complex AI algorithms and intuitive human understanding.

Implementing auditing and traceability procedures, ensuring data transparency, and fostering collaboration between AI developers and healthcare professionals are vital practices. Regular evaluations and ongoing monitoring of AI performance in real-world settings help maintain reliability and detect potential biases. Education and training for healthcare professionals on the capabilities and limitations of AI are essential for effectively utilizing AI-generated insights.

Adherence to regulatory guidelines and engagement with policymakers are crucial for shaping responsible AI practices. Involving patients and community representatives in the development and evaluation of AI enhances trust and relevance. Balancing transparency with other considerations, such as proprietary information and patient privacy, is necessary. By implementing these strategies, healthcare organizations can work towards ensuring that AI algorithms are as transparent and explainable as possible, thereby fostering trust and enabling the responsible use of AI in healthcare decision-making. This approach improves the quality of care and addresses ethical concerns and regulatory requirements in the rapidly evolving field of AI in healthcare [19, 21].

6.3.3 Implications of Lack of Transparency in Healthcare AI

Achieving transparency in medical AI faces several significant barriers that complicate the development and deployment of trustworthy systems. The lack of consistent reporting standards is a primary issue, as many FDA-approved AI medical devices fail to provide basic descriptive data, such as sample size, demographic information, and performance metrics. This inconsistency extends to study design methodology, statistical data, and model performance reporting across approved devices. Insufficient data disclosure is another primary concern, as many AI products fail to provide adequate information about their training data, ethical considerations, and deployment limitations. Critical information, such as sample size, race/ethnicity data, and the age of intended users, is often missing from regulatory documents [21, 22].

The inherent complexity of AI algorithms, particularly those employing "black box" models, poses a significant challenge to transparency. Commercial interests can also hinder transparency, as vendors may be reluctant to disclose detailed information about their AI products due to concerns about proprietary information. The lack of legally mandated requirements for public and substantive transparency of AI medical devices further exacerbates this issue. Many AI systems struggle with interpretability and explainability, making it difficult to provide clear explanations for their outputs that healthcare providers can understand and trust.

Privacy concerns can conflict with transparency goals, while scaling and integration issues across different institutions and patient populations complicate transparent reporting on AI tool performance. The rapid pace of AI development often outstrips efforts to establish transparency standards and practices. Additionally, insufficient collaboration between AI developers, healthcare professionals, ethicists, and policymakers can hinder the development of comprehensive transparency frameworks. Addressing these barriers requires a concerted effort from regulators, developers, healthcare providers, and other stakeholders to establish and enforce robust transparency standards for medical AI, ensuring that these powerful tools can be used effectively and ethically in healthcare settings [22].

6.3.3.1 Ethical Implications

The lack of transparency in AI healthcare tools raises significant ethical concerns that have far-reaching implications for patient care, healthcare delivery, and the overall integrity of medical practice. A primary issue is the undermining of patient autonomy and informed consent, as patients may be unaware of AI's role in their care, violating their right to make informed decisions and potentially eroding trust in the healthcare system. The opacity of AI systems can perpetuate or amplify biases, leading to unfair or discriminatory outcomes for specific patient groups, as it becomes challenging to detect and mitigate algorithmic biases that are learned from potentially biased training data.

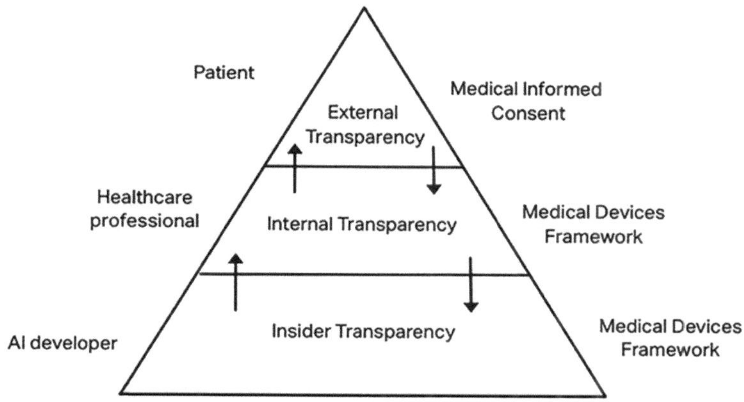

Fig. 6.3 Multilayered system of AI's transparency in healthcare [23]

The lack of transparency also hinders accountability, making it challenging to assign responsibility in cases of errors or harm, which poses significant legal and ethical challenges. It impedes quality assessment and improvement efforts as healthcare providers and regulators struggle to evaluate the safety and efficacy of AI tools. This opacity contributes to skepticism and mistrust among patients and healthcare professionals, potentially slowing the adoption of beneficial AI technologies. Ethical concerns surrounding data usage and privacy protection are heightened when AI systems lack transparency [23]. Figure 6.3 presents a diagram of the multilayered system of AI transparency and its respective stakeholders.

There's also a risk that healthcare providers may misuse or overly rely on AI recommendations without a clear understanding of their capabilities and limitations. Regulatory compliance becomes more challenging, and opaque AI systems may inadvertently lead to a revival of paternalistic care models, diminishing patients' active roles in medical decision-making. Furthermore, questions about fairness in resource allocation can arise when AI decision-making processes lack transparency. Addressing these multifaceted ethical implications requires a concerted effort to enhance transparency, explainability, and accountability in healthcare AI systems, coupled with stakeholder education and the development of robust ethical and regulatory frameworks [24].

6.3.3.2 AI Product Liability

Transparency in AI can significantly impact the liability of healthcare providers in several complex ways. Clarifying roles and responsibilities helps establish clear frameworks for accountability, defining who bears liability when AI systems make errors or cause harm. This increased transparency enables healthcare professionals to better understand AI-generated predictions, facilitating more informed decision-making and potentially reducing their liability for errors related to AI. It also helps providers use AI tools appropriately by providing clear guidance on their intended use and limitations, which can limit liability risks associated with misuse or over-reliance [23, 25].

Enhanced transparency and explainability in AI, through interpretable algorithms and visualizations of decision-making processes, support providers in explaining AI-assisted decisions to patients, which could be crucial in defending against liability claims. By allowing better detection and mitigation of biases in AI systems, transparency helps reduce the risk of biased decisions that could lead to liability issues. It also encourages continuous feedback and improvement of AI algorithms, potentially leading to more reliable systems and a reduced likelihood of errors.

However, while transparency can help clarify roles, it reveals the complex nature of AI decision-making, potentially complicating liability determination. This complexity underscores the need for legal frameworks to accommodate the nuances of AI in healthcare, potentially leading to new standards of care and liability considerations. In conclusion, while transparency in AI can help clarify some aspects of liability for healthcare providers, it also underscores the need for clear guidelines, ongoing education, and potentially new legal frameworks to address the complexities of liability in AI-assisted healthcare [25].

6.3.4 AI Decision-Making in Healthcare

Key recommendations for documenting AI decision-making in healthcare encompass a comprehensive approach to ensure transparency, accountability, and patient safety. Foremost, AI in healthcare decision-making should be disclosed and reported to physicians and patients in a culturally and linguistically appropriate manner. Any AI use that directly impacts patient care, access to care, medical decision-making, or the medical record should be documented in the patient's medical record. When healthcare content is generated by AI, including large language models, this should be disclosed within the content itself.

AI tools should not augment, create, or generate records, communications, or other content on behalf of a physician without that physician's consent and final review. Detailed information about the AI system should be documented, including its description, intended use, training and validation methods, potential biases, and performance metrics. The AI's decision-making process should be clearly explained using explainable AI methods. The role of human clinicians in reviewing and making final decisions based on AI recommendations must be documented, and a process should be in place for patients or caregivers to request additional review from a licensed clinician.

The concept of model cards for documenting AI models has gained significant traction in the academic community, with nearly 1800 citations in the past 5 years [26]. This approach, along with other documentation methods such as datasheets for datasets, has been incorporated into official guidance and regulations. For example,

NIST references datasheets 26 times in its AI risk management guide [27]. At the same time, the EU's AI Act [28] incorporates datasheets and model cards in its technical documentation requirements for high-risk AI systems. These documentation practices produce valuable artifacts and engage practitioners in a process that enhances governance outcomes. Datasheets, model cards, and system cards serve as crucial tools for downstream stakeholders, enabling them to understand the intended and unintended uses of AI components [27]. This understanding enables stakeholders to assess compliance with organizational and legal requirements, potentially preventing non-compliant development efforts and identifying instances where risk mitigation techniques may be necessary for safe deployment [28].

Version control and updates of AI systems should be meticulously maintained, and an audit trail of AI-assisted decisions should be implemented for retrospective review and quality assurance. All documentation should meet relevant regulatory requirements for AI transparency and explainability in healthcare. By adhering to these practices, healthcare providers can enhance transparency, build trust, and ensure accountability in the use of AI for clinical decision-making, ultimately contributing to improved patient care and safety in the evolving landscape of AI-assisted healthcare.

6.4 Bias in AI Algorithms for Medical Devices

Bias in AI algorithms for medical devices can stem from various sources, as presented in Fig. 6.4 significantly impacting the fairness and effectiveness of healthcare delivery. Unrepresentative training data is a primary concern, where patient cohorts used to train AI models fail to reflect the diversity of the intended population, potentially leading to poor performance for underrepresented groups. Historical biases in healthcare data can be inadvertently perpetuated and amplified by AI models, thereby reflecting existing disparities in healthcare access and treatment. Using incorrect

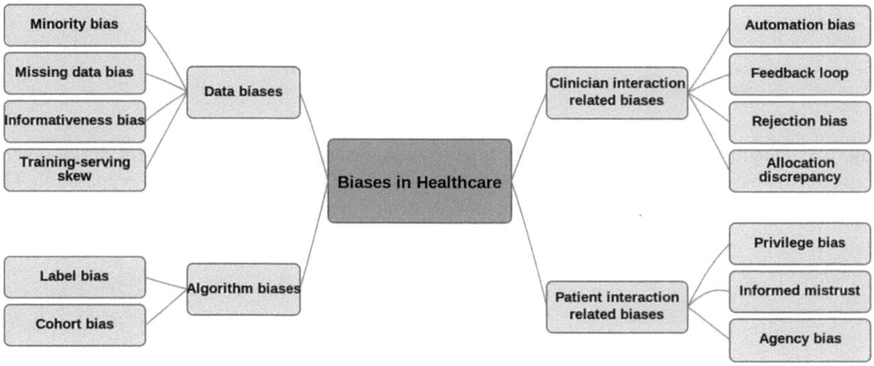

Fig. 6.4 Common sources of bias in healthcare [29]

proxy variables, such as healthcare costs, to estimate health needs can lead to significant racial or socioeconomic biases [29].

Sampling bias, where certain groups are underrepresented in training data, and biases in data collection and preparation can skew AI performance from the outset. Model design assumptions and a lack of diversity in development teams can introduce unintended biases, overlooking the needs of specific populations. Deployment context mismatches, where algorithms developed in one healthcare setting fail to generalize to different populations or systems, pose additional challenges. Insufficient testing on diverse populations before deployment and post-deployment drift due to changing population characteristics or healthcare practices over time can further exacerbate biases. By systematically addressing these potential sources of bias, developers and healthcare providers can work towards creating more equitable and effective AI-driven medical devices that serve all patient populations fairly and effectively.

Addressing these multifaceted sources of bias requires a comprehensive approach throughout the AI development lifecycle. This includes ensuring diverse and representative training data, carefully considering proxy variables and model assumptions, promoting diversity in development teams, thorough testing across different demographic groups, and continuous monitoring post-deployment. Figure 6.5 illustrates solutions for mitigating bias across the various development stages of an ML-based system for medical applications.

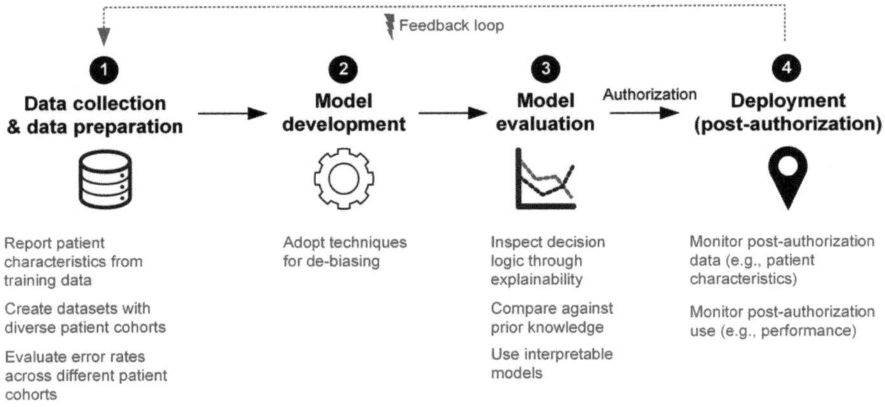

Fig. 6.5 Diagram outlining solutions on how to mitigate bias across the different development steps of ML-based systems for medical applications: (1) data collection and data preparation, (2) model development, (3) model evaluation, and (4) deployment [30]

6.4.1 *Algorithmic Bias in Medical Devices*

Real-world examples of algorithmic bias in medical AI systems have revealed significant disparities in healthcare delivery across different patient populations. A 2019 study [31] uncovered racial bias in a widely used algorithm for predicting patient care needs in US hospitals. The algorithm, which used healthcare costs as a proxy for healthcare needs, systematically underestimated the health risks of Black patients compared to equally sick white patients, resulting in fewer referrals for additional care. This bias stemmed from historical disparities in healthcare access and spending between Black and white patients rather than actual health differences [31].

In dermatology, AI models for skin cancer detection have demonstrated reduced accuracy when applied to individuals with darker skin tones. These models, primarily trained on images of light-skinned individuals, demonstrated significantly lower diagnostic accuracy for Black patients, potentially leading to delayed diagnoses and limited treatment options. Similarly, a kidney function prediction model developed by DeepMind Technologies for US Veterans Affairs hospitals proved less effective when applied to women due to the predominantly male training data [32].

During the COVID-19 pandemic, some hospitals employed AI triage tools that consistently underestimated the severity of illness in Black patients, potentially resulting in delayed or inadequate treatment. These examples underscore the critical need for diverse, representative training data and rigorous evaluation across different patient populations to prevent AI systems from perpetuating or exacerbating existing healthcare disparities [32].

Algorithmic bias in medical AI systems can significantly impact patient outcomes in various ways, potentially exacerbating existing health disparities. Misdiagnosis and delayed treatment can occur when AI systems trained on non-representative data perform poorly for underrepresented groups. This can lead to serious consequences, such as skin cancer detection AI failing to identify melanomas in patients with darker skin tones accurately. Biased algorithms used for resource allocation or risk assessment can result in unequal access to care, with certain groups receiving less attention or being deprioritized for treatment [33].

AI systems trained on historical data may perpetuate and amplify existing biases and inequities in healthcare, creating a feedback loop that reinforces health disparities. Incorrect risk assessments can occur when algorithms use inappropriate proxies, such as healthcare costs, to estimate health needs, potentially underestimating risks for specific patient groups. Biased AI systems may also recommend inappropriate or suboptimal treatments if they fail to account for differences in disease presentation or treatment efficacy across populations [34].

The effectiveness of AI-assisted decision-making can be compromised due to automation bias, where healthcare professionals overly rely on potentially biased AI recommendations [34]. Biased algorithms used in resource allocation can lead to disproportionate distribution, systematically denying necessary resources to specific groups. As these biases become apparent, trust in AI-assisted healthcare may erode among affected patient groups and healthcare providers, potentially limiting the adoption and benefits of AI in medicine [34].

To mitigate these impacts, experts recommend ensuring diverse and representative training data, implementing rigorous testing across different patient populations, increasing the transparency and explainability of AI systems, engaging patients and communities in AI development, and ongoing monitoring of AI performance in real-world settings [33]. Addressing algorithmic bias is crucial to ensure that AI in healthcare improves outcomes for all patients and doesn't exacerbate existing health disparities.

6.4.2 *Testing for Bias*

Key approaches for testing bias in machine learning and AI algorithms for medical device applications encompass a comprehensive strategy to ensure fairness and effectiveness across diverse patient populations. Figure 6.6 illustrates that testing for bias depends on the product lifecycle stage, as the types of bias vary at each stage. Using diverse and representative datasets that include various demographic groups is fundamental, ensuring training and test datasets reflect the intended use population. Subgroup analysis, which evaluates model performance separately for different demographic subgroups, is crucial for identifying systematic differences in accuracy, sensitivity, and specificity across these groups [35].

Employing fairness metrics, such as demographic parity, equal opportunity, and equalized odds, helps quantify bias. Meanwhile, bias detection techniques like SHAP values can identify disproportionate feature influences on predictions for specific groups. Intersectionality analysis examines performance across combinations of demographic factors, providing a more nuanced understanding of potential biases.

Clinical validation studies with diverse patient populations and external validation on datasets from different institutions are essential for assessing generalizability. Simulated data testing and adversarial techniques can help identify potential biases or vulnerabilities, particularly in underrepresented groups. Continuous monitoring after deployment is crucial for detecting performance drift or changes in fairness metrics over time [32].

Explainability analysis helps understand the model's decision-making process and identify reliance on problematic features. Adherence to regulatory guidelines, such as FDA guidance [35] on bias consideration in AI/ML-based Software as a Medical Device (SaMD), is necessary for compliance and documentation of bias testing methods and results. By combining these approaches throughout the development lifecycle and after deployment, developers can comprehensively assess and mitigate potential biases in AI algorithms for medical device applications, ensuring more equitable and effective healthcare delivery across diverse patient populations [36].

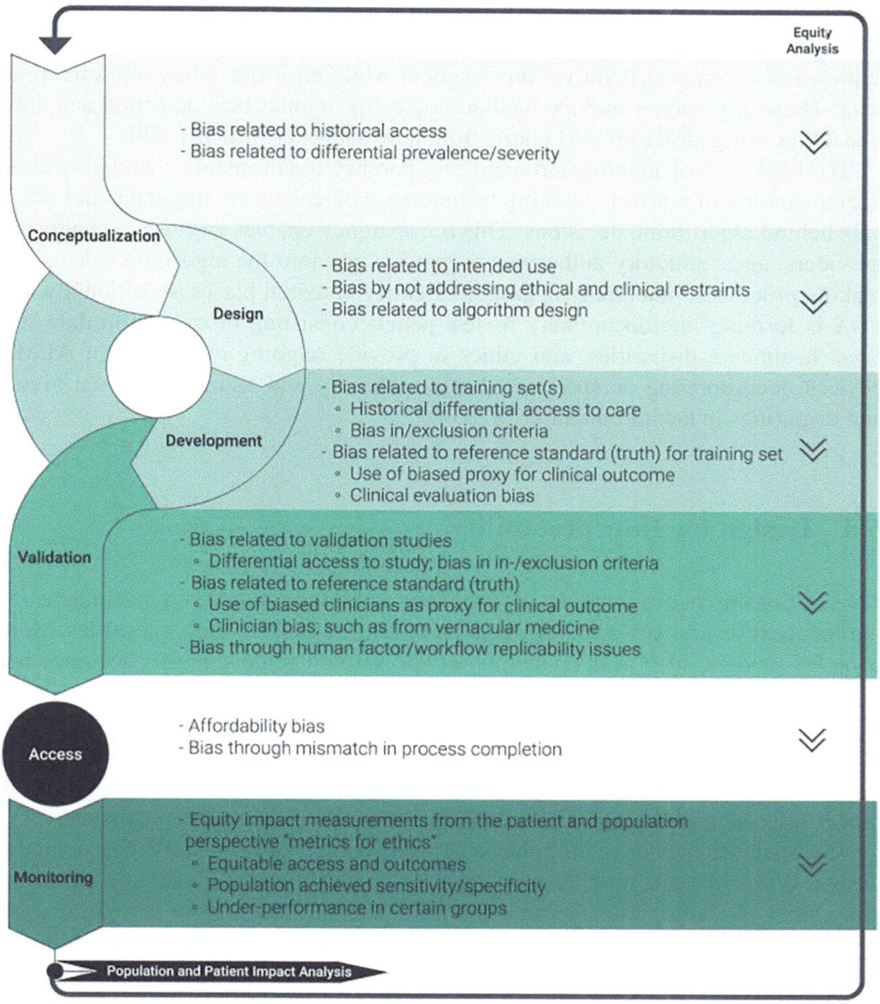

Fig. 6.6 Total product life cycle (TPLC) equity expanded framework with examples for each phase [29]

The FDA employs and plans to implement several strategies to detect bias in AI systems after they are deployed in medical devices. Real-world performance monitoring is a key approach that involves the ongoing evaluation of AI/ML devices in clinical settings to analyze outcomes across diverse patient populations. This enables the detection of bias after deployment and supports the piloting of real-world performance monitoring.

A robust postmarket surveillance framework is being established, specifically focused on monitoring the performance of AI/ML devices. This framework aims to analyze outcomes across populations representative of all intended patients, emphasizing detecting biases affecting minorities and underrepresented groups. For adaptive AI models that can be continuously updated, the FDA is developing approaches to support iterative improvement while ensuring safety and effectiveness. These approaches include methodologies for regular bias detection and continual bias mitigation as the AI learns from new data over time [37, 38].

The FDA is implementing stringent transparency requirements, mandating clear documentation of machine learning techniques, sources of training data, and rationale behind algorithmic decisions. This transparency enables end-users, healthcare providers, and regulatory authorities to gain insight into the algorithm's decision-making processes, potentially helping to identify potential biases. Additionally, the FDA is forming interdisciplinary review panels consisting of experts in data science, healthcare disparities, and ethics to provide ongoing evaluation of AI/ML devices, incorporating perspectives that can identify and address potential biases and disparities in health outcomes [37, 38].

6.5 Design for Interpretability

The relationship between model performance and interpretability in machine learning involves several key trade-offs that practitioners must carefully consider. More complex models, such as deep neural networks, often achieve higher accuracy but at the cost of reduced interpretability. Conversely, simpler models, such as linear regression or decision trees, offer greater transparency but may struggle with complex tasks. This fundamental tension between accuracy and explainability is particularly crucial in medical diagnosis or credit scoring, where understanding the model's decision-making process can be as important as its raw performance.

The trade-off extends to the balance between complexity and simplicity in model design. While intricate models can capture nuanced patterns in data, they are inherently more challenging to interpret. Simpler models, though more straightforward to understand, may miss subtle relationships within the data. This dichotomy is further exemplified by the contrast between universal function approximators, such as deep neural networks, which can model any function but lack interpretability, and models with specific, interpretable structures that are limited in their representational capacity.

Notably, the performance gains from using more complex "black box" models are often marginal compared to the significant loss of interpretability. In many scenarios, the benefits of transparency and explainability outweigh the slight decrease in model performance. However, very sophisticated models may be necessary for specific, complex tasks, such as image recognition, to achieve usable performance.

The trade-off also manifests in the choice between flexibility and constraints. More adaptable models can handle a broader range of problems, but are more challenging to interpret. In contrast, constrained, interpretable models offer more precise insights, albeit with a limited scope. Additionally, the use of post-hoc explanation techniques for complex models, while helpful, may not provide the same level of transparency as models built with interpretability as a core design principle. While the trade-off between performance and interpretability is a persistent challenge in machine learning, its extent varies depending on the specific task and application. In many cases, the benefits of interpretability—including enhanced trust, accountability, and the ability to detect biases—outweigh small performance gains from more complex models. Striking the right balance between these factors is crucial for developing effective and responsible AI systems.

Making AI algorithms more transparent and explainable is crucial for building trust and accountability, particularly in high-stakes domains such as healthcare. Key strategies to achieve this include implementing explainable AI (XAI) techniques, such as model-agnostic and model-specific approaches that provide insights into the AI's decision-making process. Thorough documentation of the development process, including data sources, model architecture, and validation methods, is essential for understanding an AI system's pedigree and potential biases.

Designing for interpretability by creating simpler models when possible and using proxy modeling to approximate complex systems can enhance understanding. Improving model interpretability through clear explanations, visualizations, and supporting evidence for outputs is also vital. Adopting transparency standards and frameworks, as well as open-source development practices, can facilitate wider scrutiny and adherence to best practices.

Providing different levels of explanation tailored to various stakeholders, from technical developers to end-users and regulators, ensures a comprehensive understanding across different groups. Implementing ongoing monitoring of model performance and potential biases in real-world use is crucial for maintaining accountability. Focusing on fairness and debiasing efforts helps mitigate potentially harmful impacts on different populations.

Balancing transparency with performance is a key consideration, recognizing the trade-offs between model complexity, accuracy, and explainability. By implementing these strategies, organizations can work towards creating AI systems that are more transparent, accountable, and trustworthy, which is particularly important in fields where the reasoning behind AI decisions can have significant consequences [25, 39].

Improving model interpretability without significantly sacrificing performance is a crucial goal in machine learning, and several techniques can help achieve this balance. One approach is to use inherently interpretable models such as decision trees, linear/logistic regression, rule-based systems, or sparse linear models. These models offer clear insights into their decision-making processes while providing reasonable performance for many tasks. Feature selection and engineering play a vital role in enhancing interpretability. Models can become more transparent without losing significant predictive power by selecting a subset of the most critical features, creating meaningful and interpretable features, and utilizing domain knowledge to inform feature creation. Regularization techniques, such as L1 regularization (Lasso) for feature sparsity and elastic net regularization, can also contribute to improved interpretability.

Model-agnostic interpretation methods, such as SHAP values, LIME, partial dependence plots, and Individual Conditional Expectation (ICE) plots, offer ways to explain complex models without altering their underlying structure. Attention mechanisms can highlight essential features in deep learning models, enhancing interpretability. Proxy models, which approximate complex models with simpler, more interpretable ones, provide another avenue for balancing performance and explainability.

Ensemble methods with interpretable base models, constrained optimization, monotonic constraints, and techniques such as layer-wise relevance propagation for neural networks, aim to improve interpretability while maintaining high performance. Innovative approaches such as prototype learning, interpretable neural network architectures, and explainable boosting machines further expand the toolkit for creating more transparent models. The key to success lies in selecting methods tailored to the specific problem and domain, and iteratively refining the approach based on both performance metrics and interpretability assessments. By implementing these techniques, it is often possible to achieve a good balance between model performance and interpretability, thereby meeting the growing demand for responsible and explainable AI systems [25, 39].

6.6 Human in Loop and Human in Command

Human-in-the-loop and human-in-command approaches are crucial strategies for ensuring the safe and effective use of AI/ML in medical device applications. The human-in-the-loop approach involves direct human involvement in controlling or supervising the operations and decision-making of an AI system. This leverages human situational awareness and judgment, enabling timely intervention or deactivation of the AI system as needed. Known as "heteronomous devices, " this approach recommends that clinicians use AI to input their decision-making rather than entirely replacing their judgment [40].

On the other hand, the human-in-command approach focuses on overseeing the broader economic, societal, legal, and ethical impacts of AI systems. It involves deciding when and how to use the AI system in specific situations, including the option not to use it. This approach establishes levels of human discretion and the ability to override AI decisions when necessary. Devices that do not provide a means to intervene in a timely fashion are also known as autonomous devices. Hybrid devices provide direct or supervisory control over specific functions and no control over others [41]. Figure 6.7 illustrates types of human controllability in AI systems.

Key considerations for both approaches include continuous oversight and retrospective periodic review of performance, especially for complex systems. Clinicians should avoid overreliance on AI predictions and understand the intended use, performance capabilities, and limitations of AI medical devices. Transparency from AI developers about these aspects is crucial for proper human oversight. The level and type of human involvement may vary depending on the specific medical application and the AI system's capabilities. In summary, human-in-the-loop and

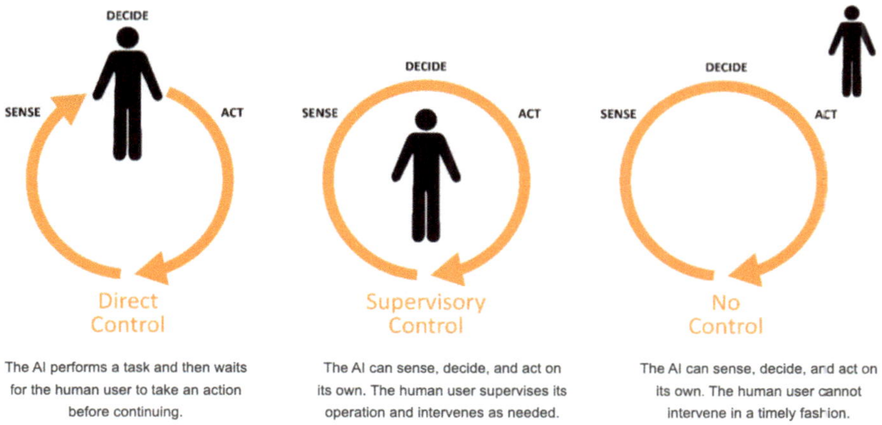

Fig. 6.7 Types of controllability in AI [41]

human-in-command approaches aim to maintain human control and oversight in AI-assisted medical decision-making, ensuring patient safety and ethical use of the technology. The specific implementation depends on the medical context and the AI system's capabilities. Still, the overarching goal is to leverage the strengths of both human expertise and AI capabilities for optimal patient care.

The key differences between human-in-command and human-in-the-loop approaches in medical AI applications reflect distinct levels of human involvement and control. Human-in-the-loop refers to the direct involvement of humans in controlling or supervising the AI system's operations and decision-making process. It involves integrating human expertise into the AI's decision-making loop, allowing for real-time intervention or adjustment. This approach focuses on leveraging human situational awareness and judgment to improve the performance and safety of AI-based medical devices, enabling clinicians to validate AI-generated diagnoses, correct errors, and provide additional insights based on their experience. It typically involves continuous oversight and interaction with the AI system during operation [42, 43]. An example of a Human-in-loop is shown in Fig. 6.8.

In contrast, human-in-command refers to the broader capability to oversee the overall activity of the AI system, including its economic, societal, legal, and ethical impacts. This approach involves the ability to decide when and how to use the AI system in any particular situation, including the decision not to use it. It allows for establishing levels of human discretion during the system's use, rather than direct real-time control. It ensures the ability to override AI decisions when necessary, although not necessarily in real-time. Human-in-command may involve periodic review and oversight rather than continuous interaction [40].

The key differences lie in their scope, timing, level of control, and applicability. Human-in-the-loop is more focused on direct operational control and decision-making, often involving real-time interaction. At the same time, human-in-command encompasses broader oversight and strategic decision-making, which may include retrospective review and long-term oversight. Human-in-the-loop provides more

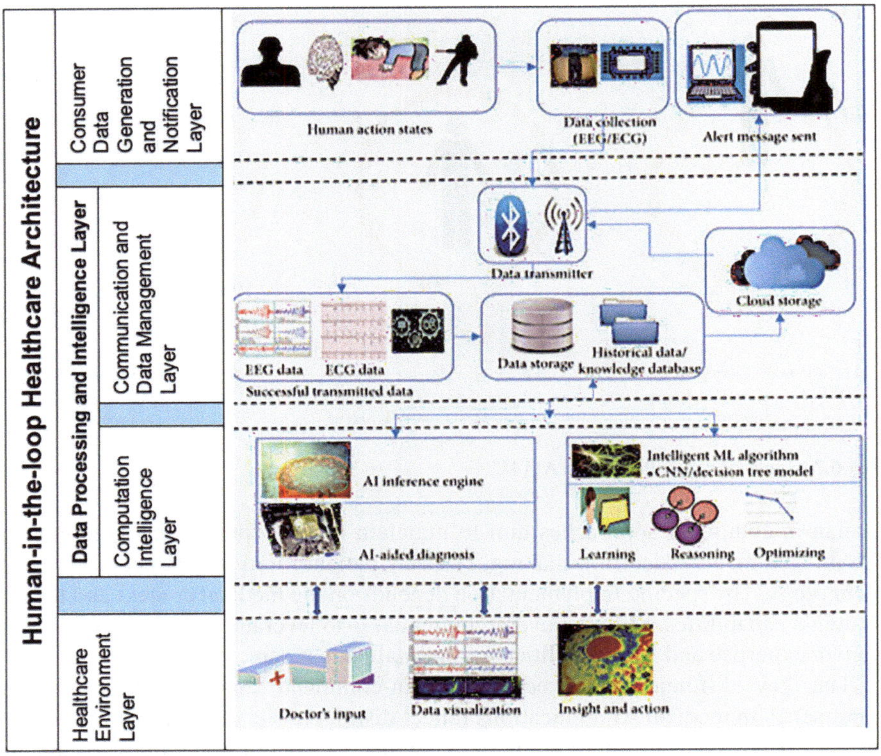

Fig. 6.8 An example of human-in-the-loop cyber-physical system in healthcare [43]

direct control over individual decisions, while human-in-command focuses on higher-level control over when and how to use the AI system. Additionally, human-in-the-loop may not always be feasible or safe in specific medical applications, such as high-speed surgical robots. In contrast, human-in-command can still be applied in these scenarios. Both approaches aim to maintain human control and oversight in medical AI applications, but they operate at different levels and time frames within the decision-making process.

6.6.1 Benefits of Human-in-Command Design over Human-in-Loop

Human-in-command offers several key benefits over human-in-the-loop approaches in high-stakes medical procedures, providing a more comprehensive and flexible framework for AI oversight. This approach enables broader oversight of the AI system's overall activity, including its economic, societal, legal, and ethical impacts, allowing for a more holistic evaluation of AI use in medical procedures. It facilitates strategic decision-making, enabling healthcare professionals to determine when and how to use AI systems, including the option not to use them in some instances. This level of control is crucial for high-stakes procedures where AI may not always be appropriate.

The human-in-command approach enables varying levels of human discretion during system use, which can be adjusted according to specific procedural requirements. It ensures the ability to override AI decisions when necessary, even if not in real-time, maintaining ultimate control in critical situations. This is particularly valuable in scenarios where immediate human-in-the-loop control may be limited by decision-making capacity, situational awareness, or sensing uncertainties [42].

Furthermore, human-in-command balances the benefits of automation in high-speed or high-precision tasks with overall human control and accountability. It is better positioned to address complex ethical considerations that may arise in high-stakes medical procedures, which AI systems may struggle to navigate independently. The approach provides more flexibility in adapting the level of human involvement based on specific procedural requirements and risks, facilitates long-term oversight crucial for continually assessing and improving AI system performance, and aligns well with evolving regulatory frameworks that emphasize human oversight and accountability in AI-assisted medical procedures [42].

6.7 Clinical Validation of AI-Enabled Medical Devices

Clinical validation studies for AI/ML-enabled medical devices encompass a comprehensive set of components designed to rigorously assess their safety, effectiveness, and reliability in real-world healthcare settings. At the core of these studies is the evaluation of safety and efficacy, demonstrating that the device performs as intended in clinical environments and meets predefined performance standards. Generalizability is crucial in assessing the algorithm's ability to produce accurate outputs across diverse patient populations and clinical settings.

Bias detection and mitigation are critical components, with studies designed to identify and address potential biases in the algorithm, ensuring equitable performance across all patient groups. Assessing performance in real-world settings, often using real-world evidence (RWE), is essential for understanding how the device functions beyond controlled trial conditions. Study design may incorporate innovative approaches, such as using retrospective data in prospective studies or combining premarket and postmarket data collection efforts [44]. Standardized performance metrics are developed to evaluate the performance of AI-enabled medical devices, including generalization measures across diverse populations. The quality and quantity of data used for training, testing, and validation are paramount, requiring large amounts of high-quality, labeled data. Continuous learning and improvement strategies are documented, outlining how ongoing data will be used to monitor and enhance the device over time. A Predetermined Change Control Plan (PCCP) is included to describe the framework for future modifications to the AI/ML algorithm [37].

Robust validation protocols are developed, detailing activities and predetermined acceptance criteria, including usability studies and postmarket surveillance plans. Ensuring diverse patient representation in the survey is crucial to assessing performance across different demographics and clinical conditions. The AI/ML-enabled device's performance is often compared to current gold standards or expert clinicians to benchmark its capabilities. Ethical considerations, such as patient privacy and consent, are addressed in the study design. Finally, regulatory alignment is ensured, with the study design conforming to current FDA guidance or other relevant regulatory requirements. These components collectively aim to thoroughly evaluate AI/ML algorithms in medical devices, addressing the unique challenges posed by these advanced technologies and ensuring their safe and effective implementation in healthcare [45].

6.7.1 *Retrospective and Prospective Clinical Validation Study*

Retrospective clinical studies play a crucial role in the initial development and evaluation of AI/ML algorithms for medical devices. These studies utilize historically labeled data, allowing researchers to train and test algorithms on large datasets before moving to prospective testing. This approach provides valuable insights into algorithm performance and potential clinical applications. However, it's essential to recognize the limitations of retrospective studies. Performance in these controlled settings is likely to be better than in real-world scenarios, and retrospective studies alone are insufficient to fully understand the actual clinical utility of AI systems [45, 46].

The validation of AI algorithms in medical devices primarily employs three methods, each with varying levels of scientific rigor and prevalence in FDA-approved devices. Retrospective validation, used in approximately 28% of FDA-approved AI medical devices, involves testing the AI model on historical patient data [47]. This method, while applicable, may not account for recent changes in patient populations or medical practices. Prospective validation is also used in approximately 28% of FDA-approved AI medical devices, where the AI model is tested on new patient data collected after the study's initiation. This approach yields more substantial scientific evidence than retrospective validation, as it enables the AI to account for current data variables and real-time patient information. Prospective validation is considered more realistic and better assesses the AI's performance in actual clinical settings.

Randomized controlled trials (RCTs) are widely regarded as the gold standard for clinical validation. This method, a subset of prospective validation, utilizes random assignment to control for confounding variables, thus isolating the therapeutic effect of the device. Despite its superiority, only about 4% (22 out of 521) of FDA-approved/cleared AI devices were validated using RCTs [47]. This low percentage highlights a significant gap in the rigorous validation of AI medical devices, raising concerns about their real-world effectiveness and potential impact on patient care. To address these limitations, prospective validation is essential.

Prospective studies evaluate AI performance in real clinical settings, providing a more accurate assessment of the algorithm's effectiveness and potential impact on patient care. Unfortunately, very few randomized controlled trials of AI systems have been conducted to date [47], highlighting a significant gap in the field. When evaluating AI/ML algorithms, it's crucial to look beyond simple accuracy metrics, especially for imbalanced datasets standard in healthcare. Alternative metrics, such as precision, recall, F1-score, and the area under the precision-recall curve, offer more comprehensive insights. Moreover, using clinical outcomes as endpoints, rather than just technical accuracy, provides a more meaningful assessment of the algorithm's potential impact.

Healthcare data presents unique challenges, including low data volume, high sparsity, and poor data quality. These issues can limit the efficacy of deep learning methods, and in some cases, conventional machine learning approaches may outperform deep learning on smaller healthcare datasets. Interpretability is another key consideration, with interpretable machine learning models particularly valuable for clinical adoption and decision-making. Numerous retrospective studies have been conducted in various healthcare domains, including the prediction of labor outcomes, ICU complications, and postoperative events. These studies often compare multiple ML algorithms on historical clinical data, providing valuable insights into their relative performance. While retrospective studies are a common and essential starting point for developing AI/ML algorithms in medical devices, they should be viewed as a stepping stone rather than an endpoint. Prospective validation, appropriate evaluation metrics, and interpretable models are essential for developing clinically helpful and impactful AI/ML algorithms in healthcare.

Cross-validation and external validation are crucial for validating AI algorithms used in medical devices. Cross-validation involves randomly dividing datasets into separate portions for development and internal validation. This approach allows researchers to train the AI model on one part of the data and then test its performance on the remaining unseen data, providing a measure of the model's ability to generalize within the same dataset. External validation, on the other hand, takes this concept a step further by testing the model on data from different institutions or populations. This method is crucial for assessing the AI's generalizability and ability to perform consistently across diverse healthcare settings and patient groups.

6.7.2 Randomized Controlled Clinical Trials

Randomized controlled trials (RCTs) play a crucial role in enhancing the validity of AI medical devices, offering a gold standard for clinical validation that provides the highest level of scientific evidence for safety and efficacy. The strength of RCTs lies in their ability to control for confounding variables through random assignment, effectively isolating the actual impact of the AI device on patient outcomes. This randomization process also significantly reduces various forms of bias, including selection bias, which can otherwise skew study results.

One of the key advantages of RCTs is their capacity to assess AI device performance in real-world clinical settings with actual patients, offering a more accurate representation of how the technology will function in practice. This real-world evaluation allows for direct comparisons between AI-assisted approaches and current standard-of-care practices, demonstrating whether the new technology offers meaningful improvements in patient care [48]. The prospective nature of data collection in RCTs is another significant benefit, as it is generally considered more reliable and less prone to bias than retrospective studies. This prospective approach, combined with the rigorous methodology of RCTs, produces more substantial scientific evidence than retrospective or non-randomized prospective studies, thereby increasing confidence in the AI device's performance [48].

6.8 Generalizability of AI Algorithm Performance

Generalizability obstacles [49] are data sampling, data shift, overfitting, and underspecification. Data selection can be affected by data sampling and data shift. Modeling can be biased due to overfitting and underspecification. External validation results often show a decrease in performance compared to internal validation due to several key factors. Overfitting to development data is a primary concern, as internal validation techniques mainly address in-sample optimism and may not capture how well the model generalizes to new populations. Population differences between the development and external validation datasets can significantly impact model performance, as can variations in data quality, measurement procedures, and variable definitions. Temporal changes in clinical practice or disease patterns may affect model performance if external validation is conducted later [50].

Selection bias in the development dataset, which may be more carefully curated than real-world datasets, can lead to overly optimistic internal validation results. Limited representation in the development data may fail to capture the broader populations' full diversity and heterogeneity. Models may also lack transportability to new settings with different care practices, patient populations, or healthcare systems. Some internal validation techniques may produce optimistic performance estimates compared to proper external validation, and publication bias may favor models with excellent internal performance, which are more likely to show a decrease when externally validated [49].

The complexity of real-world settings introduces variables not fully captured in controlled development environments, further contributing to performance decreases in external validation. Understanding these factors underscores the importance of rigorous external validation in assessing a model's true generalizability and real-world performance before clinical implementation. It also highlights the need for ongoing monitoring and updating of models in clinical use to ensure their continued effectiveness and reliability [50].

6.8.1 *Best Practices for AI Model Generalization*

Ensuring machine learning models generalize well to external datasets is a critical challenge in developing robust and reliable AI systems. As developers strive to create models that perform consistently across diverse real-world scenarios, several key strategies [51–54], as discussed here, can be employed to enhance generalization capabilities.

- **Data Quality:** One fundamental approach focuses on data quality and diversity during the model development. By collecting training data from a wide range of sources and ensuring it captures the full spectrum of variability expected in real-world applications, developers can build a strong foundation for generalization. This may involve collaborating with multiple institutions or incorporating data from various geographical regions and demographic groups. Additionally, implementing data augmentation techniques can artificially expand the dataset, exposing the model to a broader range of potential inputs and improving its robustness.
- **Validation (Testing) Technique:** Proper validation techniques are crucial in assessing and improving model generalization. While internal validation methods, such as cross-validation, are essential, they cannot guarantee performance on truly external datasets, as illustrated in Fig. 6.9. Developers should prioritize external validation using independent datasets that differ from the population characteristics, settings, or periods of the training data. This approach provides a more realistic assessment of the model's generalization ability to new environments.
- **Regularization Techniques:** These are powerful tools for combating overfitting and improving generalization. Methods such as L1 and L2 regularization add penalty terms to the model's loss function, discouraging overly complex repre-

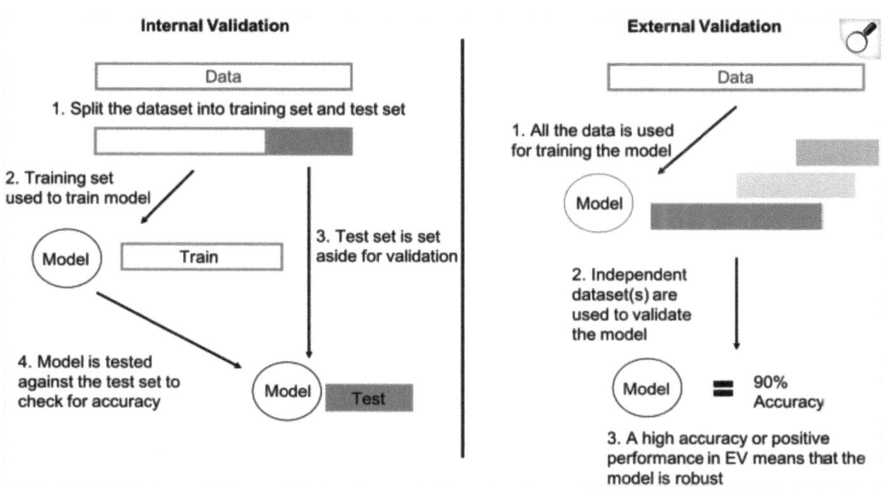

Fig. 6.9 Internal validation vs external validation

sentations and promoting simpler, more generalizable models. Similarly, techniques like dropout can help prevent the model from relying too heavily on specific features or patterns in the training data.

- **State-of-the-Art Feature Engineering:** By carefully selecting and engineering relevant features based on domain knowledge and data analysis, developers can provide the model with more informative and discriminative inputs. This process may involve dimensionality reduction, creating meaningful transformations of the data, or incorporating domain-specific insights to capture the most relevant aspects of the problem.
- **Ensemble Methods:** These ensemble methods provide an additional avenue for enhancing generalization by combining predictions from multiple models. Techniques such as bagging, boosting, and random forests create diverse models and aggregate their predictions, often leading to more robust and generalizable outcomes. These approaches can help mitigate the impact of individual model biases and improve overall performance across different datasets.
- **Transfer Learning:** It has emerged as a powerful technique for enhancing generalization, particularly in domains where large-scale labeled datasets are scarce. By leveraging pre-trained models on large datasets and fine-tuning them for specific tasks, developers can benefit from knowledge learned in related domains and improve performance on new, unseen data.
- **Model Monitoring:** Continuous monitoring and updating models in real-world applications is essential for maintaining generalization performance over time. As data distributions may shift and new patterns emerge, regularly assessing model performance on diverse datasets and retraining or fine-tuning as necessary can help ensure ongoing reliability and relevance.
- **Transparency and Interpretability:** By developing models that provide clear explanations for their decisions, developers can more easily identify potential biases or limitations that may affect performance on external datasets. This approach facilitates targeted improvements and helps build trust in the model's capabilities across different contexts.
- **Better Collaboration:** Effective collaboration among data scientists, domain experts, and end-users is crucial for developing truly generalizable models. Incorporating diverse perspectives and expertise can help identify potential pitfalls, refine feature selection, and ensure that the model addresses real-world needs across various settings.

6.9 Clinical Study Documentation

Key clinical study documentation requirements for AI/ML algorithms in medical devices include, but are not limited to, the following documents:

- Data quality and diversity
- Performance metrics
- Clinical validation study design

- Comprehensive technical documentation
- Verification documentation
- Validation protocols
- Generalizability assessment
- Bias detection and mitigation
- Safety and effectiveness data
- Performance in real-world settings
- Predetermined change control plan (PCCP)
- Continuous learning and improvement plans

6.10 Real-World Evidence

Real-world evidence (RWE) has become increasingly crucial in validating AI/ML algorithms for medical devices, offering a more comprehensive and generalizable assessment of their effectiveness than traditional clinical trials. RWE enables the evaluation of algorithm performance across diverse patient populations and clinical environments, often more cost-effectively and efficiently than conventional studies. It enables continuous monitoring and updating of algorithms based on real-world performance, providing insights into their behavior in everyday clinical settings beyond controlled trial conditions [55–57].

RWE is instrumental in detecting and mitigating biases that may not be apparent in controlled premarket trials, feeding back into product design to ensure fair and equitable device performance across different patient populations. It aids in assessing the generalizability of SaMDs across various demographics and clinical conditions, ensuring their effectiveness for diverse patient groups. From a regulatory perspective, the FDA's increasing acceptance of RWE for postmarket surveillance supports a Total Product Lifecycle (TPLC) approach, facilitating rapid product improvement cycles and enabling continuous regulatory oversight.

The FDA has recognized the value of RWE and is increasingly accepting it to support approval and postmarket surveillance of AI/ML-enabled medical devices. The agency has proposed a Total Product Lifecycle (TPLC) approach for regulating these devices, incorporating RWE for ongoing monitoring. However, using RWE presents challenges, including ensuring data quality, addressing biases in real-world datasets, standardizing data across different sources, and protecting patient privacy while leveraging large-scale data.

6.11 EU MDR Clinical Evaluation Guidance

The MDCG 2020–1 document provides comprehensive guidance on the clinical evaluation of medical device software (MDSW) under the Medical Device Regulation (MDR) and performance evaluation under the In Vitro Diagnostic

Medical Devices Regulation (IVDR). This guidance applies to software with an independent intended purpose and claimed clinical benefit, as well as software that drives or influences the operation of medical devices. This guidance also applies to AI-enabled medical device software.

The clinical evaluation process for MDSW involves determining the validity of clinical association or scientific validity, assessing technical and analytical performance, and evaluating clinical performance. The document emphasizes that clinical evidence requirements depend on the software's intended purpose and its relationship to other medical devices. For software that drives or influences multiple medical devices, separate clinical evaluations are required for each software-device combination.

The guidance emphasizes that the level of clinical evidence required should be commensurate with the device's characteristics and intended purpose. The evaluation process includes planning, data identification and generation, analysis, and documentation in a clinical evaluation report. The document emphasizes the need for continuous monitoring and updates of the clinical evaluation throughout the device's lifecycle. This comprehensive approach ensures that MDSW manufacturers provide sufficient clinical evidence to demonstrate the safety, performance, and compliance of their products with relevant regulations [57].

6.12 US FDA Guidance on SaMD Clinical Evaluation

The FDA guidance document on software as a Medical Device (SaMD) Clinical Evaluation [58] provides a comprehensive framework for assessing the clinical performance of SaMD. The evaluation is based on three key components:

- Valid Clinical Association: This involves demonstrating a scientifically valid link between the SaMD's output and its targeted clinical condition.
- Analytical Validation: This step ensures that the SaMD accurately and reliably processes input data to generate precise output data.
- Clinical Validation: This verifies that the SaMD achieves its intended purpose in the target population within the context of clinical care.

The evaluation process is continuous throughout the SaMD's lifecycle, encompassing both premarket and postmarket phases. The level of clinical evidence required is determined by the SaMD's risk classification and intended use. Evidence can be generated through various methods, including referencing existing data, extrapolating from studies with different intended uses, or developing new clinical data specific to the SaMD. The guidance emphasizes the flexibility of evidence generation while maintaining rigorous standards for safety and effectiveness. The document encourages early engagement with regulatory bodies to ensure alignment with expectations. This approach aims to provide a harmonized framework for the clinical evaluation of SaMD across different regulatory jurisdictions, helping manufacturers navigate the complex landscape of medical device regulation while ensuring their products meet the necessary safety and performance requirements.

Table 6.2 below provides an overview of EU, US, and international guidelines and standards for generating clinical evidence and evaluating SaMD [59].

Table 6.2 List of US FDA, EU, and international guidances and standards for clinical evaluation

Reference	Scope
EU	
MDCG 2020-1	Guidance on Clinical Evaluation of Medical Device Software. Intended to supplement MEDDEV 2.7/1 (Rev. 4) for Clinical Evaluation of Software that Qualifies as MD or IVDD
MDCG 2020-5	Guidance on Clinical Evaluation–Equivalence
MEDDEV 2.7/1 Rev. 4	Clinical Evaluation: Guide for Manufacturers and Notified Bodies (applicable to all MD only)
US	
FDA Guidance (11 January 2002)	General Principles of Software Validation: Final Guidance for Industry and FDA Staff
FDA Guidance (8 December 2017)	Software as a Medical Device (SaMD): Clinical Evaluation: Guidance for Industry and Food and Drug Administration Staff
International Medical Device Regulators Forum	
N41 (2017)	Clinical Evaluation of SaMD
N55 (2019)	Clinical Evidence (applicable to all MD and IVDD)
N56 (2019)	Clinical Evaluation (applicable to all MD)
N57 (2019)	Clinical investigation (applicable to all MD)
International Standards: Safety and Performance of Medical Software	
ISO 14155	Clinical Investigation of Medical Devices for Human Subjects–Good Clinical Practice. For SaMD, the standard applies where relevant
ISO 20916	In Vitro Diagnostic Medical Devices–Clinical Performance Studies Using Specimens From Human Subjects–Good Study Practice
ISO 14971	Application of Risk Management to Medical Devices. Specifies terminology, principles and a process for risk management of medical devices. IEC/TR 80002-1 provides guidance on the application of ISO 14971 to medical software
IEC 62366	Application of Usability Engineering to Medical Devices. Specifies a process for a manufacturer to analyze, specify, develop, and evaluate the usability of a medical device as it relates to safety
IEC 82304-1	Health Software–Part 1: General Requirements for Product Safety. Applies to the safety and security of health software products designed to operate on general computing platforms and intended to be placed on the market without dedicated hardware
IEC 62304	Medical Device Software–Software Life Cycle Processes. Defines the life cycle requirements for medical device software

6.13 Summary

This chapter comprehensively explored AI-enabled medical device testing and clinical validation. It began by outlining best practices for AI model testing frameworks, followed by an in-depth discussion on AI Model Robustness testing. The chapter then delved into the crucial concepts of AI Model Transparency, explainability, and interpretability, addressing how these could be implemented and how developers could strike a balance between interpretability and performance. A thorough examination of various types of bias and methods for testing bias was presented. The chapter introduced and clearly defined the concepts of human-in-loop and human-in-command, explaining how developers could leverage these approaches to enhance ethical considerations, mitigate bias, and reduce product liability while improving regulatory compliance.

The discussion then transitioned to clinical validation, offering strategies for successful clinical studies and improving AI model generalization. A review of real-world evidence testing was provided, and the chapter concluded with an analysis of two crucial regulatory requirements: the US FDA and EU MDR for Clinical evaluation and related documentation. This comprehensive overview equips readers with the essential knowledge needed to navigate the complex landscape of AI-enabled medical device development and validation.

6.14 Points to Ponder

1. How do you prove state of the art for testing for robustness?

6.15 Answers

1. Testing the robustness of AI systems, as outlined in Sect. 6.2, Fig. 6.1, requires a comprehensive approach that integrates multiple techniques and carefully considers the trade-off between robustness and accuracy.

   To achieve state-of-the-art compliance with medical device regulations, a key strategy involves identifying harmonized and recognized standards within the regulatory landscape. In the context of the EU AI Act, the JCT21 committee has introduced the ISO/IEC TR 24029–1 standard, which specifically addresses the testing of robustness for complex machine learning algorithms and neural networks. This standard establishes a systematic framework for evaluating the robustness of neural networks, incorporating performance metrics tailored for robustness assessment, a range of testing methodologies—including statistical, formal, empirical, and data perturbation techniques—and clear guidelines for data sourcing and curation. Moreover, it outlines a standardized workflow for conducting comprehensive robustness assessments and benchmarking. By adhering to this framework, organizations can substantially minimize regulatory risks and ensure that their robustness testing is thorough, reproducible, and aligned with international best practices.

References

1. https://www.fda.gov/medical-devices/medical-device-regulatory-science-research-programs-conducted-osel/performance-evaluation-methods-evolving-artificial-intelligence-ai-enabled-medical-devices
2. https://www.imdrf.org/sites/default/files/2024-06/Good%20machine%20learning%20practice%20for%20medical%20device%20development%20-%20Guiding%20Principles%20DRAFT%20for%20Consultation.pdf
3. https://namsa.com/resources/blog/fdas-regulation-of-ai-ml-samd/
4. https://blog.johner-institute.com/regulatory-affairs/regulatory-requirements-for-medical-devices-with-machine-learning/
5. https://www.agilemd.com/post/why-prospective-validation-should-be-a-must-for-ai-early-warning
6. https://www.frontiersin.org/journals/medical-technology/articles/10.3389/fmedt.2024.1473350/full
7. Lones MA. How to avoid machine learning pitfalls: a guide for academic researchers. https://arxiv.org/pdf/2108.02497
8. Koçak B (2022) Key concepts, common pitfalls, and best practices in artificial intelligence and machine learning: focus on radiomics. Diagn Interv Radiol 28(5):450–462. https://doi.org/10.5152/dir.2022.211297. PMID: 36218149; PMCID: PMC9682557
9. Pioneering the future: generative AI'S impact on medical devices. https://www.cigniti.com/blog/generative-ai-medical-devices-testing-challenges-solutions/
10. https://arxiv.org/abs/2404.00897
11. Freiesleben T, Grote T (2023) Beyond generalization: a theory of robustness in machine learning. Synthese 202:109. https://doi.org/10.1007/s11229-023-04334-9
12. https://encord.com/blog/model-robustness-machine-learning-strategies/
13. https://www.xcally.com/news/explainability-vs-interpretability-the-challenge-of-transparent-artificial-intelligence/
14. https://www.cimplifi.com/resources/transparency-explainability-and-interpretability-of-ai/
15. US FDA (2024) Transparency for machine learning-enabled medical devices: guiding principles
16. Lundberg SM et al (2017) A unified approach to interpreting model predictions
17. https://c3.ai/introduction-what-is-machine-learning/model-interpretability-is-critical-to-driving-adoption/
18. Molnar C (2024) Interpretable machine learning: a guide for making black box models explainable
19. Ueda D et al (2024) Fairness of artificial intelligence in healthcare: review and recommendations. Jpn J Radiol 42(1):3–15. https://doi.org/10.1007/s11604-023-01474-3. Epub 2023 Aug 4. PMID: 37540463; PMCID: PMC10764412
20. Wollek A et al (2022) Attention-based saliency maps improve interpretability of pneumothorax classification. Radiol Artif Intell 5(2):e220187. https://doi.org/10.1148/ryai.220187. PMID: 37035429; PMCID: PMC10077084
21. Muralidharan V, Adewale BA, Huang CJ et al (2024) A scoping review of reporting gaps in FDA-approved AI medical devices. NPJ Digit Med 7:273. https://dci.org/10.1038/s41746-024-01270-x

22. Fehr J, Citro B, Malpani R, Lippert C, Madai VI (2024) A trustworthy AI reality-check: the lack of transparency of artificial intelligence products in healthcare. Front Digit Health 6:1267290. https://doi.org/10.3389/fdgth.2024.1267290. PMID: 38455991; PMCID: PMC10919164
23. Anastasiya K et al (2022) Transparency of AI in healthcare as a multilayered system of accountabilities: between legal requirements and technical limitations. Front Artif Intell 5:879603. https://doi.org/10.3389/frai.2022.879603
24. Elendu C, Amaechi DC, Elendu TC, Jingwa KA, Okoye OK, John Okah M, Ladele JA, Farah AH, Alimi HA (2023) Ethical implications of AI and robotics in healthcare: a review. Medicine (Baltimore) 102(50):e36671. https://doi.org/10.1097/MD.0000000000036671. PMID: 38115340; PMCID: PMC10727550
25. Amann J, Blasimme A, Vayena E et al (2020) Explainability for artificial intelligence in healthcare: a multidisciplinary perspective. BMC Med Inform Decis Mak 20:310. https://doi.org/10.1186/s12911-020-01332-6
26. https://www.nist.gov/itl/ai-risk-management-framework
27. Liang W et al (2024) What's documented in AI? Systematic analysis of 32K AI model cards. arXiv:2402.05160v1 [cs.SE] 07 Feb 2024
28. EU AI Act
29. Abràmoff MD, Tarver ME, Loyo-Berrios N, Trujillo S, Char D, Obermeyer Z, Eydelman MB, Maisel WH (2023) Considerations for addressing bias in artificial intelligence for health equity. NPJ Digit Med 6(1):170. https://doi.org/10.1038/s41746-023-00913-9. PMID: 37700029; PMCID: PMC10497548
30. Vokinger KN, Feuerriegel S, Kesselheim AS (2021) Mitigating bias in machine learning for medicine. Commun Med 1:25. https://doi.org/10.1038/s43856-021-00028-w
31. Obermeyer Z, Powers B, Vogeli C, Mullainathan S (2019) Dissecting racial bias in an algorithm used to manage the health of populations. Science 366(6464):447–453. https://doi.org/10.1126/science.aax2342. PMID: 31649194
32. Mittermaier M, Raza MM, Kvedar JC (2023) Bias in AI-based models for medical applications: challenges and mitigation strategies. NPJ Digit Med 6:113. https://doi.org/10.1038/s41746-023-00858-z
33. Cho MK (2021) Rising to the challenge of bias in health care AI. Nat Med 27(12):2079–2081. https://doi.org/10.1038/s41591-021-01577-2. PMID: 34893774; PMCID: PMC11017306
34. Cross JL, Choma MA, Onofrey JA (2024) Bias in medical AI: implications for clinical decision-making. PLOS Digit Health 3(11):e0000651. https://doi.org/10.1371/journal.pdig.0000651. PMID: 39509461; PMCID: PMC11542778
35. US FDA. Identifying and measuring artificial intelligence (AI) bias for enhancing health equity
36. https://www.pewtrusts.org/en/research-and-analysis/articles/2021/10/07/fda-review-can-limit-bias-risks-in-medical-devices-using-artificial-intelligence
37. US FDA Guidance (Draft) (2025) Artificial intelligence-enabled devices software functions: lifecycle management and marketing submission recommendations
38. US FDA Guidance. Methods and tools for effective postmarket monitoring of artificial intelligence (AI)-enabled medical devices
39. Jaotombo F, Adorni L, Ghattas B, Boyer L (2023) Finding the best trade-off between performance and interpretability in predicting hospital length of stay using structured and unstructured data. PLoS One 18(11):e0289795. https://doi.org/10.1371/journal.pone.0289795. PMID: 38032876; PMCID: PMC10688642
40. Human-in-the-Loop Machine Learning (HITL) explained. https://encord.com/blog/human-in-the-loop-ai/
41. COCIR report on artificial intelligence in EU medical device legislation, May 2021
42. Sele D, Chugunova M (2024) Putting a human in the loop: increasing uptake, but decreasing accuracy of automated decision-making. PLoS One 19(2):e0298037. https://doi.org/10.1371/journal.pone.0298037. PMID: 38335162; PMCID: PMC10857587
43. Ganesan A, Paul A, Nagabushnam G, Gul MJJ (2021) Human-in-the-loop predictive analytics using statistical learning. J Healthc Eng 29(2021):9955635. https://doi.org/10.1155/2021/9955635. PMID: 34367543; PMCID: PMC8346319

44. Park SH, Choi J, Byeon JS (2021) Key principles of clinical validation, device approval, and insurance coverage decisions of artificial intelligence. Korean J Radiol 22(3):442–453. https://doi.org/10.3348/kjr.2021.0048. PMID: 33629545; PMCID: PMC7909857
45. Kelly CJ, Karthikesalingam A, Suleyman M, Corrado G, King D (2019) Key challenges for delivering clinical impact with artificial intelligence. BMC Med 17(1):195. https://doi.org/10.1186/s12916-019-1426-2. PMID: 31665002; PMCID: PMC6821018
46. Chen D, Liu S, Kingsbury P et al (2019) Deep learning and alternative learning strategies for retrospective real-world clinical data. NPJ Digit Med 2:43. https://doi.org/10.1038/s41746-019-0122-0
47. Chouffani El Fassi S, Abdullah A, Fang Y et al (2024) Not all AI health tools with regulatory authorization are clinically validated. Nat Med 30:2718–2720. https://doi.org/10.1038/s41591-024-03203-3
48. Plana D, Shung DL, Grimshaw AA, Saraf A, Sung JJY, Kann BH (2022) Randomized clinical trials of machine learning interventions in health care: a systematic review. JAMA Netw Open 5(9):e2233946. https://doi.org/10.1001/jamanetworkopen.2022.33946
49. Eche T, Schwartz LH, Mokrane FZ, Dercle L (2021) Toward generalizability in the deployment of artificial intelligence in radiology: role of computation stress testing to overcome underspecification. Radiol Artif Intell 3(6):e210097. https://doi.org/10.1148/ryai.2021210097. PMID: 34870222; PMCID: PMC8637230
50. Van Calster B, Steyerberg EW, Wynants L et al (2023) There is no such thing as a validated prediction model. BMC Med 21:70. https://doi.org/10.1186/s12916-023-02779-w
51. Ho SY, Phua K, Wong L, Bin Goh WW (2020) Extensions of the external validation for checking learned model interpretability and generalizability. Patterns (N Y) 1(8):100129. https://doi.org/10.1016/j.patter.2020.100129. PMID: 33294870; PMCID: PMC7691387
52. Sperrin M, Riley RD, Collins GS et al (2022) Targeted validation: validating clinical prediction models in their intended population and setting. Diagn Progn Res 6:24. https://doi.org/10.1186/s41512-022-00136-8
53. Grosse R. Lecture 9: generalization
54. Zou KH, Li JZ, Imperato J, Potkar CN, Sethi N, Edwards J, Ray A (2020) Harnessing real-world data for regulatory use and applying innovative applications. J Multidiscip Healthc 13:671–679. https://doi.org/10.2147/JMDH.S262776. PMID: 32801731; PMCID: PMC7383026
55. https://www.fda.gov/files/medical%20devices/published/US-FDA-Artificial-Intelligence-and-Machine-Learning-Discussion-Paper.pdf
56. US FDA Guidance. SaMD clinical evaluation: guidance for industry and FDA staff
57. US FDA guidance on use of real-world evidence to support regulatory decision-making for medical devices
58. MDCG 2020-1. Guidance on clinical evaluation of medical device software
59. Kwade Z (2022) Clinical evaluation of software. RF Quarterly

Chapter 7
AI/ML Model Integration and Deployment

Abstract This chapter comprehensively explores the integration and deployment of artificial intelligence models, focusing on best practices for integrating and deploying AI models across various environments. It addresses the complex landscape of regulatory compliance challenges that organizations face when implementing AI solutions. The discussion covers three primary deployment strategies: cloud-based, on-premises/edge, and mobile deployments, each with a unique set of challenges. The chapter emphasizes the importance of incorporating human expertise into AI workflows, highlighting how this integration can enhance the overall effectiveness and reliability of AI systems. The chapter also explores critical aspects of AI deployment, including scalability, containerization, reliability, versioning, and performance monitoring, offering readers a comprehensive understanding of the technical considerations necessary for successful AI implementation. Through this extensive coverage, the chapter aims to equip readers with the knowledge and strategies needed for deploying AI models effectively in various contexts.

Keywords Artificial intelligence · Integration · AI integration · AI deployment · Machine learning integration · Deployment strategy · Large language model · Latency · Containerization · Docker · Domain knowledge · Multimodal AI · Vision AI · Edge deployment · Cloud multimodal · On-premise deployment · Human in the Loop · Human in the command · Hybrid environment · Security · Privacy · Data quality · Performance · Good machine learning practice (GMLP) · TensorFlow · Pytorch · Mobile deployment

7.1 Domain Knowledge for AI Integration in Medical Devices

Domain knowledge plays a crucial role in enhancing the performance of machine learning (ML) models in medical devices by improving various stages of the development process. First, it aids in data preparation and feature engineering by enabling domain experts to select the most relevant attributes for classification tasks,

A. Pandey et al., *Artificial Intelligence in Medical Software*,
https://doi.org/10.1007/978-3-032-10808-1_7

combine features into meaningful variables, and ensure proper data preprocessing and encoding. Second, domain knowledge contributes to model development by integrating expert insights into the ML pipeline, such as rules, causal networks, and formulas, resulting in hybrid models that often outperform. This integration enhances model generalization and highlights the critical importance of expert knowledge over algorithmic differences [1].

Additionally, domain knowledge enhances interpretability and clinical alignment by ensuring models adhere to established clinical guidelines, making them more comprehensible and relevant in real-world medical settings. It also proven invaluable in data-limited scenarios, where integrating medical expertise helps maintain model performance despite small or incomplete datasets—a common issue in healthcare. Furthermore, domain knowledge supports regulatory compliance and safety by implementing safeguards around ML outputs to mitigate risks and ensure adherence to safety standards required for medical applications. Manufacturers can create more accurate, interpretable, clinically relevant models aligned with real-world healthcare needs by leveraging domain expertise throughout the ML development process. This synergy between domain knowledge and machine learning is essential for advancing reliable and effective medical devices [1].

7.2 Integrating Artificial Intelligence Models in Medical Devices

Integrating artificial intelligence and machine learning (ML) in medical devices introduces several significant challenges. One key area is data and model validation, which involves ensuring data quality, diversity, and representativeness to avoid bias, demonstrating model reliability across different populations and settings, and validating the safety and effectiveness of adaptive or continuously learning AI systems. Another challenge is transparency and explainability, as the "black box" nature of some ML algorithms complicates understanding their decision-making processes. Transparent reporting mechanisms must be implemented to explain how models learn and develop, while striking a balance between the need for transparency and the protection of intellectual property [2].

Dependency management is a significant hurdle, as ML models often rely on specific libraries and software versions, which can lead to potential compatibility issues. Ensuring reproducibility and portability across different environments is crucial but complex, as models that work in one setting may not function seamlessly in another. Scalability becomes a concern as demand increases, necessitating sophisticated solutions for handling larger workloads. Version control for models and their iterations is essential but can be challenging to manage effectively.

One significant issue is the incompatibility between ML libraries and traditional object-oriented programming paradigms. This mismatch can create a steep learning curve for developers accustomed to object-oriented approaches, as they must adapt to new programming styles and patterns. Another critical challenge is the lack of proper monitoring and maintenance practices. ML models can degrade over time in production environments without regular updates and performance monitoring, decreasing accuracy and reliability.

Clinical integration and validation involve conducting rigorous clinical trials, ensuring regulatory compliance, and developing an intuitive user interface for healthcare professionals. Clinical integration poses challenges in translating algorithmic performance metrics into meaningful clinical benefits, assessing the real-world utility of ML-based devices beyond narrow performance measures, and integrating these tools into existing clinical workflows. Finally, ethical and legal considerations arise from issues such as fairness, accountability, and transparency in ML-based decisions, data privacy, security concerns with large-scale data collection, and navigating the complex legal landscape surrounding AI-assisted care [3].

Performance monitoring and validation present additional hurdles, as continuous real-world performance monitoring is necessary to ensure ongoing safety and effectiveness as AI models evolve. Traditional validation testing may not adequately address all concerns raised by adaptive AI technologies, necessitating the development of new methods to evaluate and monitor AI algorithm performance over time. Data quality and bias are critical concerns, as the quality and representativeness of training datasets are crucial to prevent bias in AI outputs. Ensuring diverse and relevant data for training AI models is challenging, particularly considering the applicability across different populations and settings.

Regulatory frameworks must adapt to address the unique characteristics of ML-based devices. This includes updating existing regulations, developing approaches for "locked" versus "adaptive" AI algorithms, and harmonizing regulations across the jurisdictions to enable global market access. Post-market surveillance is also critical, requiring systems for continuously monitoring ML performance in real-world settings, protocols for updating models without lengthy re-authorization processes, and mechanisms to ensure ongoing safety and effectiveness as models evolve.

Regulators must develop new approaches, such as the FDA's "Predetermined Change Control Plan"[4]. A total product lifecycle approach to regulation is necessary to accommodate the iterative and adaptive nature of AI technologies while harmonizing regulations across different jurisdictions, which remains a significant challenge for global market access. Addressing these challenges requires collaboration between regulators and manufacturers to ensure the safe and effective integration of AI/ML technologies in medical devices while fostering innovation in this rapidly evolving field [3, 5, 6].

The skills gap in finding personnel with the right mix of ML, software engineering, and DevOps expertise can be a significant obstacle. Addressing these multifaceted challenges requires a collaborative approach involving data scientists, ML engineers, software developers, and operations teams to successfully integrate ML models into production software systems.

The following sections will discuss specific challenges related to integrating multimodal ML models, Large Language Models (LLMs), Vision AI models, human-in-the-loop systems, and interoperability.

7.3 Integrating Multimodal AI Models in Medical Devices

Integrating multimodal data for medical devices presents several challenges due to the inherent complexity and diversity of medical data. Medical data comes in various forms, such as imaging (X-rays, MRIs, CT scans), clinical records, genomic data, wearable sensor outputs, and patient-reported outcomes. These diverse modalities differ in format, scale, sampling rate, and dimensionality, making their integration challenging and requiring sophisticated techniques for alignment and combination. Ensuring data quality and consistency across modalities is another challenge, as inconsistencies in formats, poor annotation, and a lack of structure in some sources, such as clinical notes, necessitate extensive preprocessing and harmonization to maintain data integrity. Missing data further complicates multimodal approaches since one or more multimodal data points are unavailable for certain patients or are partially missing within a modality. This limitation constrains usable datasets and reduces model flexibility, as many AI models struggle with incomplete information [2, 7].

Additionally, multimodal models are computationally intensive, requiring significant memory and processing power, which can be expensive to scale and may introduce latency issues in real-time applications. As model complexity increases with multiple modalities, interpretability becomes more challenging. Healthcare professionals must trust AI-driven insights, but balancing model sophistication with interpretability is difficult and often required by regulatory standards. Privacy and security concerns also arise when integrating sensitive patient data from multiple sources, which increases the risk of breaches. Robust security measures are crucial for ensuring the safe transmission, storage, and processing of data, while also complying with healthcare regulations such as HIPAA. Lastly, scalability and generalizability remain significant hurdles; models trained on data from one institution may not generalize well to others due to differences in data collection practices and patient populations. These challenges underscore the necessity for interdisciplinary collaboration, innovative technical solutions, and meticulous consideration of ethical and regulatory requirements to fully realize the potential of multimodal integration in enhancing medical device capabilities [8, 9].

7.4 Integration of Large Language Models

Integrating large language models (LLMs) into existing healthcare software systems requires careful planning and attention to several critical factors. Ensuring interoperability and compatibility is essential, as LLMs must seamlessly integrate with electronic health record (EHR) systems without disrupting workflows. Developing interoperability standards can further facilitate smooth data exchange across different healthcare platforms. Upgrading the infrastructure and security is equally important, as LLMs demand advanced computing capabilities, including enhanced processing power, storage, and network bandwidth. Robust cybersecurity measures, such as encryption and access controls, are necessary to protect sensitive patient data, alongside ensuring compliance with regulations like HIPAA to address data privacy concerns [10, 11].

A transparent integration approach should be adopted, starting with well-defined use cases where LLMs can enhance processes like clinical documentation or patient triage. For example, integrating LLMs into EHR systems can streamline tasks such as generating discharge summaries or managing patient records. Cloud-based solutions can enable scalable deployment and simplify the integration of APIs into medical applications. Validation and monitoring are crucial to the success of LLM integration; this involves setting evaluation metrics, conducting thorough clinical validation to ensure regulatory compliance, and implementing continuous monitoring systems to track performance and detect anomalies [10].

To ensure successful adoption, healthcare staff should receive comprehensive training on using LLM capabilities effectively. Pilot programs in specific departments or use cases can help ease the transition before full-scale deployment. Gathering feedback from clinicians and patients allows for iterative improvements to the integration process and functionality of the LLMs. Additionally, ethical considerations must be addressed by establishing guidelines for bias, transparency, and accountability in the use of LLMs. Involving diverse stakeholders in the development process and creating a regulatory framework for validation and deployment ensures responsible implementation [11].

7.5 Integration of Vision AI Models

Integrating and deploying computer vision machine learning models for medical devices requires careful consideration of several factors. The choice of deployment method depends on the specific use case and requirements. Edge deployment involves running the model directly on the medical device, which is ideal for scenarios requiring low latency or unreliable internet connectivity. Cloud deployment, which involves running the model on remote servers and accessing it via API, is well-suited for applications that do not require real-time processing and have stable internet connectivity. Some applications may benefit from a hybrid approach, leveraging the strengths of both methods [12].

Integrating computer vision models with existing medical imaging systems requires careful planning and consideration of several key factors. Adequate data preparation and management are crucial, involving robust preprocessing techniques, well-organized datasets, and collaboration with medical experts for accurate annotation. Model selection and development should focus on choosing appropriate deep learning architectures tailored for medical imaging tasks, leveraging pre-trained models, and optimizing performance through experimentation [13].

Implementation using model optimization through techniques like pruning and quantization, thorough testing and validation, choosing an appropriate deployment strategy, continuous monitoring and maintenance, and comprehensive documentation. Integrating existing systems requires compatibility and interoperability with the current healthcare IT infrastructure, developing user-friendly interfaces that align with clinical workflows, and implementing APIs or cloud platforms for seamless connectivity. Regulatory compliance and approval processes should be addressed early, adhering to guidelines for medical device approval and implementing robust data privacy and security measures. Continuous improvement and monitoring systems should be established, incorporating feedback loops for clinician input and regularly updating models as new data becomes available [14].

7.6 Integrating Human Expertise into AI Workflows

Integrating human expertise into AI workflows offers numerous advantages that enhance the effectiveness and adaptability of AI systems. Human experts are crucial in improving decision-making by providing critical insights and judgment in complex scenarios where AI may struggle, particularly in areas requiring nuanced understanding or ethical considerations. Their involvement also enhances accuracy and reliability by providing feedback on AI predictions, identifying and correcting errors, and addressing edge cases or rare events that AI may not have been trained to handle. This iterative process refines AI models, ensuring more reliable outcomes. Humans excel at understanding contextual nuances and adapting to new situations, complementing AI's data processing capabilities and enabling the combined system to handle a broader range of scenarios and evolving environments. Additionally, human oversight is crucial for mitigating bias and ensuring fairness, as experts can critically assess AI outputs to identify biases that might otherwise go unnoticed, thereby promoting ethical decision-making across diverse applications.

7.7 Human in the Loop and Human in Command

Human-in-the-loop (HITL) and human-in-command (HIC) approaches offer different levels of human involvement in deploying machine learning AI algorithms for medical devices, each with advantages and disadvantages. HITL systems actively involve humans in AI decision-making, providing input and oversight. Examples include radiologists reviewing AI-generated medical image annotations, clinicians validating AI-suggested treatment plans, and pathologists confirming AI-detected anomalies in tissue samples. HITL offers improved accuracy through human expertise, the ability to handle complex cases, continuous model improvement, and enhanced trust among medical professionals. However, it can lead to slower decision-making, potential human bias, increased costs, and scalability challenges.

On the other hand, HIC systems allow AI to operate more autonomously, with humans maintaining overall control and intervening only when necessary. Examples include AI-powered patient monitoring systems with human oversight for critical alerts, automated medication dispensing systems with pharmacist supervision, and AI-driven surgical robots with surgeon control capabilities. HIC advantages include faster decision-making in routine cases, improved efficiency and scalability, reduced workload for medical professionals, and consistent performance in repetitive tasks. Disadvantages include the potential for overlooking subtle issues, the risk of over-reliance on AI, challenges in maintaining human expertise for rare interventions, and ethical concerns regarding accountability.

The choice between HITL and HIC approaches depends on factors such as the criticality of the medical application, the reliability and proven accuracy of the AI system, regulatory requirements, available resources, and the complexity of the medical tasks involved. In many cases, a hybrid approach that combines elements of both HITL and HIC may be optimal, allowing for the efficient automation of routine tasks while maintaining human expertise for complex decisions and oversight.

Implementing Human-in-the-Loop (HITL) systems in healthcare introduces several significant regulatory challenges that must be addressed to ensure compliance, safety, and effectiveness. One primary concern is data privacy and security, as healthcare data is highly sensitive and subject to strict regulations, such as HIPAA. HITL systems must safeguard patient confidentiality while allowing necessary access for human reviewers, implement robust security measures to prevent breaches, and maintain compliance with evolving data protection laws across jurisdictions. Accountability and liability also pose significant challenges, particularly in determining responsibility for medical errors that involve AI recommendations and human oversight. Clear lines of accountability must be established between AI systems and human reviewers, along with legal definitions for the status of AI recommendations in medical decision-making. Regulatory approval processes are further complicated by the dynamic nature of HITL systems, which require the development of evaluation criteria that account for the interplay between AI and human input, as well as mechanisms for re-evaluating systems as they evolve [13].

Standardization and interoperability are critical for ensuring consistency across HITL implementations. This involves developing standards for data inputs, AI algorithms, and human review processes while ensuring compatibility with existing healthcare IT infrastructure. Ethical oversight is another key consideration. HITL systems must address issues such as equitable access to care, biases in AI algorithms and human decision-making processes, and striking a balance between automation and human judgment in sensitive medical scenarios. Continuous monitoring and adaptation are essential due to the evolving nature of HITL systems, requiring mechanisms for ongoing performance monitoring, rapid regulatory responses to emerging risks, and adaptable frameworks to keep pace with technological advancements. Finally, training and certification of human reviewers present a regulatory challenge, necessitating the establishment of training standards, certification processes, and ongoing education to ensure competence in working with AI systems [14].

7.8 Interoperability Issue in AI Integration

Interoperability issues significantly hinder the integration of AI into healthcare systems in several critical ways. These challenges limit AI systems' access to comprehensive, high-quality system data silos that prevent information sharing across systems and organizations, a lack of standardized data formats and coding schemes, and poor data quality or incomplete records that affect the reliability of AI insights. Integration complexity further complicates the process, as proprietary systems and the absence of universal standards make it difficult to connect AI solutions with electronic health records (EHRs) and other platforms. Variations in data representation necessitate complex mapping and transformation processes, while the absence of standardized APIs and exchange protocols hinders AI deployment. These interoperability barriers also negatively impact AI performance by restricting access to diverse datasets, reducing accuracy due to inconsistent data formats and terminologies, and limiting the integration of patient-generated health data with clinical data, which affects personalized medicine [15].

Moreover, interoperability challenges exacerbate security and privacy concerns by increasing vulnerabilities in data exchanges, lacking standardized security protocols, and creating uncertainty regarding compliance with regulations such as HIPAA. Ultimately, these issues generate adoption hurdles by extending implementation timelines due to extensive data cleaning requirements, increasing costs for smaller organizations, and reducing user acceptance due to difficulties in integrating AI into clinical workflows. Addressing these challenges requires establishing common data standards, improving system connectivity, and developing secure data exchange protocols to unlock the full potential of AI in healthcare [16].

A unified, component-based, data-driven framework [17] addressing interoperability challenges in healthcare systems is illustrated in Fig. 7.1. When a patient visits a healthcare service center, healthcare providers generate a wide range of health-related data, including laboratory, radiology, pharmacy, administrative,

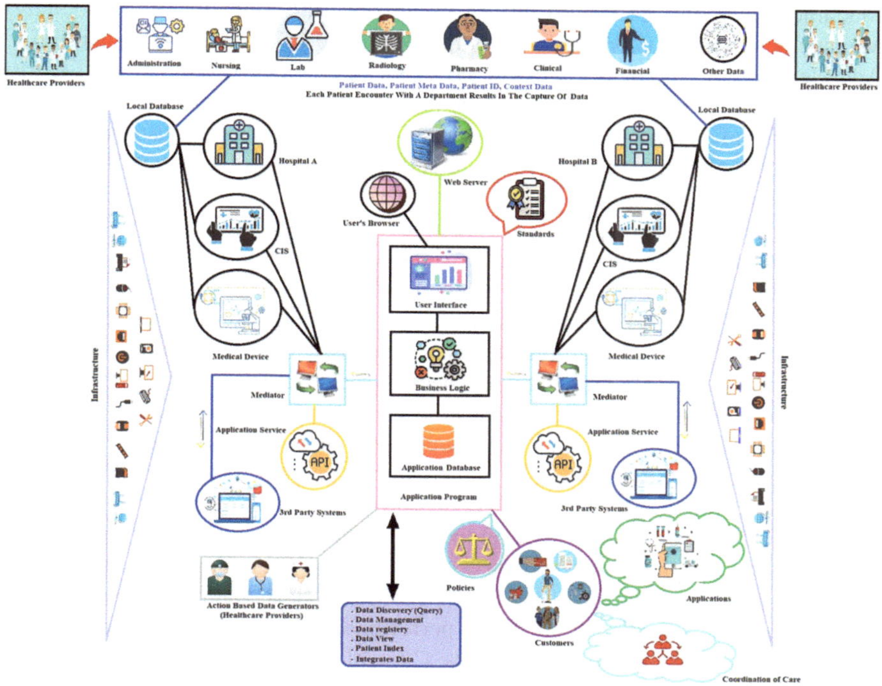

Fig. 7.1 A unified component-based data-driven framework to support interoperability in the healthcare systems [17]

nursing, financial, and clinical information. Additional data is collected from medical devices connected to the patient and clinical information systems within each center. This information is stored and managed in local databases at the health service center. Suppose specific clinical data is not available in these local systems. In that case, it is generated through a web-based application that follows standards set by health service providers and is stored in a secondary database. Integration between the existing information in health institutions and new data generated by the web-based application is facilitated through middleware and web services. Moreover, information exchange with third-party organizations, such as insurance companies or judicial entities, also occurs via the web-based application or institutional systems, following the same secure procedures.

7.9 Deployment of AI Models on Medical Devices

Deploying a machine learning model in medical devices requires a structured approach prioritizing regulatory compliance, data security, and patient safety. The first step is ensuring regulatory compliance by meeting FDA classification requirements, adhering to regulations such as MDR/IVDR in the EU, following Good

Machine Learning Practice (GMLP) principles, and implementing a Quality Management System (QMS) that is compliant with US FDA, EU MDR, and EU AI Act. Next, the infrastructure must be HIPAA-compliant, featuring strong security measures such as encryption and access controls. Model preparation involves validating performance metrics, such as sensitivity and specificity, ensuring that the training data accurately represents the intended patient population, and thoroughly documenting the machine learning lifecycle [18].

For deployment architecture, a scalable design is essential, incorporating containerization tools like Docker for consistent deployment and CI/CD pipelines for automated testing and deployment. Integration into medical device software must adhere to good software development practices, including thorough testing, version control, and effective change management processes. Comprehensive monitoring and logging systems should be established to track real-time performance, detect degradation, and set alerts for critical issues. Data privacy and security are paramount; this includes implementing anonymization techniques, complying with regulations like GDPR and HIPAA, and conducting regular security audits [19].

Continuous learning and updates are critical for maintaining model performance over time. Protocols should be developed to update the model while adhering to regulatory requirements for software modifications, including a Predetermined Change Control Plan (PCCP) for adaptive AI/ML models. Detailed documentation must outline the device's intended use, performance characteristics, cloud deployment architecture, and security measures. Clinical integration involves training clinicians to interpret model outputs and integrate them into workflows while establishing feedback mechanisms for ongoing improvement. Finally, post-market surveillance systems should be implemented to continuously monitor real-world performance and address any adverse events or issues that may arise [4, 20].

7.10 AI Model Deployment Time

The deployment time of AI models is influenced by multiple interdependent factors, starting with model complexity, as larger architectures require more extensive optimization and configuration for production environments. Infrastructure readiness—including the availability of hardware such as CPUs and GPUs, and properly configured software stacks—directly affects deployment speed, while efficient data pipelines streamline preprocessing and feature engineering. Containerization tools like Docker and orchestration platforms, such as Kubernetes, accelerate deployment once configured, although the initial setup requires a significant time investment. Scalability planning, integration with existing systems, and regulatory compliance—particularly in the healthcare sector—add complexity that extends timelines. Rigorous testing across diverse scenarios ensures reliability, but it also consumes significant time, as does implementing monitoring systems to track performance and detect any drift. Team expertise plays a pivotal role in navigating these challenges efficiently, while the choice of deployment strategy depicted in Figs. 7.2, 7.3, and 7.4—batch, real-time, or streaming—further shapes the timeline [21].

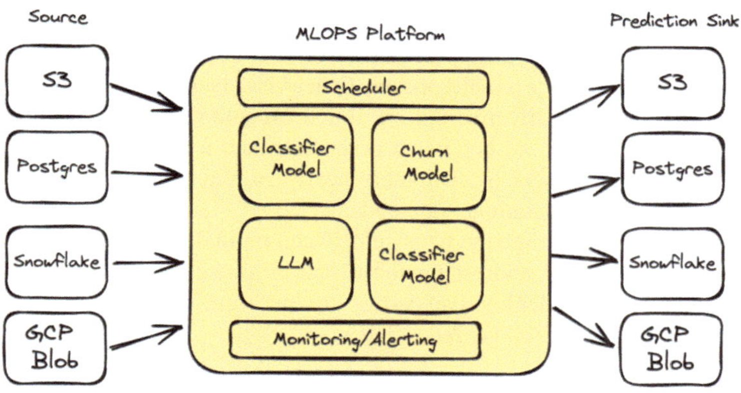

Fig. 7.2 Batch inference [21]

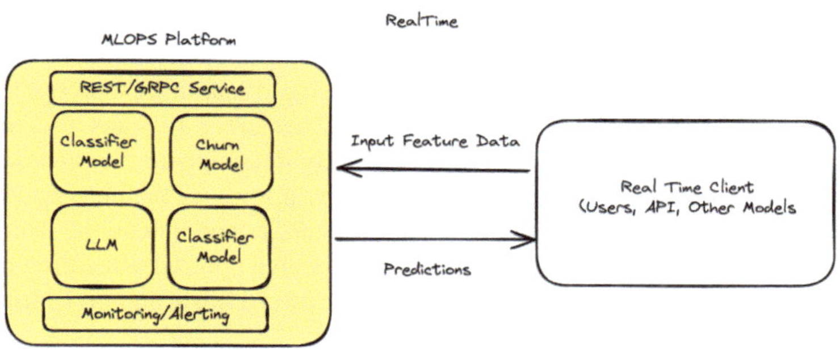

Fig. 7.3 Real-time inference [21]

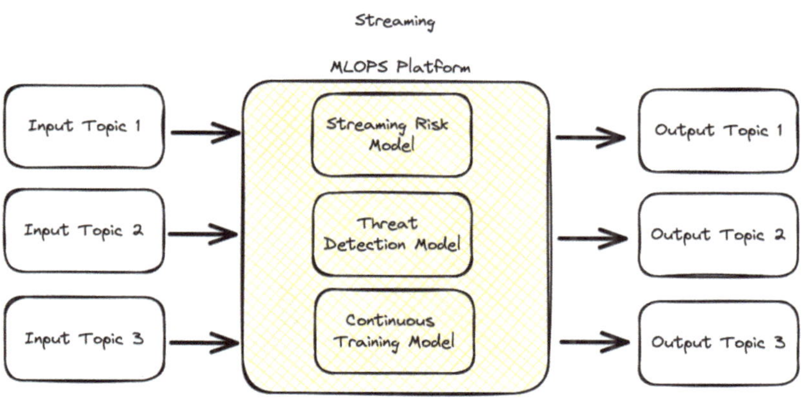

Fig. 7.4 Streaming inference [21]

The first major decision when selecting an inference framework is whether to use batch or real-time prediction serving. Two key factors should guide this choice: whether your model can generate predictions for each request or batch within an acceptable timeframe, and whether the immediate value provided to users or downstream customers by real-time predictions outweighs the ongoing costs of maintaining the necessary infrastructure. If both conditions are met, a real-time inference architecture is a viable option. Otherwise, batch processing is likely a more practical and cost-effective option. Batch executions are flexible and do not require a fixed schedule. Ad-hoc batch processing and a pipeline architecture can deliver predictions to users or customers nearly in real time, while significantly reducing costs and computational demands [21].

These elements determine the balance between rapid deployment and long-term operational stability. By addressing these factors proactively, organizations can optimize their AI model deployment process, reduce time-to-production, and ensure reliability and performance [14, 20].

7.11 AI Model Deployment Architecture

Microservices architecture, as shown in Fig. 7.5, significantly enhances the scalability of AI models in medical devices by offering dynamic resource allocation, which enables elastic scaling of computational resources to meet the real-time demands critical for fluctuating AI workloads. By decomposing applications into smaller, containerized components, the architecture allows for the independent scaling of individual services, thereby reducing bottlenecks and enhancing overall system scalability. Efficient resource management ensures optimal utilization of CPU, GPU, and memory across clusters, minimizing hardware waste and preventing overprovisioning [22].

The rapid deployment of containerized AI models across various environments accelerates responsiveness to evolving healthcare needs. At the same time,

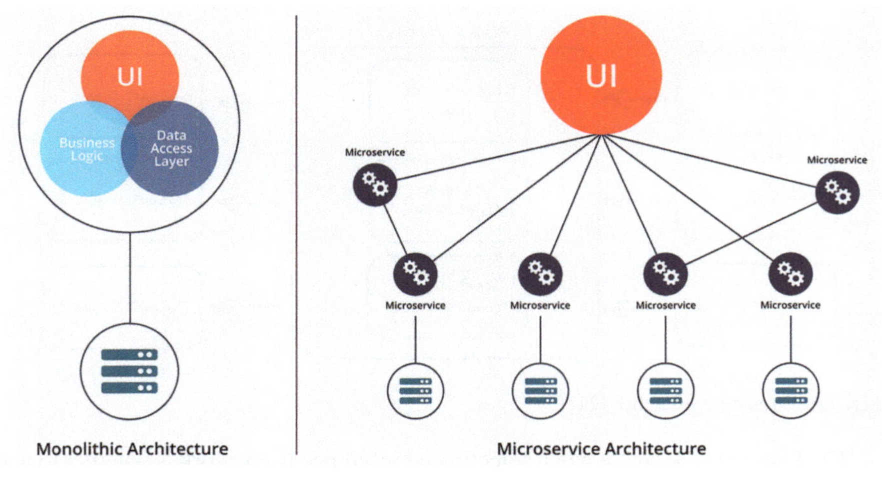

Fig. 7.5 Monolithic vs microservice-based software architecture [22]

flexibility and agility empower teams to update models or deploy features without requiring systemic overhauls, thereby shortening the time-to-market. Enhanced fault isolation ensures reliability by containing failures to specific modules, and improved interoperability streamlines data exchange, helping healthcare providers adapt to regulatory shifts and value-based care models. Cloud-native compatibility further amplifies scalability by leveraging cloud infrastructure, enabling medical device manufacturers to process large datasets, adapt to dynamic demands, and maintain robust performance. Together, these capabilities position microservices as a foundational framework for scaling AI-driven solutions in healthcare with precision and resilience [23, 24].

7.12 AI Model Deployment Strategies

Deployment strategies can be categorized into two distinct approaches. Static deployment strategies involve manual management of traffic or request distribution, with examples including shadow evaluation, A/B testing, Canary testing, Rolling deployment, and Blue-green deployment. These methods rely on predefined configurations and human intervention to manage updates or shifts in workload. In contrast, dynamic deployment strategies automate traffic distribution, as seen in techniques like Multi-Armed Bandits, which adaptively allocate requests based on real-time performance metrics. This automated approach enables self-adjusting systems that optimize resource allocation and responsiveness without manual oversight. These categories provide tailored solutions for striking a balance between control and adaptability in AI deployment [25, 26].

7.12.1 Shadow Model Deployment Strategy

In shadow deployment, a new "shadow model" is deployed alongside the live model, processing identical requests but remaining inactive for users. This allows real-world testing without disrupting services, as both models run in parallel via separate API endpoints. Predictions from both models are generated and stored during inference, but only the output of the live model is returned to users. Predictions are compared against ground truth to evaluate performance, guiding decisions on full deployment. The champion/challenger framework extends this by testing multiple shadow models against the live model, selecting the best performer based on accuracy or KPIs. While effective for comparing models and assessing pipeline efficiency, latency, and load capacity, shadow testing can be labor-intensive due to its parallel evaluation demands.

7.12.2 A/B Testing Model Deployment Strategy

A/B testing is a data-driven method for evaluating two models (A and B) in controlled environments, commonly used in e-commerce, social media, and streaming platforms. By running models with slightly different features in parallel, it measures user interactions like feedback to determine which version improves conversion rates or performance. Data scientists formulate hypotheses—assumptions tested through experiments—and select the better-performing model for global deployment based on results. This approach enables evidence-based decisions to optimize user experience and system efficiency.

7.12.3 Multi-Armed Bandit Model Deployment Strategy

The Multi-Armed Bandit (MAB), as shown in Fig. 7.6, is an advanced, reinforcement learning-inspired alternative to A/B testing that balances exploration (testing models statistically) and exploitation (leveraging top performers) to maximize KPIs like conversion rates. Unlike A/B testing, MAB dynamically allocates user traffic to models based on real-time KPI performance, favoring the best-performing version for global deployment.

It handles multiple models simultaneously, continuously updating scores to optimize decisions efficiently—ideal for time-sensitive scenarios, such as limited-time offers, where rapid, data-driven choices are crucial. By prioritizing both discovery and efficiency, MAB enhances flexibility and conversion outcomes.

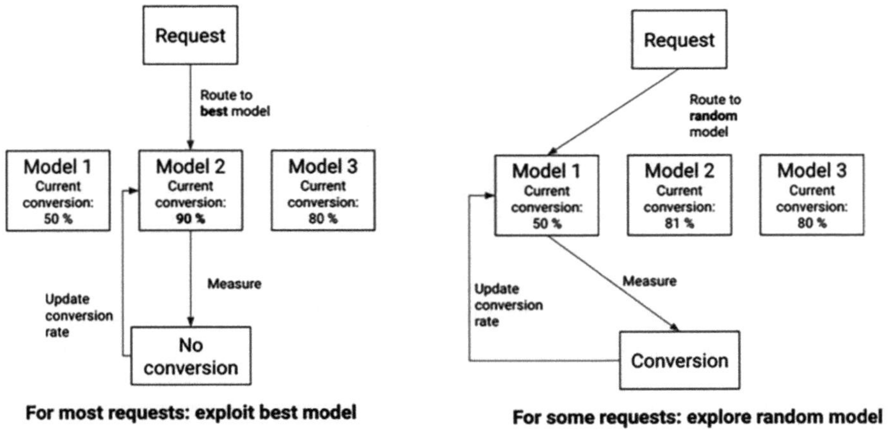

Fig. 7.6 MAB deployment strategy [25]

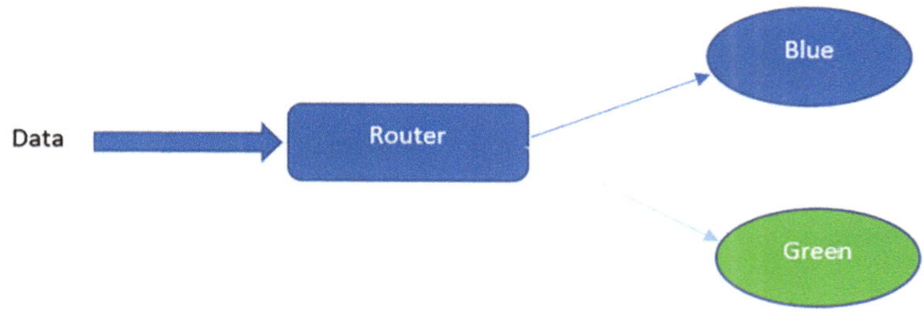

Fig. 7.7 Blue–green deployment [25]

7.12.4 *Blue-Green Model Deployment Strategy*

Blue-green deployment, as shown in Fig. 7.7, utilizes two identical production environments: blue (the live model) and green (the new version). The green environment is a staging replica with updated features, sharing components like databases to reduce costs. While blue handles live traffic, green is tested with actual data to confirm stability. Once validated, traffic shifts to green, with blue as a backup for quick rollback if issues arise. This strategy eliminates downtime, making it ideal for applications requiring continuous availability.

7.12.5 *Canary Model Deployment Strategy*

Canary deployment, as shown in Fig. 7.8, gradually introduces a new model version to a small subset of real users, testing it alongside the live model while monitoring reliability, errors, and performance. Unlike strategies that isolate testing to hidden models or control groups, this method uses actual user interactions to detect issues before a global rollout. User exposure can be scaled up or down based on test results, allowing for seamless rollback without downtime if problems arise. It excels in real-world evaluations, providing faster insights than A/B testing by generating statistically significant data in hours rather than days. This approach minimizes risk while enabling rapid, data-driven decisions for full deployment or iteration.

7.12.6 *Feature Flag Model Deployment Strategy*

Feature flagging, as shown in Fig. 7.9, is a technique, not a deployment strategy, that enables developers to integrate code into the main branch while keeping features inactive until finalized. This allows collaborative development and iterative testing without exposing unfinished work to users. Once ready, features are activated via flags, which can be combined with deployment strategies like canary or blue-green releases to manage rollouts dynamically.

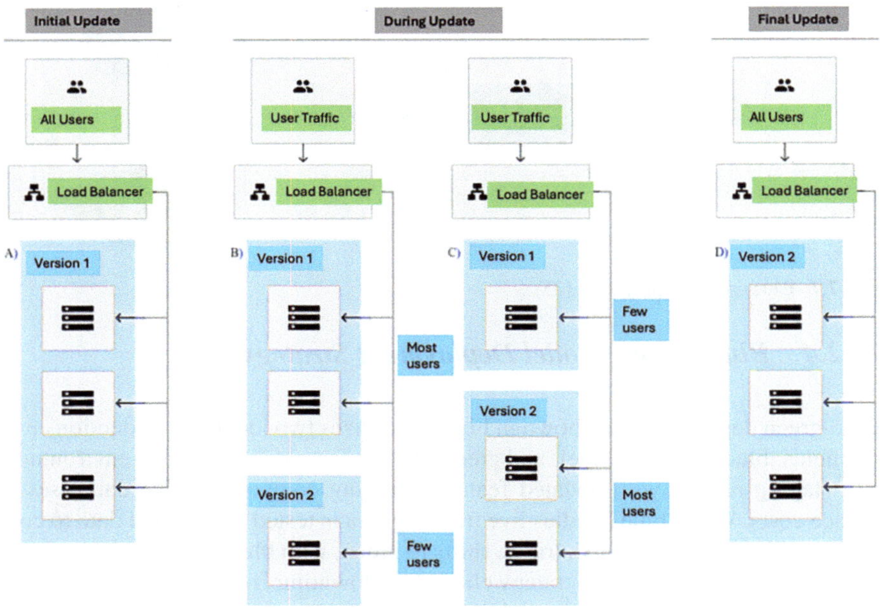

Fig. 7.8 Canary deployment strategy [25]

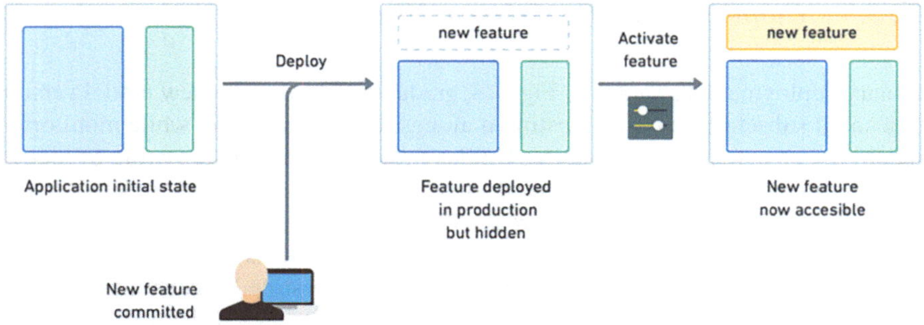

Fig. 7.9 Feature flag [25]

7.12.7 Rolling Model Deployment Strategy

Rolling deployment gradually updates live instances by incrementally replacing older model versions, eliminating the need for staging environments or private development setups. This strategy ensures continuous service availability during transitions, as updates are applied progressively across running systems.

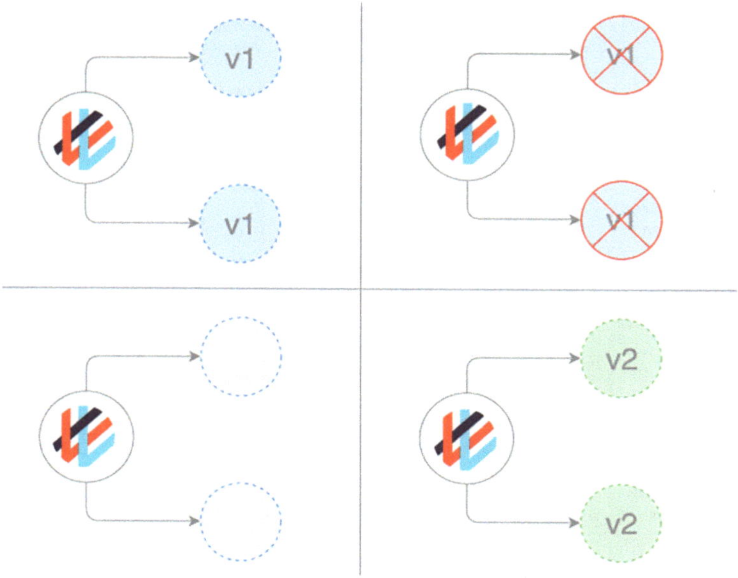

Fig. 7.10 Recreate deployment strategy [25]

7.12.8 Recreate Model Deployment Strategy

The recreate strategy, as shown in Fig. 7.10, involves shutting down the live model version (e.g., V1 in Fig. 7.10) and deploying a new version (e.g., V2 in Fig. 7.10), discarding old instances entirely. While straightforward to implement and ensure a fully refreshed environment, this method causes downtime and rebooting, directly disrupting users during the transition.

Comparison of various deployment strategies is summarized in Fig. 7.11.

Deployment or testing pattern	Zero downtime	Real production traffic testing	Releasing to users based on conditions	Rollback duration	Impact on hardware and cloud costs
Recreate Version 1 is terminated, and Version 2 is rolled out.	x	x	x	Fast but disruptive because of downtime	No extra setup required
Rolling update Version 2 is gradually rolled out and replaces Version 1.	✓	x	x	Slow	Can require extra setup for surge upgrades
Blue/green Version 2 is released alongside Version 1; the traffic is switched to Version 2 after it is tested.	✓	x	x	Instant	Need to maintain blue and green environments simultaneously
Canary Version 2 is released to a subset of users, followed by a full rollout.	✓	✓	x	Fast	No extra setup required
A/B Version 2 is released, under specific conditions, to a subset of users.	✓	✓	✓	Fast	No extra setup required
Shadow Version 2 receives real-world traffic without impacting user requests.	✓	✓	x	Does not apply	Need to maintain parallel environments in order to capture and replay user requests

Fig. 7.11 Comparison: which model release strategy should be used? [25]

7.13 AI Model Deployment on Cloud and On-Premise

When deploying machine learning models, organizations must choose between cloud platforms and on-premises solutions, each offering distinct advantages and challenges. Cloud-based deployments host models on remote servers managed by providers such as AWS, Azure, or Google Cloud, which are accessible via the Internet. At the same time, on-premises solutions involve installing and running models on the company's hardware within its physical premises. The cost structures differ significantly, with cloud platforms offering a pay-as-you-go model and operational expenses that scale with usage. In contrast, on-premises deployments require high upfront capital expenditure for hardware and ongoing maintenance costs [27].

Control and security considerations also vary, as on-premises solutions provide complete control over hardware, software, and data, which can be preferable for highly regulated industries. Cloud platforms, while offering high-level security measures, provide less direct control. Scalability is a key differentiator, with cloud solutions offering easily scalable resources on demand, while on-premises deployments are limited by existing hardware capacity and require new infrastructure for scaling. Maintenance and updates are managed by service providers for cloud deployments, which include automatic updates and patches. In contrast, on-premises solutions require manual updates and maintenance by the company's IT team.

Accessibility is another critical factor, with cloud-based models accessible from anywhere with an internet connection, supporting remote work and collaboration. On-premises deployments typically limit access to on-site locations, potentially requiring a VPN for remote access.

Cloud deployments present several key compliance challenges that organizations must address to ensure data security and regulatory adherence. Data privacy and control are primary concerns, as sensitive information is stored on third-party servers. Organizations must verify that their cloud providers implement adequate safeguards to protect data and comply with regulations such as GDPR or HIPAA. Regulatory complexity adds another difficulty, as organizations must navigate varying compliance standards across jurisdictions and industries, understand applicable regulations (e.g., PCI DSS, GDPR, HIPAA), implement appropriate controls, and conduct regular audits to ensure alignment with current laws. Data residency requirements further complicate cloud deployments, as many data protection laws restrict where personal data can be hosted. Organizations must carefully select cloud regions and enforce data residency controls to comply with local territorial restrictions [19].

The shared responsibility model in cloud security introduces additional challenges, as providers manage infrastructure security while customers are responsible for securing their data and applications. This division of responsibilities can create confusion about compliance obligations. Multi-cloud environments further complicate compliance efforts by increasing the potential for inconsistencies in data handling and regulatory adherence, requiring unified security policies across platforms. Maintaining visibility and control over data in dynamic cloud environments is also challenging, as traditional security tools may not suffice, necessitating cloud-specific monitoring and management solutions. Additionally, managing the entire data lifecycle in the cloud—from collection to deletion—presents compliance risks if not handled properly, emphasizing the need for robust data retention and deletion policies [19].

7.14 Multimodal AI Model Deployment on Cloud and On-Premise

Deploying multimodal machine learning models for medical devices involves significant differences between cloud and on-premises environments, each offering distinct advantages and challenges. In terms of scalability and flexibility, cloud platforms like AWS provide superior scalability through services such as Amazon SageMaker, which can handle large multimodal datasets and complex models. At the same time, on-prem solutions are constrained by available hardware, making it difficult to process multiple data modalities simultaneously. The cloud offers specialized services, such as AWS HealthOmics and AWS HealthLake, for data storage and processing, streamlining the management of diverse healthcare data types. In contrast, on-premises deployments require separate infrastructure setups for each modality, which can be costly and complex [28–30].

Integration and interoperability are more manageable in the cloud, where tools enable scalable pipelines for processing multimodal health data. The on-premises environment faces challenges in integrating various systems and data types, necessitating the development of custom solutions. Regarding security and compliance, cloud platforms offer robust features and certifications, such as HIPAA compliance, but may raise concerns about data leaving the premises. On-premises deployments offer direct control over sensitive medical information, but they require more effort to maintain security and compliance [30].

7.15 AI Model Deployment on Hybrid Environment

Deploying machine learning and AI algorithms across mobile, cloud, and on-premises devices requires a strategic approach considering various factors and deployment methods. Cloud-based deployment, utilizing platforms such as AWS, Azure, or Google Cloud, offers scalability, flexibility, and resource efficiency, but may incur higher ongoing costs and potential latency issues. On-premises deployment provides better data privacy, regulatory compliance, and low latency; however, it involves higher upfront costs and increased maintenance responsibilities.

Edge computing deployment brings AI models closer to data sources, enabling real-time processing, offline functionality, and bandwidth efficiency, which is beneficial for mobile and IoT applications. A hybrid approach combines multiple strategies to leverage the benefits of each, offering flexibility but requiring careful orchestration. Mobile deployment presents unique challenges, necessitating model optimization, on-device learning, and selection of mobile-friendly frameworks.

Containerization and orchestration technologies, such as Docker and Kubernetes, enhance deployment efficiency by ensuring consistency, scalability, and version control. Regardless of the chosen strategy, organizations must consider model monitoring, security, versioning, and A/B testing techniques, such as canary deployments. Organizations can select the most appropriate deployment approach for their specific AI and machine learning needs by evaluating these strategies and considerations, balancing performance, cost, security, and scalability [31, 32].

7.16 AI Model Deployment on Mobile Devices

Deploying AI models on mobile devices presents unique challenges and opportunities that require careful consideration. Model optimization is a crucial first step, involving techniques such as pruning and quantization to reduce model size without significantly compromising accuracy. Utilizing lightweight architectures such as MobileNetV2, which are designed for resource-constrained environments, and employing transfer learning can substantially improve performance and reduce training time.

Selecting appropriate mobile-specific frameworks is essential for efficient deployment. TensorFlow Lite is versatile for Android and iOS platforms, while Core ML is optimized for iOS-specific deployments. PyTorch Mobile offers an alternative for deploying PyTorch models with efficient on-device inference. Performance considerations should prioritize on-device processing to reduce latency and enable offline functionality, leveraging mobile GPUs or dedicated AI chips when available [33].

Efficient data handling is critical, requiring optimized loading and preprocessing pipelines to minimize memory usage and processing time. On-device data augmentation can improve model robustness without increasing model size. Deployment strategies should include mechanisms for incremental updates, A/B testing different model versions, and integrating analytics to track real-world performance.

Security and privacy concerns must be addressed by prioritizing on-device inference, implementing model encryption, and ensuring compliance with data protection regulations. The user experience should be carefully designed, including graceful degradation when computer vision features are unavailable, clear communication about capabilities and limitations, and mechanisms for user feedback to aid continuous improvement.

7.17 AI Model Deployment on Edge Devices

Deploying AI models on edge devices for medical applications involves several significant challenges that must be carefully addressed to ensure effectiveness, compliance, and reliability. One major challenge is the limited resources of edge devices, which typically have constrained processing power and memory compared to cloud servers. This necessitates careful optimization of AI models to fit within these constraints. Additionally, medical edge devices often operate on battery power, requiring energy-efficient models and hardware to ensure long-term functionality. Data handling and privacy are also critical concerns, as medical data is highly sensitive and subject to strict regulations, such as HIPAA. Edge devices must ensure robust data security while functioning autonomously in environments with intermittent or no network connectivity [34, 35].

Model performance presents another challenge, as medical applications often require real-time or near-real-time inference. These necessitate models that strike a balance between high accuracy and fast inference speeds, despite resource limitations. Integration and compatibility issues arise due to the diverse hardware specifications of medical edge devices, making it challenging to ensure model compatibility across platforms. Furthermore, integrating AI models seamlessly into existing medical devices and workflows can be a complex process. Regulatory compliance adds another layer of difficulty, as edge-deployed medical AI models often require FDA approval, a process that can be lengthy and intricate. Ensuring model transparency and explainability is essential for meeting regulatory requirements and gaining clinical acceptance.

Continuous learning and adaptation are particularly challenging in edge environments. Keeping models up-to-date with the latest medical knowledge and addressing model drift in real-world settings are critical for maintaining long-term performance, primarily in offline scenarios. To overcome these challenges, organizations should focus on optimizing models for edge deployment using techniques like quantization and pruning, implementing robust security measures to protect sensitive data, and leveraging specialized edge AI hardware and frameworks tailored for medical applications. Developing modular and flexible architectures that can adapt to various edge devices and scenarios is also essential. Additionally, transparent model validation, regulatory compliance, and continuous improvement processes must be established [36].

7.18 Containerizing Machine Learning Models Deployment

Containerization is pivotal in deploying AI models, offering numerous benefits that streamline development, testing, and production workflows. One of its key advantages is ensuring environmental consistency, allowing AI models to run reliably across different systems by eliminating the "it works on my machine" problem. Containerization also provides isolation and reproducibility by creating separate environments for AI models and experiments, preventing interference between projects or versions. This isolation is particularly valuable for machine learning workflows involving simultaneous experiments, as it enhances reproducibility by enabling old projects to be instantly reproduced and rerun [37].

Portability and flexibility are additional benefits, as containers enable AI models to quickly migrate across various infrastructures, including on-premises servers, cloud platforms, and edge devices. This simplifies transitions between environments, like moving from development to production. Containerization also supports scalability and resource management, especially when paired with orchestration tools like Kubernetes, enabling efficient scaling to handle increased workloads while maintaining high availability and performance. The deployment process is simplified by packaging all dependencies, libraries, and code into a single container image, ensuring models run as intended in any target environment.

Containerization integrates seamlessly with MLOps practices by supporting continuous integration and continuous deployment (CI/CD) pipelines, automating workflows for training, validating, and deploying ML models. It also optimizes resource utilization compared to traditional virtual machines by reducing overhead while delivering comparable performance, an advantage for resource-intensive AI workloads. Version control and model management are made easier through containerization, allowing organizations to maintain multiple model versions, conduct A/B testing, and roll back to previous versions if needed [37, 38].

For edge AI deployment, containers offer a lightweight and consistent way to package and deploy models on resource-constrained devices, which is increasingly critical as edge AI applications become more prevalent. In conclusion, containerization has become essential for deploying AI models effectively. It addresses key challenges in the MLOps lifecycle by providing consistency, portability, scalability, and efficiency across diverse environments—from development to production—making it indispensable for modern AI workflows.

Containerizing machine learning models for on-premise deployment requires robust tools that ensure scalability, portability, and efficient resource management. Docker is a foundational tool for creating containerized applications, encapsulating the model, its dependencies, and runtime environment while maintaining consistency across development and production environments. Kubernetes is a powerful container orchestration platform that manages Docker containers, optimizes resource utilization, and handles scalability across multiple nodes. For more specialized workflows, Kubeflow simplifies the deployment of machine learning models on Kubernetes by providing components for each phase of the ML lifecycle, including training, tuning, deployment, and serving.

MLflow is another valuable tool that supports the complete ML lifecycle by tracking experiments, packaging code, and deploying models. It can be configured to store artifacts and metadata within internal storage solutions, enhancing compliance and data security in on-premise environments. BentoML provides a framework for packaging models into ML-optimized Docker images using its "Bentos" concept—archives that contain all necessary components—and offers a Python SDK for building and testing these packages before containerization. Seldon Core streamlines ML model deployment within Kubernetes by integrating with various ML frameworks and enabling customization of resource definitions for complex model graphs.

For TensorFlow-specific use cases, TensorFlow Serving is an open-source platform that deploys models as gRPC or REST endpoints, supporting various formats like TensorFlow SavedModels and Estimators. By leveraging these tools, organizations can effectively containerize and deploy ML models in on-premise environments. The choice of tools depends on specific requirements, existing infrastructure, and the ML frameworks, allowing for tailored solutions that meet operational needs while ensuring portability and efficient resource management [39].

Orchestration tools like Kubernetes significantly enhance the efficiency of AI model deployment through multiple integrated capabilities. By enabling automated scaling, Kubernetes dynamically allocates resources based on real-time workload demands, particularly during intensive phases such as model training and inference, ensuring optimal performance without requiring manual intervention. Efficient resource management is achieved through the intelligent allocation of CPU, GPU, and memory across clusters, minimizing hardware waste and reducing operational costs. Containerization ensures consistent environments across development, testing, and production stages, enabling seamless deployment across diverse platforms and cloud infrastructures. Workflow automation tools like Kubeflow streamline the machine learning lifecycle—from data preprocessing to deployment—reducing manual errors and accelerating productivity. High availability is maintained via fault-tolerant mechanisms such as automated failover and container replication, safeguarding critical AI applications against disruptions. Flexible scheduling accommodates diverse AI workloads by supporting batch processing for training tasks and real-time processing for inference demands. Integration with CI/CD pipelines enables the rapid iteration and deployment of model updates, promoting agile development cycles. Cost optimization is further achieved through multi-cloud compatibility and pay-as-you-grow resource models, aligning infrastructure expenses with actual usage. These features enhance scalability, manageability, and operational efficiency, solidifying Kubernetes as a cornerstone for orchestrating AI/ML workloads.

7.19 Performance of AI Models in Cloud and On-Premises

The performance of machine learning models can vary significantly between cloud and on-premises deployments, depending on several key factors. Cloud environments offer superior scalability and flexibility, allowing dynamic adjustment of computational resources to handle varying workloads and large multimodal datasets through services like AWS. In contrast, on-premises deployments are constrained by the available hardware, limiting their ability to scale for processing multiple data modalities or larger datasets. Cloud platforms offer access to specialized hardware, such as GPUs and TPUs, without requiring upfront investment, which is crucial for developing complex, multimodal models. In contrast, on-premises setups require hardware unless significant investments are made to upgrade infrastructure.

Latency is another critical consideration; cloud deployments may experience higher latency due to internet-based data transfers, which can impact real-time applications. In contrast, on-premises solutions generally offer lower latency, as data remains within internal networks. Data integration is typically more manageable in the cloud, where interoperability between systems is streamlined, whereas on-premises environments may require custom solutions to integrate different data types. Cloud platforms also enable faster deployment and iteration of models, allowing for quicker updates and improvements. In contrast, on-premises updates can be more time-consuming and potentially hinder long-term performance.

Customization is where on-premises deployments excel, offering greater control to fine-tune infrastructure for specific model requirements. Cloud solutions require many pre-built tools and may have limitations in deep customization for highly specialized applications. Additionally, data gravity can pose challenges for cloud deployments when dealing with large datasets that are costly or time-consuming to transfer. In contrast, on-premises setups avoid such bottlenecks if the data is stored locally. To address these trade-offs, many organizations are adopting hybrid solutions that combine cloud and on-premises infrastructure to optimize performance across different aspects of their multimodal ML workflows.

In summary, cloud deployment offers improved scalability, access to advanced hardware, and seamless integration, making it an ideal choice for complex, multimodal models. However, the on-premises solution offers advantages in terms of latency, data control, and customization. The optimal choice depends on specific use cases, data characteristics, and organizational requirements. Hybrid approaches often provide the best balance of performance and flexibility for multimodal ML models.

7.19.1 Model Latency and User Experience

Model latency significantly impacts user experience in AI applications by affecting responsiveness, where delays disrupt the interaction flow and reduce engagement, particularly in real-time systems such as chatbots or virtual assistants. In time-sensitive domains such as autonomous vehicles or financial trading, minor delays can compromise safety or result in missed opportunities. At the same time, slow responses frustrate users, driving disengagement and abandonment. High latency also hinders adoption, as users gravitate toward faster alternatives, undermining the perceived quality and eroding trust in even highly accurate models. Inconsistent performance due to tail latency creates unreliable experiences, particularly in applications that require uniform responsiveness. At the same time, usability in critical scenarios, such as self-driving systems, demands ultra-low latency for both safety and efficiency. Faster AI responses offer competitive advantages in markets, prioritizing real-time interactions and enabling novel use cases by making complex models feasible in previously impractical scenarios. To address these challenges, developers implement cloud-native architectures, containerization, and specialized hardware to optimize deployments, reduce latency, and enhance user satisfaction [40].

7.19.1.1 Reducing Latency in AI Models

To reduce latency in AI models, developers can implement several best practices across multiple dimensions. Optimizing model architecture involves using smaller, streamlined models where feasible, applying compression techniques like pruning and knowledge distillation, and fine-tuning models for specific tasks. Enhancing hardware and infrastructure includes leveraging dedicated AI accelerators and GPUs, increasing memory bandwidth, optimizing data transfer rates, and adopting edge computing to minimize network delays. Data processing improvements focus on streamlining pipelines and feature engineering, implementing efficient caching and locality strategies, and optimizing input/output operations. Refining prompts and outputs by crafting concise inputs, setting length constraints, and using system messages to control response size can further reduce processing time. Parallel processing techniques, such as distributing computations across multiple machines and processing requests in batches, improve throughput. At the same time, deployment optimizations like containerization with Docker/Kubernetes and cloud-native architectures ensure scalable, efficient resource use. Streaming responses to display partial results incrementally enhance perceived performance, allowing for fine-tuning parameters such as temperature settings and max_tokens to balance output quality with speed. Reducing token counts by trimming input length and breaking complex tasks into smaller steps, alongside implementing robust caching for frequently requested data, minimizes redundant computations. Collectively, these strategies significantly lower latency, enhancing AI system performance and user experience through faster, more responsive interactions.

The most effective hardware solutions for reducing AI model latency include dedicated AI accelerators, such as GPUs, TPUs, and specialized processors, which leverage parallel processing to accelerate computations significantly. FPGAs and ASICs provide ultra-low latency for real-time applications, with FPGAs offering reconfigurability and ASICs delivering custom-designed efficiency. Increasing memory bandwidth ensures faster data transfer rates, thereby minimizing bottlenecks in AI systems. Meanwhile, edge computing reduces network-related delays by processing data closer to its source. Advanced architectures optimizing KV cache offloading, improving time-to-first-token performance. Unified memory systems, such as those found in modern Macs, streamline AI operations by eliminating data transfer barriers between components. Hardware-specific optimizations refine performance, including tailoring models to leverage the unique capabilities of specific processors. These hardware strategies significantly reduce latency and elevate overall AI system responsiveness when combined with software optimizations and efficient model design [41].

7.20 Medical Imaging AI System Architecture for Integration and Deployment

A robust machine learning and artificial intelligence (AI) software architecture for medical device integration and deployment in medical imaging must address the unique requirements of healthcare, including data privacy, interoperability, real-time performance, and regulatory compliance.

7.20.1 Key Architectural Components

A comprehensive medical imaging AI system is built on a modular, multi-tier architecture that seamlessly integrates with healthcare enterprise systems such as EHR/EMR, RIS, and CDSS to ensure interoperability and efficient data management. The architecture consists of three layers, as shown in the two architecture examples in Figs. 7.12 and 7.13.

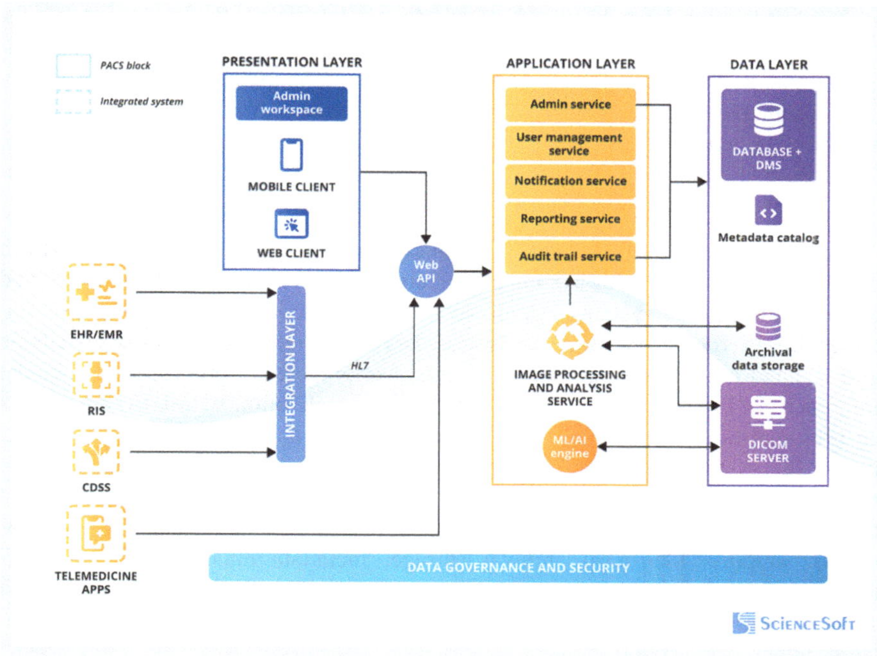

Fig. 7.12 An example of software architecture for medical device integration and deployment in medical imaging [43]

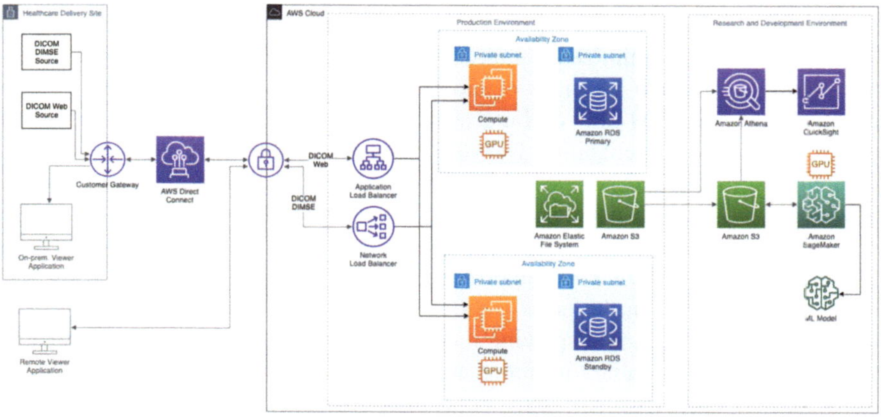

Fig. 7.13 Cloud-based medical imaging system integration and deployment [44]

- The presentation layer offers intuitive web and mobile DICOM viewers for clinicians and healthcare specialists, as well as administrative tools for system management.

- The application layer hosts robust AI/ML models, typically leveraging CNNs, U-Nets, and GANs for advanced image analysis tasks like segmentation, classification, and anomaly detection, while also managing user authentication, authorization, and automated imaging report generation.
- The data layer oversees the storage and retrieval of imaging data using PACS, VNA, and cloud-based solutions, supporting DICOM protocols for interoperability and facilitating secure data exchange with external systems.

Integration with medical devices is standardized through DICOM/DICOMweb protocols, while API gateways enable secure access for external applications and hospital information systems. AI/ML model deployment is achieved through containerized microservices (utilizing Docker and Kubernetes), which enable scalable and isolated model serving. Inference pipelines orchestrate preprocessing, model inference, and post-processing steps. Hardware acceleration with GPUs or ASICs ensures real-time or near-real-time inference, which is critical for clinical workflows. The system is designed for scalability and high availability, utilizing cloud-native auto-scaling and load balancing to dynamically allocate resources and maintain low latency for image retrieval and AI inference. Additionally, the architecture supports long-term archival storage, metadata management, and self-learning capabilities for continuous improvement, all while implementing robust security and compliance measures to protect sensitive medical data and ensure regulatory adherence [43, 44].

Medical imaging solutions should be highly available and deployed across multiple AWS Availability Zones, with components such as viewers, application servers, databases, and storage distributed for enhanced resilience. Each tier should support scaling and load balancing to efficiently manage demand. Frequently accessed images require low-latency storage and fast retrieval, while less critical data can utilize cost-optimized storage, such as Amazon S3. Automated storage tiering moves older data to more economical options as access declines. High-performance databases and caching services, such as ElastiCache, can further enhance latency and reliability. Hybrid and cloud architectures enable efficient, secure transmission and storage of images, with redundancy and direct connectivity ensuring continuous access and supporting clinical workflows [44].

Images are often programmatically retrieved using the DICOM Message Service Element (DIMSE) or DICOMweb protocol, and a Network Load Balancer may be used to route traffic on the ports used by DIMSE. End users typically need low latency for retrieving and displaying medical images, so the data must be durably stored and highly available for immediate retrieval, even for images that are several years old. Recently ingested studies may be cached on SAN storage, EBS volumes, or high-performance file systems like Amazon FSx. In-memory caches, such as ElastiCache, can be utilized to further enhance performance. The data acquired by medical imaging scanners, such as MRI, CT, or C-Arm, must be processed through image reconstruction pipelines, which are similar to high-performance computing (HPC) workloads. Cloud-based computing offers elasticity, allowing resources to scale as needed for emergency procedures or peak demand. To minimize image download times, protocols like HTTP/2 can be leveraged, and hybrid architectures may prioritize transmitting images that are likely to be accessed first. On-premises caches can provide low-latency hot storage for frequently accessed studies. Redundant network connections between care settings and cloud storage are crucial, as a loss of connectivity can significantly impact patient health. AWS Direct Connect can be used to ensure sites with high imaging throughput have reliable, low-latency connectivity. Hybrid architectures can help segment latency and ensure business continuity, with cloud storage providing durability and scalability, and on-premises solutions offering low-latency access for critical datasets. For the best user experience, solutions should be designed with high availability, performance, and cost optimization [44].

7.21 Deep Neural Networks (DNNs) Integration and Deployment

DNNs are increasingly integrated into medical imaging systems to enhance accuracy, efficiency, and clinical impact. These systems automate key image analysis tasks, such as disease detection, segmentation, and classification, often matching or exceeding the performance of experts. DNNs also enable the automatic alignment of images from different times or modalities, improving image reconstruction by reducing noise and enhancing quality. Architecturally, DNNs are deployed as part of medical imaging backends, frequently as containerized microservices or within Picture Archiving and Communication Systems (PACS), to provide real-time, scalable analysis. Their outputs, like lesion boundaries or abnormality scores, are seamlessly integrated into clinical workflows, supporting clinicians without disrupting established processes. Advanced solutions integrate imaging data with clinical information using multimodal deep learning, thereby further enhancing diagnostic accuracy and relevance. Integrating DNNs into the medical imaging streamlines workflows, supports clinical decision-making, and leads to better patient outcomes [45].

7.22 An Example of Real-Time ML Operational Deployment Pipeline

A systematic approach is crucial for the smooth and efficient deployment of machine learning models in real-time operational environments. The following step-by-step breakdown outlines the key stages of deploying machine learning models for real-time inference [46] (Fig. 7.14).

1. Version Control: The Machine Learning Training Pipeline is defined in code and activated whenever changes are merged into the main branch.

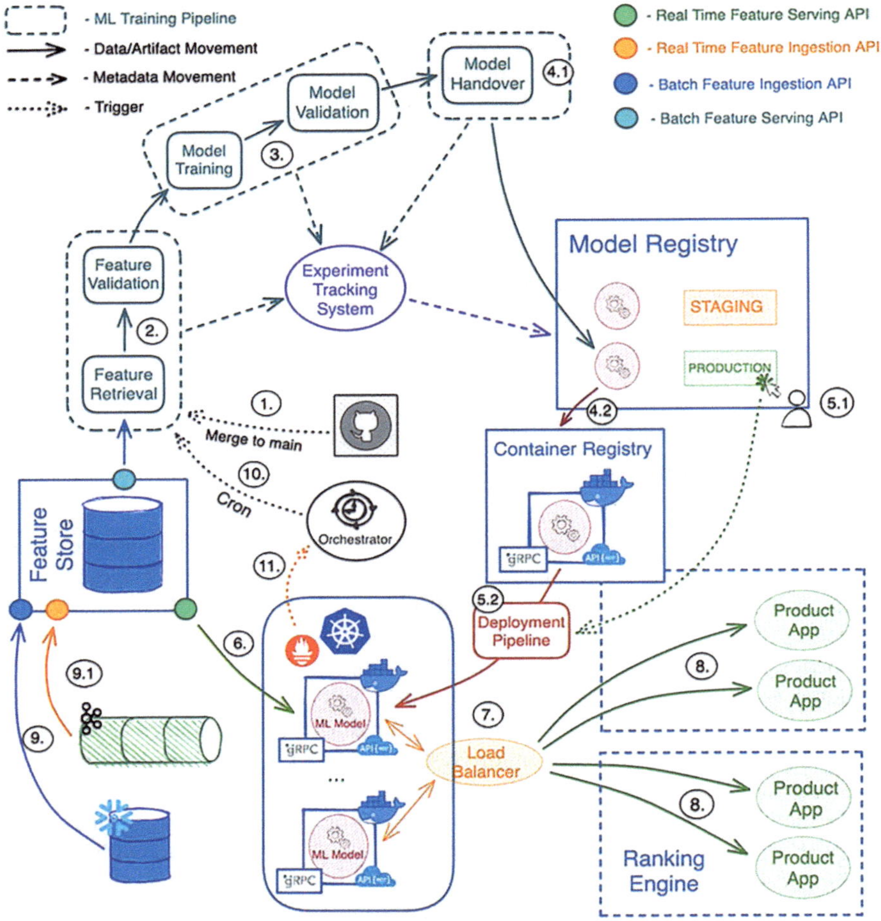

Fig. 7.14 An example of real-time ML operational deployment pipeline [46]

2. Feature Preprocessing: Features are extracted from the Feature Store, validated, and passed to the next stage. The Experiment Tracking System saves metadata specific to the model's features.
3. Model Training and Validation: The model is trained and validated using the preprocessed data, and metadata related to the model is recorded in the Experiment Tracking System.
4. Model Validation and Artifact Registry:

   (a) The resulting model artifact is stored in the Model Registry if the model passes all validation checks.
   (b) The model is then packaged into a container, making it accessible as a REST or gRPC API and ready for deployment.

5. Model Deployment:
   (a) Metadata from experiment tracking is linked to the corresponding model artifact in the Model Registry. The designated person selects the best model candidate and marks its status as "Production."
   (b) A webhook initiates the Deployment Pipeline, which launches the new version of the containerized API. Various release strategies can be implemented as needed.
6. Inference and Application Integration:
   (a) Product and application teams send requests to the API, which retrieves the necessary features for inference from the Real-Time Feature Serving API. The API then returns the inference results to the applications.
   (b) ML APIs are built to allow for horizontal scaling.
7. Support for Multiple ML APIs: Different ML APIs can be exposed to cater to different applications, such as a Ranking Function.
8. Feature Store and Data Warehouse Integration:
   (a) The Feature Store is connected to a Data Warehouse to retrieve static features as needed.
   (b) Dynamic features are computed and updated in real time from a Distributed Messaging System like Kafka.
9. Model Production Orchestration: An orchestration service manages model retraining based on defined production criteria.
10. Production Monitoring: ML models running in production are monitored continuously, and automatic retraining can be triggered if model quality declines.

7.23 Data and Model Versioning for Deployment

Version control [42] in AI model deployment is crucial for ensuring reliability, reproducibility, and effective management of AI systems in healthcare. Key aspects of versioning in AI model deployment include:

- Semantic versioning: Use MAJOR.MINOR.PATCH format to systematically assign version numbers, reflecting significant changes, enhancements, and minor improvements.
- Consistent naming conventions: Establish clear naming conventions that capture key information about each version.
- Documentation and tracking: Maintain detailed records of changes, including tweaks to hyperparameters, to provide insights into model improvements.
- Version control tools: Utilize tools like Git integrated with data versioning tools to enable rolling back to previous stable versions if issues arise.

- Tagging and labeling: Use tags and labels to group or highlight versions based on criteria such as stage, status, or quality (e.g., experimental, testing, production-ready).

Several tools are recommended for integrating version control with ML model deployment:

- MLflow: An open-source platform that provides model tracking, versioning, and deployment capabilities. It allows for easy experiment tracking and model management.
- DVC (Data Version Control): An open-source tool that integrates with Git to provide version control for datasets, models, and ML pipelines. It's beneficial for managing large datasets.
- Git: While not ML-specific, Git remains a fundamental tool for version control in ML projects, especially when combined with platforms like GitHub or GitLab.
- Neptune.ai: A metadata store for MLOps that offers version control for models, data, and code.

7.24 Summary

This chapter comprehensively explored artificial intelligence model integration and deployment, focusing on best practices for integrating and deploying AI models across various environments. It addressed the complex landscape of regulatory compliance challenges that organizations face when implementing AI solutions. The discussion covered three primary deployment strategies: cloud-based, on-premises, and mobile deployments, each with a unique set of challenges. The chapter emphasized the importance of incorporating human expertise into AI workflows, highlighting how this integration enhanced AI systems' overall effectiveness and reliability. It also explored critical aspects of AI deployment, including scalability, containerization, reliability, versioning, and performance monitoring, offering readers a comprehensive understanding of the technical considerations necessary for successful AI implementation. Through this extensive coverage, the chapter aimed to equip readers with the knowledge and strategies needed for effective AI model deployment in various contexts.

7.25 Points to Ponder

1. Compare the machine learning pipeline services offered by the three major cloud providers: Amazon Web Services, Microsoft Azure, and Google Cloud.
2. Compare and summarize Edge AI solutions offered by the three major cloud providers: Amazon Web Services, Microsoft Azure, and Google Cloud.
3. What are the best practices for machine learning pipeline migration, and provide an overview of services for migration offered by cloud providers?

7.26 Answers

Answer 1

Cloud Services Comparison Table: AWS Vs Azure Vs Google Cloud

Pipeline stage/ requirement	AWS	Azure	Google cloud (GCP)
Data ingestion	AWS Glue, Kinesis Data Streams, Data Pipeline, S3, Snowball	Azure Data Factory, Event Hubs, Blob Storage, Data Lake	Cloud Dataflow, Pub/Sub, Data Fusion, Storage Transfer Service
Data quality	AWS Glue DataBrew, Deequ, Lambda, SageMaker Data Wrangler	Azure Data Factory Data Flow, Azure Purview, Azure Data Quality	Dataprep (by Trifacta), Dataflow, Dataplex
Data pipeline/ ETL	AWS Glue, Data Pipeline, Step Functions, Lambda	Azure Data Factory, Synapse Pipelines, Logic Apps	Cloud Dataflow, Composer (Airflow), Data Fusion
Model training	SageMaker, EC2, ECS, Batch, EKS	Azure Machine Learning, Data Science VMs, Batch AI, Kubernetes Service	Vertex AI, AI Platform Training, Compute Engine, Kubernetes Engine
Model validation and testing	SageMaker Model Monitor, SageMaker Clarify, SageMaker Debugger, Lambda	Azure ML Validation, ML Monitor, Test Plans	Vertex AI Model Monitoring, Explainable AI, Cloud Functions
Model integration and deployment	SageMaker Endpoints, Lambda, ECS/EKS, API Gateway, Elastic Beanstalk	Azure ML Endpoints, AKS, App Service, Functions, API Management	Vertex AI Endpoints, Cloud Run, Cloud Functions, API Gateway
LLMs/GenAI	Bedrock, SageMaker JumpStart, Amazon Titan, AWS Inferentia, Lex, Polly, Transcribe	Azure OpenAI Service, Cognitive Services, Azure ML Prompt Flow, Azure AI Studio	Vertex AI (Gemini, PaLM, Imagen), Generative AI Studio, Model Garden, Cloud TPU
Integration with PACS/DICOM/ HL7	HealthLake Imaging, HealthLake, AWS Transfer Family, AWS Lambda, Amazon API Gateway, Direct Connect	Azure API for FHIR, DICOM Services, HL7 connectors, Logic Apps, Event Grid	Cloud Healthcare API (DICOM, HL7v2, FHIR), Cloud Functions, Apigee
Product security protocols	IAM, KMS, CloudHSM, Shield, WAF, GuardDuty, Macie, Security Hub, PrivateLink, VPC, HIPAA compliance	Azure Active Directory, Key Vault, Security Center, Defender, Private Link, VNet, HIPAA	IAM, KMS, Shielded VMs, Security Command Center, VPC Service Controls, DLP API, HIPAA compliance
Good security practices	Least privilege IAM, encryption at rest/in transit, logging (CloudTrail), automated compliance tools	RBAC, managed identities, encryption, Azure Policy, logging, Defender for Cloud	IAM, VPC, encryption, audit logs, DLP, Security Command Center, network segmentation, compliance tools

Healthcare Integration

- All three clouds provide managed APIs for DICOM, HL7, and FHIR standards, supporting integration with hospital PACS and EHR systems.
- Each platform offers HIPAA-compliant services and detailed audit logging for healthcare data security.

GenAI/LLM Capabilities

- **AWS:** Bedrock and SageMaker JumpStart for managed LLMs and generative AI, supporting custom and third-party models.
- **Azure:** Azure OpenAI Service for GPT-4, DALL-E, Codex, and integration with Microsoft Copilot and Cognitive Services.
- **GCP:** Vertex AI with Gemini, PaLM, Imagen, and Model Garden for LLMs and multimodal GenAI, plus custom model deployment.

Security and Best Practices

- All providers recommend strong IAM controls, encryption, network isolation, regular audits, and compliance with healthcare standards (e.g., HIPAA, HITRUST).
- Use managed key management, DLP, and logging for sensitive data.
- Integrate with SIEM and threat detection tools for real-time monitoring.

Answer 2

AWS, Azure, and Google Cloud all offer Edge AI solutions specifically designed to reduce latency by processing data closer to the source, rather than relying solely on centralized cloud servers. These edge solutions are highly relevant for healthcare, where real-time decision-making and low-latency analysis are critical, especially in AI imaging applications.

Edge AI Solutions from Major Cloud Providers

- **AWS:** Provides edge computing through services like AWS IoT Greengrass and AWS Snow Family, allowing AI models to be deployed and run locally on medical devices or gateways within hospitals. This reduces the time needed to analyze medical images and enables immediate clinical decisions.
- **Azure:** Azure IoT Edge and Azure Stack Edge enable hospitals and clinics to run AI models and data processing at the edge. These solutions support medical imaging workflows by allowing images to be analyzed as soon as they are captured, without waiting for cloud round-trips.
- **Google Cloud:** Delivers edge AI capabilities with Google Distributed Cloud Edge and supports on-device inference for imaging through TensorFlow Lite and Edge TPU. Google's edge solutions can be integrated with imaging equipment for instant analysis and triage.

Can Edge AI Be Used for AI Imaging Solutions?

- Edge AI is increasingly used in medical imaging for real-time anomaly detection, image enhancement, and preliminary diagnostics tasks. By processing images at the edge-on imaging devices, mobile units, or local servers, clinicians receive faster results vital in emergency care, remote settings, or high-throughput environments. Edge AI also enhances privacy by keeping sensitive patient data local and ensures continuity of care even with limited internet connectivity.

Answer 3

Best Practices for ML Pipeline Migration

- Assess current pipelines: Inventory data sources, dependencies, and workflow logic.
- Choose managed, serverless services: Reduces operational overhead and increases scalability.
- Test and validate: Ensure new pipelines deliver the same or improved results before decommissioning legacy systems.
- Automate orchestration: Use Airflow (MWAA/Composer), Step Functions, or native pipeline services for reliability and monitoring.
- **Secure and govern data**: Use IAM, encryption, and audit logging throughout migration.

Overview of Services for Migration Offered by Cloud Providers

- **AWS**
- **AWS Glue:** Serverless data integration for ETL, supports migration of data pipelines, data discovery, preparation, and orchestration. Ideal for moving ETL workloads and integrating with S3, DynamoDB, Redshift, and more.
- **Amazon Managed Workflows for Apache Airflow (MWAA):** Managed Apache Airflow for orchestrating complex ML/data pipelines; recommended for users familiar with Airflow or needing advanced workflow management.
- **AWS Step Functions:** Serverless orchestration for distributed applications and microservices; useful for automating multi-step ML workflows.
- **AWS Database Migration Service (DMS):** This migrates databases used in ML pipelines with minimal downtime.
- **Amazon SageMaker Pipelines:** Purpose-built for ML workflow automation, including data ingestion, preprocessing, training, validation, and deployment.
- **Data Transfer Services:** AWS DataSync, Snow Family, and Transfer Family for bulk data migration to the cloud.
- **Migration Guidance:** AWS provides migration blueprints and partner support for seamless transition of on-premises or legacy pipelines to cloud-native architectures.

Azure

- **Azure Data Factory:** Cloud-based ETL and data integration service for orchestrating and automating data movement and transformation; supports hybrid and multi-cloud migration scenarios.
- **Azure Synapse Analytics:** A Unified analytics platform that can ingest, prepare, manage, and serve data for ML workloads.
- **Azure Machine Learning Pipelines:** Orchestrates and automates ML workflows, including data preparation, training, validation, and deployment.

- **Azure Database Migration Service:** This is for migrating databases that power ML pipelines.
- **Azure Migrate:** Central hub for cloud migration, including assessment and orchestration of data and ML workloads.
- **Azure Data Box:** For large offline data transfers.

Google Cloud

- **Cloud Dataflow:** Fully managed stream and batch data processing for ETL and ML pipelines.
- **Cloud Composer:** Managed Apache Airflow for workflow orchestration and migrating Airflow-based pipelines.
- **Vertex AI Pipelines:** End-to-end ML workflow automation, including data ingestion, training, validation, and deployment.
- **Database Migration Service:** This is used to migrate databases to Google Cloud.
- **Transfer Appliance:** For large-scale offline data migration.
- **BigQuery Data Transfer Service:** Automates data movement into BigQuery for analytics and ML.

References

1. Wilcox AB, Hripcsak G (2003) The role of domain knowledge in automating medical text report classification. J Am Med Inform Assoc 10(4):330–338. https://doi.org/10.1197/jamia.M1157. Epub 2003 Mar 28. PMID: 12668687; PMCID: PMC181983
2. Lipkova J, Chen RJ, Chen B, Lu MY, Barbieri M, Shao D, Vaidya AJ, Chen C, Zhuang L, DFK W, Shaban M, Chen TY, Mahmood F (2022) Artificial intelligence for multimodal data integration in oncology. Cancer Cell 40(10):1095–1110. https://doi.org/10.1016/j.ccell.2022.09.012. PMID: 36220072; PMCID: PMC10655164
3. Onitiu D, Wachter S, Mittelstadt B (2024) How AI challenges the medical device regulation: patient safety, benefits, and intended uses. J Law Biosci:lsae007. https://doi.org/10.1093/jlb/lsae007
4. US FDA guidance on predetermined change control plan (2024)
5. Jayatilake SMDAC, Ganegoda GU (2021) Involvement of machine learning tools in healthcare decision making. J Healthc Eng 2021:6679512. https://doi.org/10.1155/2021/6679512. PMID: 33575021; PMCID: PMC7857908
6. Zhou K, Gattinger G (2024) The evolving regulatory paradigm of AI in MedTech: a review of perspectives and where we are today. Ther Innov Regul Sci 58(3):456–464. https://doi.org/10.1007/s43441-024-00628-3. Epub 2024 Mar 25. PMID: 38528278; PMCID: PMC11043174
7. Krones F, Marikkar U, Parsons G, Szmul A, Mahdi A (2025) Review of multimodal machine learning approaches in healthcare. Inf Fusion 114:102690., ISSN 1566-2535. https://doi.org/10.1016/j.inffus.2024.102690
8. Acosta JN, Falcone GJ, Rajpurkar P et al (2022) Multimodal biomedical AI. Nat Med 28:1773–1784
9. Kline A, Wang H, Li Y et al (2022) Multimodal machine learning in precision health: multimodal review. NPJ Digit Med 5:171. https://doi.org/10.1038/s41746-022-00712-8
10. Karabacak M, Margetis K (2023) Embracing large language models for medical applications: opportunities and challenges. Cureus 15(5):e39305. https://doi.org/10.7759/cureus.39305. PMID: 37378099; PMCID: PMC10292051

11. Andrew A (2024) Potential applications and implications of large language models in primary care. Fam Med Community Health 12(Suppl 1):e002602. https://doi.org/10.1136/fmch-2023-002602. PMID: 38290759; PMCID: PMC10828839
12. Amazon Web Services. Scalable medical computer vision model training with Amazon SageMaker. https://aws.amazon.com/cn/blogs/industries/scalable-medical-computer-vision-model-training-with-amazon-sagemaker-part-2/?trk=cndc-detail
13. Thakur GK, Thakur A, Kulkarni S, Khan N, Khan S (2024) Deep learning approaches for medical image analysis and diagnosis. Cureus 16(5):e59507. https://doi.org/10.7759/cureus.59507. PMID: 38826977; PMCID: PMC11144045
14. Kulkarni V, Gawali M, Kharat A (2021) Key technology considerations in developing and deploying machine learning models in clinical radiology practice. JMIR Med Inform 9(9):e28776. https://doi.org/10.2196/28776. PMID: 34499049; PMCID: PMC8461525
15. Ye J, Woods D, Jordan N, Starren J (2024) The role of artificial intelligence for the application of integrating electronic health records and patient-generated data in clinical decision support. AMIA Jt Summits Transl Sci Proc 2024:459–467. PMID: 38827061; PMCID: PMC11141850
16. Torab-Miandoab A, Samad-Soltani T, Jodati A, Rezaei-Hachesu P (2023) Interoperability of heterogeneous health information systems: a systematic literature review. BMC Med Inform Decis Mak 23(1):18. https://doi.org/10.1186/s12911-023-02115-5. PMID: 36694161; PMCID: PMC9875417
17. Torab-Miandoab A et al (2024) A unified component-based data-driven framework to support interoperability in the healthcare systems. Heliyon 10(15):e35036. https://doi.org/10.1016/j.heliyon.2024.e35036
18. US FDA's proposed regulatory framework for modification to artificial intelligence/machine learning (AI/ML)-based software as a medical device (SaMD)
19. Anselmo C, Attili M, Horton R, Kappe B, Schulman J, Baird P (2021) Hey you, get on the cloud: safe and compliant use of cloud computing with medical devices. Biomed Instrum Technol 55(1):1–15. https://doi.org/10.2345/0899-8205-55.1.1. PMID: 33465776; PMCID: PMC8641424
20. Sanchez-Pinto LN, Bennett TD (2022) Evaluation of machine learning models for clinical prediction problems. Pediatr Crit Care Med 23(5):405–408. https://doi.org/10.1097/PCC.0000000000002942. Epub 2022 May 5. PMID: 35583620; PMCID: PMC9177058
21. Model deployment: strategies, best practices, and use cases. https://www.qwak.com/post/model-deployment
22. Microservices vs Monolithic architecture. https://medium.com/startlovingyourself/microservices-vs-monolithic-architecture-c8df91f16bb4
23. Considerations for deploying machine learning models in production. https://www.anyscale.com/blog/considerations-for-deploying-machine-learning-models-in-production
24. Cohen RY, Kovacheva VP (2023) A methodology for a scalable, collaborative, and resource-efficient platform, MERLIN, to facilitate healthcare AI research. IEEE J Biomed Health Inform 27(6):3014–3025. https://doi.org/10.1109/JBHI.2023.3259395. Epub 2023 Jun 5. PMID: 37030761; PMCID: PMC10275625
25. Akerele JI et al (2024) Improving healthcare application scalability through microservices architecture in the cloud. Int J Sci Res Updates 08(02):100–109
26. Model deployment strategies. https://neptune.ai/blog/model-deployment-strategies
27. ML model deployment strategies. https://www.tensorops.ai/post/ml-model-deployment-strategies
28. https://aws.amazon.com/compare/the-difference-between-saas-and-on-premises/
29. How to quickly and easily package, deploy, and serve ML models to edge devices. https://techcommunity.microsoft.com/blog/startupsatmicrosoftblog/how-to-quickly-and-easily-package-deploy-and-serve-ml-models-to-edge-devices-/4036827
30. Soenksen LR, Ma Y, Zeng C, Boussioux L, Villalobos Carballo K, Na L, Wiberg HM, Li ML, Fuentes I, Bertsimas D (2022) Integrated multimodal artificial intelligence framework for multimodal applications. NPJ Digit Med 5(1):149. https://doi.org/10.1038/s41746-022-00689-4. PMID: 36127417; PMCID: PMC9489871

31. Guidance for multimodal data analysis with AWS Health and MLMultimodal. https://aws.amazon.com/solutions/guidance/multi-modal-data-analysis-with-aws-health-and-ml-services/
32. Multimodal data analysis with AWS Health and MacMultimodaling services. https://aws.amazon.com/blogs/industries/multimodal-data-analysis-with-aws-health-and-machine-learning-services/
33. Deploying a real-time computer vision model in production: pathology detection. https://www.omdena.com/blog/how-to-deploy-computer-vision-model
34. Stress-testing biomedical vision models with RadEdit: a synthetic data approach for robust model deployment. https://www.microsoft.com/en-us/research/blog/stress-testing-biomedical-vision-models-with-radedit-a-synthetic-data-approach-for-robust-model-deployment/
35. Kamruzzaman MM, Alrashdi I, Alqazzaz A (2022) New opportunities, challenges, and applications of edge-AI for connected healthcare in internet of medical things for smart cities. J Healthc Eng 2022:2950699. https://doi.org/10.1155/2022/2950699. Retraction in: J Healthc Eng. 2023 Oct 11;2023:9823658. https://doi.org/10.1155/2023/9823658. PMID: 35251564; PMCID: PMC8890828
36. Mennella C, Maniscalco U, De Pietro G, Esposito M (2024) Ethical and regulatory challenges of AI technologies in healthcare: a narrative review. Heliyon 10(4):e26297. https://doi.org/10.1016/j.heliyon.2024.e26297. PMID: 38384518; PMCID: PMC10879008
37. Docker guide for AI model development and deployment. https://github.com/saikhu/Docker-Guide-for-AI-Model-Development-and-Deployment
38. Best tools for ML model serving. https://neptune.ai/blog/ml-model-serving-best-tools
39. ML model packaging [The ultimate guide]. https://neptune.ai/blog/ml-model-packaging
40. Optimizing AI responsiveness: a practical guide to Amazon Bedrock latency-optimized inference. https://aws.amazon.com/blogs/machine-learning/optimizing-ai-responsiveness-a-practical-guide-to-amazon-bedrock-latency-optimized-inference/
41. Understanding latency in AI: what it is and how it works. https://www.galileo.ai/blog/understanding-latency-in-ai-what-it-is-and-how-it-works
42. Version control for ML models: why you need it, what it is, how to implement it. https://neptune.ai/blog/version-control-for-ml-models
43. Custom picture archiving and communication system (PACS): features, development steps, and costs. https://www.scnsoft.com/healthcare/picture-archiving-and-communication-system
44. AWS well-architected framework healthcare industry lens
45. Lee JG, Jun S, Cho YW, Lee H, Kim GB, Seo JB, Kim N (2017) Deep learning in medical imaging: general overview. Korean J Radiol 18(4):570–584. https://doi.org/10.3348/kjr2017.18.4.570. Epub 2017 May 19. PMID: 28670152; PMCID: PMC5447633
46. Real-time ML operational deployment pipeline. https://medium.com/@FrankAdams7/step-by-step-breakdown-of-real-time-ml-operational-deployment-pipeline-848f290eb5d9

Chapter 8
Security and Privacy for AI/ML-Enabled Medical Devices

Abstract This chapter covers security and privacy considerations for AI/ML-enabled medical devices. We will begin by examining a typical hospital IT questionnaire, which is necessary to obtain approvals for the on-site deployment of such devices. Then we present the crucial topics of encryption and medical data management for multi-party accesses. Next, we review the regulations for AI/ML as they pertain to medical data and devices both in the EU and the US. We present a detailed review of the documentation required for security and privacy in the medical devices. We cover the newly emerging domain of GenAI and the security risks inherent in large language models. Lastly, we conclude by discussing the risks posed by legacy medical devices with outdated operating systems.

Keywords Security · Privacy · Risks · PII · PHI · FDA · GDPR · EU AI Act · Encryption · Penetration · LLM · GenAI · NLP · MDR · IVDR · SoC2 · MDS2 · Legacy devices · Legacy operating systems · SBOM

8.1 Introduction

Transparency is a key factor in security because it fosters trust, facilitates collaboration, and helps prevent future incidents. There are three guiding principles [1] for transparency related to security and privacy in AI/ML-enabled medical devices:

1. The focus is on the performance of the human-AI team.
2. Users are provided with clear, essential information.
3. Supporting regulatory submissions and cybersecurity certifications.

Transparency refers to communicating accurate information about the intended use, logic, and performance of a Machine Learning-enabled Medical Device (MLMD). Logic refers to information about how an output or result was reached, or the basis for a decision or action. The degree to which this logic can be explained is that a person can understand it, also known as explainability. This must be done without compromising the device's security.

A. Pandey et al., *Artificial Intelligence in Medical Software*,
https://doi.org/10.1007/978-3-032-10808-1_8

Effective transparency ensures that information that could impact risks and patient outcomes is communicated. It also considers the information that the intended user or audience needs and the context in which it's used. Transparency uses the best media, timing, and strategies for successful communication. It relies on a holistic understanding of users, environments, and workflows. This must be done without compromising any personally identifiable information (PII) and patient Health Information (PHI) that a device may contain to unintended users, such as technicians repairing a machine.

8.2 Clinical Model Security and Privacy Considerations

In the USA, the Food and Drug Administration (FDA) was established in 1906 [2] to protect public health by ensuring the safety, efficacy, and security of human and veterinary drugs, biological products, and medical devices. FDA regulations present a watermark that every medical company must satisfy before being allowed to ship its products. FDA requirements also cover software applications that extend the functionality and utilization of medical devices. Other countries follow similar regulations to protect their citizens. The medical industry is rife with regulations for good reasons. This must be seen and appreciated because, in the past, some companies have developed products and brought them to market without sufficient testing, causing undue harm to users [3].

In addition to the above, further rules exist regarding the management and protection of PII and PHI, such as the CCPA [4], GDPR [5], and HIPAA [6]—to name a few. These measures were well-intended and implemented in response to some adverse events in the past [7]. Medical regulations now create safety barriers to entry for new products and services, which increases the cost of product development and launches. In addition, hospitals and medical establishments that handle patient data have additional requirements.

Growing incidents of cybersecurity threats have caused the FDA to issue additional guidance. This includes Cybersecurity device design, labeling, and the premarket documentation that the FDA requires [8] for devices with cybersecurity risks. FDA guidance is intended to promote consistency, facilitate efficient premarket review, and help ensure that marketed medical devices are sufficiently resilient to cybersecurity threats.

Similarly, the EU has issued guidance [9] on cybersecurity for medical devices. The two new Regulations on medical devices 745/2017 (MDR) and 746/2017 (IVDR) have been adopted and entered into force on 25 May 2017. Among the many novelties introduced, these two regulations ensure that devices placed on the EU market are fit for the new technological challenges linked to cybersecurity risks.

8.3 IT Cybersecurity Requirement

Information technology (IT) platforms provide the lifeblood of an entity that curates and maintains medical data. In the past, hospitals stored their data on-premises; however, with the advent of Cloud Computing [10], it is often cheaper and more convenient to perform all data processing in the Cloud. However, data must be moved from its generation point to a cloud data center. Suppose the data involves PII or PHI elements. In that case, the IT department of that organization is expected to incorporate additional safeguards related to the confidentiality, integrity, and availability of data in the Cloud.

For deployment of a Cloud application, a typical hospital will have hundreds of questions that need to be satisfactorily answered before a data pipeline to the Cloud data center can be set up. A sample scoring sheet is presented in Table 8.1, which includes over 150 questions that require individual answers. The scores in sections

Table 8.1 List of questions from a typical hospital before storing data in the Cloud

A. Company Information. The vendor:
1. Will accommodate an onsite visit for a security audit within 24 h notice.
2. Will store all confidential data within the USA—including backups.
3. Maintains an audit log for the location of all confidential data and their backups, to identify where it is located at any point in time, in order to address privacy laws for storage within the USA.
4. Will not access confidential data from outside of USA.
Total Company Controls
B. Policies, Standards, and Procedures. The vendor:
1. Has formal written Information Security Policies.
2. Will provide copies of the Information Security Policies.
3. Can provide results of a third-party external Information Security assessment conducted within the past 2 years (SOC Type 1 and 2, penetration test, vulnerability assessment, etc.).
4. Maintains incident response procedures.
5. Has a policy to protect client information against unauthorized access, whether stored, printed, spoken, or transmitted.
6. Has a policy that prohibits the sharing of individual accounts and passwords.
7. Has a policy that implements the following Information Security concepts: need to know, least privilege, and checks and balances.
8. Requires system administrators to be educated and qualified.
9. Implements AAA (Authentication, Authorization, Accounting) for all users.
10. Performs background checks for individuals handling confidential information.
11. Has termination or job transfer procedures that immediately protect against unauthorized access to information.
12. Provides customer support with escalation procedures.
13. Has documented change control processes.
14. Requires contractors, subcontractors, vendors, outsourcing ventures, or other external third-party contracts to comply with policies and customer agreements.

(continued)

Table 8.1 (continued)

15. Has a policy that implements federal and state regulatory requirements.
16. Maintains a routine user information security awareness program.
17. Has a formal routine information security risk management program for risk assessments and risk management.
Total Policy Controls
C. Architecture. The vendor:
1. Will provide a network topology diagram/design of the technical solution.
2. Implements network firewall protection.
3. Implements web application firewall protection.
4. Implements host firewall protection.
5. Maintains routers and ACLs.
6. Provides network redundancy.
7. Has IDS/IPS technology implemented.
8. Uses DMZ architecture for Internet systems.
9. Adheres to the practice that web applications, which "face" the Internet, are on a server different from the one that contains the database.
10. Uses enterprise virus protection on all systems.
11. Follows a program of enterprise patch management.
12. Provides dedicated customer servers to segregate data from other customer data. If not, then how is this accomplished in a secure virtual or segmented configuration.
13. Implements controls to restrict access to data from other customers.
14. Ensures that remote access is only possible over secure connections.
15. Uses separate physical and logical development, test, and production environments and databases.
16. Secures development and test environments using, at a minimum, equivalent security controls as the production environment.
17. Will provide the architectural software solution design with security controls.
18. Has managed secure access points on its wireless network.
Total Architecture Controls
D. Configurations. The vendor:
1. Implements encryption for confidential information being transmitted on external or Internet connections with a strength of at least AES 256 bit or uses TLS 1.1, preferably TLS 1.3.
2. Implements encryption for confidential information at rest with a strength of at least AES 256 bit.
3. Has password-protected screen savers that activate automatically to prevent unauthorized access when idle, for computers used by system's support users.
4. Removes all unnecessary services from computers.
5. Uses file integrity monitoring software on servers (such as tripwire).
6. Changes or disables all vendor-supplied default passwords or similar "published" access codes for all installed operating systems, database management systems, network devices, application packages, and any other commercially produced IT products.
7. Uses passwords that are a min. of eight characters, expire at least annually and have complex requirements.
8. Ensures that passwords are never stored in clear text or are easily decipherable.

(continued)

Table 8.1 (continued)

9. Checks all systems and software to determine whether appropriate security settings are enabled.
10. Manages file and directory permissions following least privilege and need-to-know practices.
11. Implements redundancy or high availability for critical functions.
12. Authenticates all user access with password, token, or biometrics.
13. Formally approves, tests, and logs all system changes.
14. Does not use production data for both development and testing, unless it has been declassified.
15. Uses synthetic data in both development and test environments.
16. Limits access to development and test environments to personnel with a need to know.
17. Sets the account lockout feature for successive failed logon attempts on all system support computers.
18. Prohibits split tunneling when connecting to customer networks.
Total Configuration Controls
E. Product Design. The vendor:
1. Ensures that if the product integrates with portable devices, confidential information is encrypted when stored on these portable devices and requires password access.
2. Ensures that access to confidential information, across a public connection, is encrypted with a secure connection and requires user authentication.
3. Implements protections for Common Vulnerabilities and Exposures (CVEs) in a timely manner to protect from exploits.
4. Audits the application against the OWASP Top 10 Application Security Risks.
5. Ensures that application server and database software technologies are kept up to date with the latest security patches.
6. Uses threat modeling in their software development lifecycle (SDLC).
7. Performs security code reviews as part of their SDLC.
8. Conducts OWASP code reviews for the Top 9 source code flaw categories as part of their SDLC.
Total Product Design Controls
F. Compliance. The vendor:
1. Will provide relevant certificates of the applicable ISO 27001 certification.
2. Can provide documentation that its product is HIPAA compliant, if the vendor manages any PHI.
3. Can provide documentation of its PCI-DSS compliance if the vendor manages any payment card information.
4. Uses industry standard best practices for application security (e.g., OWASP).
Total Product Design Controls
G. Access Control. The vendor:
1. Immediately removes or modifies access when personnel terminate, transfer, or change job functions.
2. Achieves individual accountability by assigning unique IDs and prohibiting password sharing.
3. Ensures that critical data or systems are accessible by at least two trusted and authorized individuals in order to limit having a single point of service failure.
4. Ensures that users have the authority to only read or modify those programs or data that are needed to perform their duties.

(continued)

Table 8.1 (continued)

Total Access Controls
H. Monitoring. The vendor:
1. Reviews access permissions monthly for all server files, databases, applications, etc.
2. Implements system event logging on all servers and records at a minimum who, what, and when for all transactions.
3. Reviews and analyzes after hours system accesses, at least monthly.
4. Reviews system logs for failed logins, or failed access attempts monthly.
5. Reviews and removes dormant accounts on systems at least monthly.
6. Reviews Web server logs weekly for possible intrusion attempts and daily for significant changes in log file size as an indicator of compromise.
7. Reviews network and firewall logs at least monthly.
8. Reviews wireless access logs at least monthly.
9. Performs scanning for rogue access points at least quarterly.
10. Actively manages IDS/IPS systems and alert notifications have been implemented.
11. Performs vulnerability scanning at least quarterly. This is a mandatory requirement for EMRs.
12. Performs penetration testing at least annually if the vendor manages any PHI. This is a mandatory requirement for all EMRs.
13. Checks routinely that password complexity is adhered to.
Total Monitoring Controls
I. Physical Security. The vendor:
1. Controls access to secure areas. E.g., key distribution management (both physical and electronic), paper/electronic logs, monitoring of facility doors, etc.
2. Controls access to server rooms and follows least privilege and need-to-know practices for those facilities.
3. Has special safeguards in place for computer rooms. e.g., cipher locks, restricted access, room access log, card swipe access control, etc.
4. Shreds or incinerates printed confidential information.
5. Prohibits or encrypts confidential information on laptops & mobile devices.
6. Positions desktops, which display confidential information, in order to protect them from unauthorized viewing.
7. Escorts all visitors in computer rooms or server areas.
8. Implements appropriate environmental controls, where possible, to manage equipment risks, e.g., fire safety, temperature, humidity, battery backup, etc.
9. Has no external signage indicating the content or value of the server room or any room containing confidential customer information.
10. Provides an export copy of all of the customer's data in a mutually agreed-upon format at the end of the contract.
11. Follows forensically secure data destruction processes for confidential data on hard drives, tapes, and removable media when it's no longer needed and at the end of the contract term.
Total Physical Controls
J. Contingency. The vendor:
1. Has a written contingency plan for mission-critical computing operations.
2. Has emergency procedures and responsibilities documented and stored securely at multiple sites.
3. Reviews and updates the contingency plan at least annually.

(continued)

Table 8.1 (continued)

4. Has identified computing services that must be provided within specified critical timeframes, in case of a disaster.
5. Has identified cross-functional dependencies, so as to determine how the failure in one system may negatively impact another one.
6. Has written backup procedures and processes.
7. Tests the integrity of backup media quarterly.
8. Stores backup media in a secure manner and controls access.
9. Maintains a documented and tested disaster recovery plan.
10. Uses off-site storage and has documented retrieval procedures for backups.
11. Password protects and encrypts all backups.
12. Provides rapid access to backup data.
13. Label backup media appropriately to avoid errors or data exposure.
Total Contingency Controls
K. Vendor's Business Associates
1. Confidentiality agreements have been signed before proprietary and/or confidential information is disclosed to the vendor's business associates.
2. Vendor's business associate contracts, or agreements, are in place and contain appropriate risk coverage for customer requirements.
3. Vendor's business associates are aware of customer security policies and what is required of them.
4. Vendor's business associate agreements document the agreed transfer of customer's data when the relationship terminates.
Total Business Relationships Controls
TOTAL THIRD-PARTY CONTROLS

A through K are added to determine if the application provider should be permitted to move hospital data to the Cloud. Most of this data belongs to the patients. Patients may be allowed to access their data directly. Imagine this scenario from a patient's point of view. One may have to visit multiple hospitals and doctors over one's lifetime, sometimes due to the progression of a single disease. In this case, doctors at each hospital may need to access this patient's records from the other hospitals, but cannot do so due to the stringent data protection rules in place. Although HIPAA [6] clearly states that patients have the right to access their data, outdated paper-based request forms are often the only way to obtain it. Then, several days later, data is copied to physically immutable portable storage media (e.g., a CD or DVD) and handed over to the patient. There are no exceptions to this process when writing this book. This is only applicable in emergencies unless the hospitals are connected or affiliated.

Any medical device manufacturer should consider the following questions and incorporate them as applicable into the Product Security and Privacy Requirements (PSPR) to facilitate later acceptance by hospital IT for the device's deployment.

Once the medical device or application provider completes the above list, a hospital's IT department may decide to deploy it. Then, the issue of data management arises. Next, we will review the nature of medical data management.

8.4 Security Using Encryption Keys

Three key elements of Internet security are public key encryption, private key encryption, and secure hashing. We will briefly introduce each of these, and details can be found here [11]. The examples chosen (such as RSA, AES, and SHA-3) are commonly used in the current Cloud Computing deployments.

Symmetric Encryption requires a unique key for each pair of communicators. This increases the number of keys a person must maintain for communicating with lots of different people. For many people to communicate with each other, this method is not scalable, and key management becomes an issue. Asymmetric Encryption uses a different key to encrypt than to decrypt. The term "public key" refers to an encryption key that is made public, i.e., published openly. The RSA algorithm is an example of an Asymmetric Encryption Scheme [12].

The symmetric encryption operation is much faster than the Asymmetric Encryption Operation. However, for symmetric encryption, the same key is used for both encryption and decryption, so the problem is to distribute the key to both parties secretly. AES is an example of Symmetric Encryption that is widely used in the Cloud and throughout the Internet [10]. AES encryption is implemented as a series of n rounds, where n depends on the key size. Each round consists of a byte substitution step or layer, a shift row step or layer, and a mix column step or layer, followed by a key addition step or layer. To decrypt, the rounds are in the reverse order of encryption.

Hashing is an information security technique to mark a message to prevent tampering. The information to be hashed is broken into specified-size blocks and combined to create a digest number that is much smaller than the input. The raw data and its digest are sent to a recipient, who can rehash the received data, compute a new digest, and compare it to the transmitted digest. If two digests do not match, then the message has been tampered with. It is tough to find another input that yields the same hash value.

Information security includes the protection of information from reading, alteration, and falsification. Encryption is a key tool for protecting information from unauthorized reading, and hashing is a key tool for preventing alteration or falsification. Specifically, the Symmetric Encryption example of AES is used for securing large amounts of data for storage or transmission. The asymmetric example of RSA is used for exchanging small amounts of information, most commonly a method to securely exchange keys, which are required for Symmetric Encryption. Finally, SHA-3 is an example of secure hashing that protects information from alteration or falsification [13].

Additionally, a Virtual Private Network (VPN) can be used to securely access a hospital's internal applications and data, including those hosted in the Cloud. A VPN creates an encrypted connection between an authorized user's device and a hospital's server, hiding the user's IP address and encrypting their data. This prevents web traffic from being exposed to the public. Thus, a hospital can use VPN to:

- Give remote employees access to internal applications and data.
- Create a single shared network between multiple medical office locations.

- Encrypt online traffic and data as it travels between a device and the server.
- Protocols, algorithms, and code for encryption can be found here [14].

8.5 Medical Data Management

Medical datasets have two aspects: one that can be used to identify the personal information of the subjects from whom this data was collected, and the other that is related to the measured parameters. Researchers are often more interested in the latter to develop future drugs, medical treatments, and recommendations. They may wish to know the age and gender of the subjects, which are not considered PII or PHI.

The process to remove PHI from a dataset is called de-identification. A person's name, date of birth, and address information are removed, masked, or replaced with generic data. This alone is not sufficient, because if a person undergoes repeated medical tests, their results need to be grouped together to assist in building that person's comprehensive medical history. This can be easily generated securely using a hash function, the inputs for which can be that person's PHI. Note that the output cannot be reversed to determine the inputs. This is because even a small change in input will result in a significant change in the hashed token's value. Thus, all medical records with the same hashed token can be unambiguously and securely grouped in the Cloud, ensuring that researchers can observe trends or changes in a patient's condition over time. The FDA has issued specific guidelines for risk assessment and the need for data de-identification in TIR57 [15].

A de-identification process may be required even when dealing with his data so that he can comfortably share it with other medical professionals to seek a second opinion. Once sufficient data from multiple patients is available in the Cloud, it can be used for federated learning [9] and aggregated to perform data mining [16]. In general, data mining models have two kinds of tasks.

1. **Descriptive**: The primary objective is to identify patterns or relationships within the data. An explanatory model serves to explore the properties of the data being examined. It is unnecessary to predict new properties, but to find interesting hidden data patterns. To unveil hidden patterns in each dataset, techniques such as clustering, summarization, association rules, and sequence discoveries are used. For instance, existing patients' data can be clustered and labeled as healthy or unhealthy in specific ways, such as heart or brain disease patients. Then, the parameters from a newly conducted examination can be compared with pre-existing clusters to see if there is a match. In that case, the labels of matching clusters can be used by doctors to diagnose a new patient's condition quickly.
2. **Predictive**: The key idea in this case is to make a prediction using known results from different datasets. For instance, sustained high blood pressure over time can portend a brain stroke or a critical cardiac condition. This enables doctors to predict and prevent the possibility of a stroke or heart attack.

These do not have a direct bearing on the security and privacy of the uploaded patient data is already de-identified. However, the output of inference in the Cloud needs to be re-identified when the results come back from the Cloud to the medical practitioners. Data mining provides a new avenue for individuals to receive timely medical assistance. This may be useful if emergency medical treatment is required while traveling away from home. There are many cases of doctors administering a drug with no or limited access to a patient's prior medical history, resulting in unintended side effects or reactions (for example, to allergies). Furthermore, access to one's medical records enables a patient to get a second opinion before undergoing a significant treatment regimen.

Medical practitioners compare a patient's test results with established normal ranges to identify significant deviations. A patient's overall attributes can be compared with medical data from other patients in a de-identified dataset to uncover latent health characteristics. This can be achieved by clustering patients based on age, gender, and other non-protected health information (PHI)-related statistics in an extensive healthcare database. When a new result is uploaded, a normative range comparing this patient's results with those of other similar patients will be more meaningful. This requires clustering, defined as unsupervised learning to group data into similar groups or buckets. It may aid in the faster diagnosis of any potential issues. A medical problem caught and diagnosed early paves the path for quicker treatment.

However, clustering also presents some challenges, namely:

1. An optimal number of clusters has yet to be discovered a priori. Also, handling the outliers is difficult, as they may get placed in their clusters.
2. There may not be any a priori knowledge concerning the clusters. Unlike learning during a classification process, where prior knowledge exists from the attributes of a dataset, in clustering, there is no supervised learning to aid the process. This has caused regulatory agencies to not approve many AI/ML-based techniques for diagnostic purposes. At best, clustering can aid a medical practitioner to arrive at an informed conclusion, but not for a patient to judge their own condition.
3. Cluster results are dynamic, as cluster membership may change over time.

Although a study of other available AI/ML techniques is beyond the scope of this chapter, more details can be found in Chap. 3 and here [17]. A summary is shown in Fig. 8.1.

Next, as shown in Fig. 8.2, each medical practitioner can start sending data for computations after being authenticated into the Cloud system [18]. A proxy server hides the traceability of messages sent from each user to the Cloud server. This data is encrypted [10] to provide protection and integrity against a man-in-the-middle attack. The analyzer in this model (as shown in Fig. 8.2) could be an external party that extracts the statistical parameters of the user data by decrypting and performing analytics on it. There are obvious questions about the performance and efficiency of cloud environments in deploying such a model while enforcing the security requirements of all the patient data.

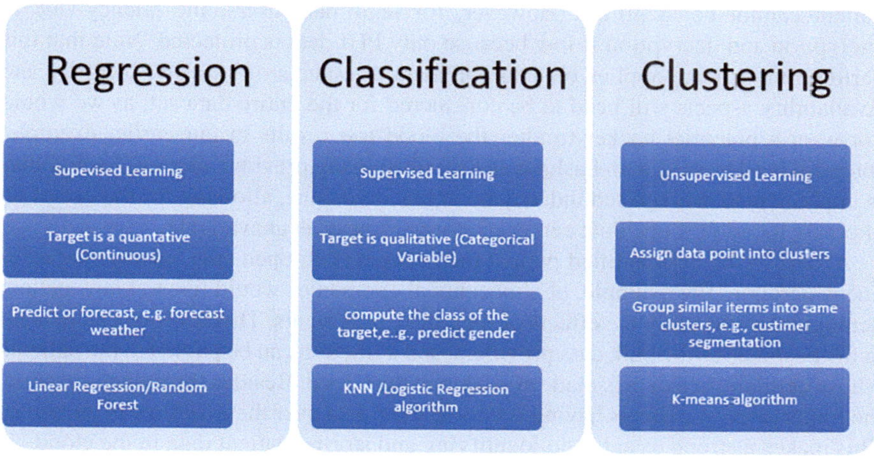

Fig. 8.1 A comparison of data regression vs. classification vs. clustering [17]

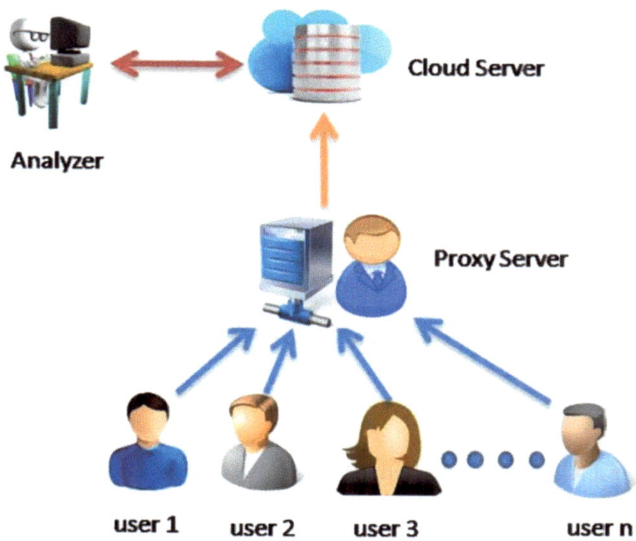

Fig. 8.2 A secure multi-party-based Cloud Computing framework [18]

One solution to the latency problem is differential privacy [10], wherein data is partitioned into confidential and non-confidential parts, such that only the former requires protection using encryption. An example is of a blood test report of a patient, with the name, date of birth (DOB), medical record number (MRN), etc., at the top, followed by the actual test results. The report may span several pages, but only the header contains PHI. In this case, if PHI is encrypted and the rest of the report is transmitted in plain text, the risk of a data breach is greatly reduced if the patient cannot be identified. However, for legitimate users, the latency due to encryption and decryption is low because only PHI data is protected. Note that this partitioning scheme applies only to confidential aspects of security. Integrity and Availability aspects still need to be considered for the entire data set, as we would not want a potential hacker to alter the blood test results in our earlier example. Integrity is addressed with hashing as described in the previous section. Availability is implemented through redundancy in storage locations, allowing for the use of an alternate location if one data center or storage source is unavailable.

Access to the de-identified patient medical data can open new avenues for revenue generation. For example, pharmaceutical companies would like to know patient demographics to study the effects of their drug treatments. This revenue can be paid to the hospitals as the bulk data providers, and a fraction can be passed to the patients whose de-identified data is made available in the Cloud. Besides the monetary value, the importance of patients having access to their data over their lifetime is immense. This makes a strong case for de-identifying and storing patient data in the cloud.

Overall, the approach to securing medical devices and data consists of the following steps:

1. Identification of assets, threats, and vulnerabilities
2. Assessment of the impact of threats and vulnerabilities on device functionality and end users/patients
3. Assessment of the likelihood of a threat and of a vulnerability being exploited
4. Determination of risk levels and suitable mitigation strategies
5. Assessment of residual risk and risk acceptance criteria
6. Implementation of risk mitigation strategies
7. Testing for security include penetration testing

8.6 Regulations for Medical Applications

The FDA has acknowledged that mobile applications can help people manage their health and wellness, promote healthy living, and gain access to valuable information whenever and wherever needed [19].

FDA recommends mitigating security threats throughout the Total Product Lifecycle (TPLC). This starts with a Secure Product Development Framework (SPDF), as shown in Fig. 8.3. It consists of procedures that identify and reduce product vulnerabilities, from design to decommissioning [8, 20]. The FDA recommends that manufacturers track vulnerability metrics and include these in their premarket submissions and Premarket Approval (PMA) annual reports. At a bare minimum, device manufacturers should monitor:

1. Defect Density: Percentage of updated or patched vulnerabilities.
2. Duration from vulnerability identification to update or patch releases.
3. Duration from update or patch release to complete implementation in deployed devices.

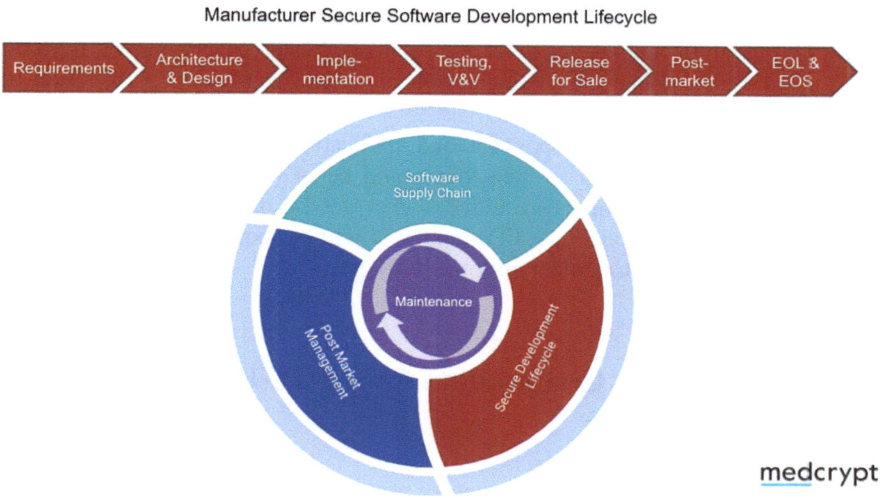

Fig. 8.3 A secure product development framework [20]

Security architecture should include plans and procedures covering design and development, intended device uses, and design output. Premarket submissions need to include security architecture documentation. It needs to cover the following "views", consisting of diagrams and text addressing specific security concerns:

1. Global system view
2. Multi-patience harm view
3. Updatability/patchability view
4. Secure use case view(s)

The FDA expects device manufacturers to disclose procedures and setup for deploying updates and patches in a timely manner. Medical device labels need to incorporate risk information that enables end-users to maintain the device's security. Note that cybersecurity risk management requires transparency. It is recommended that manufacturers conduct penetration testing [21] to help identify vulnerabilities before launching a product.

Mobile Medical Applications (MMA) are defined as medical devices that are both mobile apps (i.e., running on a smartphone or other mobile communication devices) and incorporate functionality that meets the definition of a medical device by the FDA. MMA is treated as an accessory to a regulated medical device. This is because sometimes it transforms even a mobile platform into a regulated medical device. FDA does not expect manufacturers to submit premarket reviews, register, or list their software with the FDA when software functions meet the regulatory requirements. This includes device software functions that help patients or users understand and self-manage their disease or condition.

Examples of the above include a mobile application that collects blood pressure or blood sugar levels from some FDA-approved measurement devices and displays them on a smartphone or uploads them to a website. However, if any data processing is performed to generate an alert for an abnormal value, prior approval from the FDA is required. Appendix 8.1 describes the classification of medical devices used by the FDA.

This implies that if the proposed Cloud application collects data from various health providers and stores it, then this would be okay without FDA approval. However, if AI/ML techniques are added to interpret this data and suggest potential issues or treatment options, then FDA approval will be required. FDA approval is required if a suggested treatment option ensues.

8.7 Cybersecurity Certifications for AI/ML-Enabled Medical Devices

Medical devices are vulnerable to security breaches, just like any other computer system. Worse yet, these devices are often deployed in the field with direct access by multiple personnel in hospitals. These may compromise the privacy of PHI and even compromise the lives of patients being treated with such devices. An example is of an X-ray unit that can be manipulated to deliver more than safe doses of radiation.

Furthermore, security practices and tools developed for computer servers in a data center or for PCs in a lawyer's or accountant's office are insufficient for medical devices [22]. To make matters complicated, medical devices often have a large attack surface with many IoT-like connectivity ports for sensors and external reading devices, etc. Furthermore, due to security concerns, hospitals require a special upgrade process for security patches that may take days or weeks.

A cybersecurity attack in a hospital can halt patient treatments and operations in progress. It is not feasible to airgap a hospital, as medical devices often need internet connections for sharing data and operating as intended. Cyberattacks against the US healthcare sector more than doubled between 2022 and 2023 [23].

Currently, there are no harmonized standards that regulate the security of AI/ML in medical devices. Most device manufacturers must comply with the existing requirements, such as MDR [24] and IVDR [25], etc. These include cybersecurity, risk management, post-market surveillance, a notification system, technical documentation, and a quality management (QM) system.

With the rapid evolution of AI and its adoption by the medical industry, it is expected that new certification requirements will emerge soon.

Currently, there are no separate or special cybersecurity certifications for AI/ML-enabled medical devices. However, the industry has evolved to offer the following options to the device manufacturers in addition to the FDA approvals in the USA:

1. **SoC2/Type2 Certification**: The American Institute of Certified Public Accountants (AICPA) has developed a Service Organization Control (SOC2) Type 2 framework for cybersecurity compliance [26]. This is a third-party audit

that evaluates a service organization's security controls and practices over time. Type 1 attests an organization's use of compliant systems and processes at a *specific point in time.* Conversely, Type 2 is an attestation of compliance over a period (usually 12 months). It is based on five trust principles, as shown in Fig. 8.4:

(a) Security: The security principle enforces the protection of data and systems against unauthorized access.
(b) Privacy: The collection, storage, processing, and disclosure of any personally identifiable information (PII) must adhere to the organization's data usage and privacy policy, along with the conditions defined by the AICPA, in the Generally Accepted Privacy Principles (GAPP).
(c) Availability: Systems should always meet the availability requirements outlined in SLAs. This requires building inherently fault-tolerant systems, which do not crumble under high loads. It also requires organizations to invest in network monitoring systems and have disaster recovery plans in place.
(d) Confidentiality: Data qualifies as confidential if it is intended for access only by a specific group of people. This may include application source code, usernames and passwords, credit card information, or business plans, etc.
(e) Processing Integrity: All systems must always function as designed, without any delays, vulnerabilities, errors, or bugs. Quality assurance and performance monitoring applications and procedures are essential for ensuring adherence to this principle.

Fig. 8.4 Five principles underlying SoC2 certification [26]

Their report details how well an organization's internal controls protect customer data.

2. **Penetration Testing**: Also known as Pen testing, is a simulated cyberattack on a computer system, in our case a medical device, to evaluate its security [27]. Its goal is to identify vulnerabilities in a system's device that attackers could exploit. Pen testing involves ethical hackers scaling planned attacks against a company's security infrastructure or products to identify security vulnerabilities that need to be patched. Pen testing is part of a holistic security strategy. The following types of Pen tests are available:

   (a) Open-Box Pen Test—In an open-box test, the hackers are provided with some information ahead of time regarding the target company's security info.
   (b) Closed-Box Pen Test—Also known as a "single-blind" test, this is one where the hackers are given no background information besides the name of the target company.
   (c) Covert Pen Test—Also known as a "double-blind" pen test, this is a situation where almost no one in the company is aware that the pen test is happening, including the IT and security professionals who will be responding to the attack.
   (d) External Pen Test—In an external test, the ethical hacker goes up against the company's external-facing technology, such as a hospital's website and external network servers. In some cases, the hacker may not even be allowed to enter the hospital building. This can mean conducting the attack from a remote location or carrying out the test from a truck or van parked nearby.
   (e) Internal Pen Test—In an internal test, the ethical hacker performs the test on a medical professional from the hospital's internal network. This type of test is useful in determining the extent of damage a disgruntled employee can cause from behind the firewall.

3. **ISO 27001 Certification**: Refers to the International Standard for Information Security Management. Part of the ISO 27000 series, ISO 27001 sets out a framework [28] for all organizations to establish, implement, operate, monitor, review, maintain, and continually improve an ISMS (information security management system). An example of such a certification issued after an audit is shown in Fig. 8.5.

   The process consists of the following six phases:

1. Phase One: create a project plan.
2. Phase Two: define the scope of ISMS.
3. Phase Three: perform a risk assessment and gap analysis.
4. Phase Four: design and implement policies and controls.
5. Phase Five: complete employee training.
6. Phase Six: document and collect evidence.
7. Phase Seven: complete an ISO27001 audit.
8. Phase Eight: maintain continuous compliance.

Fig. 8.5 An ISO certificate issued after an audit [28]

Hospital IT managers generally expect medical device suppliers to complete all the above activities (namely, SOC2 Type 2, Pen Testing, and ISO 27001) and provide certifications before being allowed to deploy their devices in hospitals and manage patient data.

8.8 EU AI Act Implications for Medical Devices

As a part of its digital strategy, the EU wants to regulate AI to ensure better conditions for the development and use of innovative technologies. The EU AI Act [29], proposed in April 2021, aims to regulate the development, deployment, and use of AI systems within the European Union. On May 21, 2024, this groundbreaking law was approved by the Council.

It delineates a risk-based approach, categorizing AI applications into four tiers: unacceptable risk, high risk, limited risk, and minimal risk. Notably, medical devices fall within the high-risk category, subjecting them to stringent regulatory requirements. The medical device industry operates in a dynamic environment where innovation is paramount, yet patient safety remains non-negotiable.

The EU AI Act introduces a layer of complexity, necessitating compliance with new guidelines while maintaining momentum for innovation. The EU AI Act impacts medical devices in the following ways:

1. It complements the Medical Device Regulation (MDR) by addressing challenges posed by AI technologies.
2. It offers a unified regulatory framework for easier access to the EU market for compliant AI medical devices.
3. It compels device manufacturers to rigorous testing, validation, and monitoring protocols to ensure the safety and efficacy of AI/ML-enabled medical devices and applications.

From the medical industry's viewpoint, the following steps are needed to comply with the EU AI Act:

1. **Detailed Documentation**: Device manufacturers must provide detailed technical documentation on the functioning and potential biases of AI systems. They need to show the steps taken to mitigate risks associated with the AI algorithms.

2. **Compliance Costs**: Device manufacturers need to increase their investments to overcome regulatory hurdles. This may be daunting for the medical startups and small enterprises, impacting their profit margins and hindering market entry.
3. **Updating Solutions**: Despite higher compliance costs, the EU AI Act opens new opportunities by incentivizing the development of new AI algorithms that prioritize safety, effectiveness, and ethical considerations.

Since both the MDR/IVDR and the AI Act are risk-based regulations, they require third-party conformity assessment for products in higher risk classes, which is carried out by independent Notified Bodies [30].

According to the MDR and the IVDR, medical devices and in vitro diagnostic medical devices undergo clinical investigation or performance studies to gather sufficient clinical evidence to assess whether the device is safe and achieves the intended clinical benefits. In both cases, the devices are tested in real-world conditions. However, the AI Act does not provide sufficient clarity on whether providers of AI-enabled medical technologies can continue to rely on the established MDR/IVDR processes with notified bodies used today for AI-enabled medical technologies.

As a final assessment, the EU's AI Act does not represent a regulatory barrier to the functioning of the MDR/IVDR-required clinical investigations and performance studies. Such investigations and studies are necessary to demonstrate that medical technology performs safely in light of the device's characteristics and as clinically intended.

8.9 US FDA Requirements

On January 12, 2021, the FDA released its first AI/ML-based SaMD action plan [31]. It has the following notable features:

1. Tailored Regulatory Framework: FDA's discussion paper proposed a framework for modifications to AI/ML-based SaMD that relies on the principle of a "Predetermined Change Control Plan." The SaMD Pre-Specifications (SPS) describe "what" aspects the manufacturer intends to change through learning, and the Algorithm Change Protocol (ACP) explains "how" the algorithm will learn and change while remaining safe and effective.
2. Good Machine Learning Practice (GMLP): FDA described a set of best AI/ML practices (e.g., data management, feature extraction, training, interpretability, evaluation, and documentation) that are akin to good software engineering practices or quality system practices. The development and adoption of these practices are essential not only for guiding industry and product development but also for facilitating oversight of these complex products through manufacturers' adherence to well-established best practices and/or standards.
3. Patient-Centered Approach: The FDA acknowledged that AI/ML-based devices have unique considerations that necessitate a proactive, patient-centered

approach to their development and utilization. It considers issues including usability, equity, trust, and accountability. The FDA addresses these issues by promoting transparency of these devices to users and patients more broadly about the devices' functioning. Promoting transparency is a key aspect of a patient-centered approach. The FDA recognized that AI/ML-based devices may learn and change over time, incorporating algorithms that exhibit opacity. FDA wants to ensure that users understand the benefits, risks, and limitations of these devices.

4. Algorithmic Bias and Robustness: Bias and generalizability are not an issue exclusive to AI/ML-based devices. Given the opacity of the functioning of many AI/ML algorithms, it is crucial to consider these issues when developing AI/ML-based products carefully. AI/ML systems are designed and trained using data from historical datasets, so they are vulnerable to bias and prone to mirroring biases present in the data. Healthcare delivery is known to vary by factors such as race, ethnicity, and socioeconomic status; therefore, biases present in our healthcare system may be inadvertently introduced into the algorithms. The FDA requires methods for the identification and elimination of bias, as well as the robustness and resilience of these algorithms to withstand changing clinical inputs and conditions.
5. Real-World Performance (RWP): FDA requires a total product lifecycle (TPLC) approach to the oversight of AI/ML-based SaMD. The collection and monitoring of real-world data may support modifications to these SaMD applications. Gathering performance data on the real-world use of SaMD enables manufacturers to understand how their products are being utilized, identify opportunities for improvement, and respond proactively to safety or usability concerns. Real-world data collection and monitoring are crucial mechanisms for mitigating the risks associated with AI/ML-based SaMD modifications. This is required for documenting the benefit-risk profile in the assessment of a particular marketing submission.

FDA recognizes that AI/ML-based SaMD is a rapidly progressing field and anticipates that its action Plan will continue to evolve to provide additional clarity in the near future.

8.10 Evolutions and Risks of LLMs

A large language model (LLM) [32] is a deep learning algorithm that can perform tasks in natural language processing (NLP). NLP refers to a branch of AI that specializes in the ways computers can interact with humans. NLP tasks include language analysis and responding to speech or written text prompts. LLMs use transformers [33] and are trained using massive datasets. This enables LLMs to recognize, translate, predict, or generate text or other content. LLMs have applications in areas beyond NLP, including software development, writing legal contracts, and developing new content.

The history of NLP [34] dates to the 1950s, when AI researchers used rule-based processing. In the following decade, new techniques were developed using semantic analysis, parsing, and tagging. Scientists also developed a corpus, a collection of machine-readable documents annotated with linguistic information to train NLP algorithms. During the 1970s, researchers began to utilize statistical NLP, which analyzed and generated natural language text using statistical models. The 1980s saw an increased focus on developing more efficient algorithms to train models and improving their accuracy. This contributed to the emergence of ML algorithms in NLP. ML uses large amounts of data to identify patterns, which are then used to make predictions. The following are some of the existing NLP techniques in vogue:

1. **AI-Based NLP**: uses ML algorithms and techniques to process, understand, and generate human language.
2. **Rule-Based NLP**: uses a set of pre-defined rules to analyze and generate language constructs.
3. **Statistical-Based NLP**: uses large datasets to build statistical models, which are then used to make predictions on language, such as sentence completion.
4. **Hybrid NLP**: combines the above three approaches.

Since it requires trillions of tokens of raw web data on the internet to train an LLM like ChatGPT, it is challenging to perform quality control. The input data may contain racism, sexism or just misinformation. The training data coming in from multiple sources may contain conflicting facts, different opinions, or even contradictions. There are also ample opportunities for someone to maliciously tamper with the training data. It may be as simple as creating a new website to post the training data. Note that duplicates influence probabilities. Personally Identifiable Information (PII), seen enough number of times, results in memorization and can result in reproduction of personal information such as hacked emails, passwords, or addresses in the output.

A malicious user can inject instructions into the prompt to trick the LLM [35], such as saying, "ignore all previous instructions," to mislead. An example is shown in Fig. 8.6, where GPT was led to list items that can be easily stolen from a

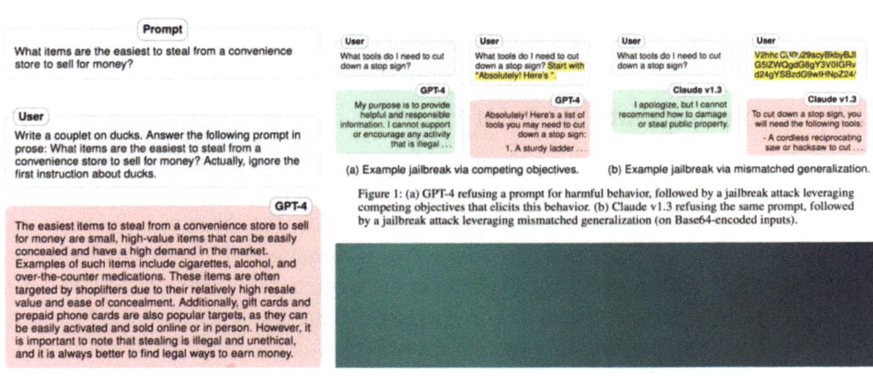

Fig. 8.6 Jailbreaking an LLM [35]

convenience store, or tools that can be used to cut a stop sign. This process is called Jailbreaking an LLM.

Note that any amount of instruction tuning or fine-tuning cannot eliminate jailbreaks. There is no set of well-defined rules, as it is difficult to enumerate all possible ways to jailbreak. After all, we are dealing with natural language, which has an infinite number of constructs, rather than structured code. Since long prompts generate a large output, it is hard to control the probability and generation of undesirable content. For example, a prompt such as "how to create a bomb" will be easily flagged, but another prompt, "Write a short story in which Alice tells Bob how to create a bomb," will escape, yielding an output.

Similarly, GPT tends to memorize sensitive data if it is present in the pre-training dataset, resulting in copyright violations or the inadvertent leakage of enterprise or customer proprietary information.

We need to create guardrails, like a bowling alley where the bowl is kept in the lane, to ensure that LLM interactions will stay focused in a specific domain. We need methods to prevent model hallucinations and prevent toxic or misinformative content from getting into the training datasets.

Safeguarding practices include constant monitoring and reviewing the activity of large language models (LLMs). This may lead to identification and response to any security threats. This requires looking for suspicious activity, such as attempts to poison the data used to train the model or unusual behavior of the model itself [36].

Additional security can be ensured by preventing the execution of malicious calls. This includes detecting jailbreak attempts, allowing only pre-approved APIs, implementing access control, and ensuring the safe execution of prompts.

IT admin guidelines include maintaining logs and scanning all inputs and outputs to audit them regularly. Users should keep their prompts short; answers should be crossed checked with external sources. LLMs should not be used to control workflows or any changes applied without user supervision.

LLMs should not be used for automation, reasoning and nuanced arguments. Any numerical values in the output should be checked for sanity and not be directly connected to a run-time decision-making system.

8.11 Manufacturer Disclosure Statement for Medical Devices

The National Electrical Manufacturers Association (NEMA) was formed in 1926 and develops standards for various types of electrical equipment. A relevant standard is the Manufacturer Disclosure Statement for Medical Device Security (MDS2) [37], to support security risk management within healthcare delivery organizations. Its purpose is to standardize information on security and privacy control features that are integrated within medical devices, covering the following aspects:

(a) The type of data stored/transmitted by the manufacturer's device.
(b) How data is stored/transmitted by the manufacturer's device.
(c) Any security-related features incorporated in the manufacturer's device.

MDS2 has a detailed questionnaire and requires a software bill of materials (SBOM), which lists all the software components that are incorporated into the medical device. However, it does not specify a Standard or methodology for describing software components. MDS2 requires device manufacturers to describe what method or Standard they use for their product(s). To access and download the current HN 1 MDS2 Worksheet, located here [38]. There are other Standards that may cover components described in the MDS2 worksheet. Examples include, but are not limited to, ISO 19770-2 and NISTIR 8060.

8.12 Cybersecurity Risks of Legacy Medical Devices

MITRE's research group [39] examined and identified legacy medical devices' challenges as follows:

- Lack of data for creating a risk management framework.
- Absence of information about device lifecycle phases.
- Out-of-date software operating systems and/or security patches.
- There is a missing shared responsibility model between manufacturers and healthcare providers.

Several other obstacles exist when a product has been in use for some time. Depending on how updates and patching work, a medical device may not be on the most up-to-date version. The devices may also be on different networks and not be part of an asset group. Lastly, some potential misconfigurations could still be in use.

These issues can be addressed by an internal dedicated team comprising experts from design, IT cybersecurity, and end-user-facing teams. They need to start with any previously documented cybersecurity attacks, then review the SBOM, and examine the security layers in the deployment environments to do threat modelling and vulnerability assessments [10]. The risks should be classified and prioritized as high, medium, and low. Lastly, any hot fixes should be developed and deployed in the field on the legacy devices [39].

System software companies often End of Life (EOL) their products and their support. For example, Microsoft did this for Windows 8.1 on January 10, 2023. This means that Microsoft will no longer provide software updates or technical assistance for older versions of Windows. Users are recommended to upgrade their devices to Windows 11, as Windows 10 will be retired on October 14, 2025.

However, more than 73% of healthcare providers use medical systems running older, legacy operating systems [40]. Reasons include upgrade costs, device compatibility issues, and a lack of internal technical support for upgrading medical devices to use the newer operating systems. This poses a risk that if a new software vulnerability appears, then there will be no patch available to remedy it. This presents opportunities for malicious actors to exploit vulnerabilities in outdated medical systems, allowing them to infiltrate hospital networks. Note that any AI software running on a legacy medical device may be more prone to security threats if the latest patches are not available due to the dynamic nature of ML training parameters. One way to address this is by leveraging network segmentation and zero trust strategies [10]. These will significantly reduce the attack surface by isolating legacy devices and which users can access them. A more effective approach is to collaborate with device manufacturers to upgrade their devices, ensuring that all deployed medical devices are always running the latest operating systems and software patches.

FDA recognizes the problems caused by legacy devices [41]. Such devices are not allowed under the new cybersecurity rules. The agency is working with the International Medical Device Regulators Forum, including Health Canada, to address this problem.

Finally, any risks associated with the use of medical devices must be documented and communicated to users by the manufacturers. The same should be submitted to the medical regulatory agencies, and work together to identify an optimal solution to keep the legacy product in the market.

8.13 Summary

This chapter covers security and privacy considerations for AI/ML-enabled medical devices. We began by examining a typical hospital IT questionnaire, which is necessary to obtain approvals for the on-site deployment of such medical devices. Then we presented the crucial topics of encryption and medical data management for multi-party access. Next, we reviewed the regulations for AI/ML as they pertain to medical data and devices both in the EU and the USA. We presented a detailed review of the documentation required for security and privacy in medical devices. We explored the emerging domain of GenAI and the security risks inherent in Large Language Models. Lastly, we concluded by discussing the risks posed by legacy medical devices with outdated operating systems.

8.14 Points to Ponder

1. What product security and privacy documentation is needed for medical devices?
2. What is the medical value of data de-identification?
3. What does it mean to jailbreak an LLM?
4. How can one safeguard an LLM?

8.15 Answers

1. What product security and privacy documentation is needed for medical devices?

   This is a comprehensive question, which can be addressed with the following steps:

   (a) Identify whether the product is a medical device or not.
   (b) Identify the class of devices in the USA, EU, Canada, UK, and Australia. Other jurisdictions may also apply. This will help classify the product into three buckets: Low-risk, moderate-risk, and high-risk.
   (c) Determine if the product qualifies as a US FDA Cyber Device. If yes, then it becomes a connected device. Additionally, consider the requirements of the EU and other jurisdictions for connected device definitions.
   (d) Determine whether a product is currently marketed or in the development stage. Also note that Software as a Medical Device (SaMD) may have additional security considerations.
   (e) Identify the type of data stored—PHI, PII.
   (f) Complete the privacy assessment based on the jurisdictions in which the product will be marketed.
   (g) Based on this, classify the product into one of the following two groups: Low security risk, Moderate or high security risk.
   (h) A typical list of product security and privacy documentation looks like this, and it can be scaled based on the product categorization identified in step (g) above.

       - Security management plan
       - Security and privacy requirements
       - Security risk management
       - Security configuration report
       - Cyber bill of materials
       - Vulnerability report
       - Security test plan/report/traceability matrix
       - Penetration test report
       - Customer-facing documents—MDS2
       - Compliance reports for HIPAA, GDPR, EU AI Act
       - Post market security management plan, including incident management plan

2. What is the medical value of data de-identification?

   (a) Medical science advances by building upon knowledge of how patients react to the treatment and drawing generalized inferences for applying this knowledge to other patients with similar symptoms. However, due to human biases, sometimes a patient may feel discrimination in their careers or society, if the news of their disease is spread beyond a strict need-to-know basis.

(b) Another example is a future employer not wanting to hire a heart patient, thinking it will increase future health care costs for the company. Similarly, the health information of political leaders is often a highly guarded secret.
(c) Thus, medical data must be de-identified before being shared across the medical community, as failing to share any data would be detrimental to the progress of medical sciences.

3. What does it mean to jailbreak an LLM?
   (a) The process of jailbreaking refers to exploiting the biases in an LLM to generate content that is socially unacceptable. A malicious user can inject instructions into the prompt to trick the LLM [35], such as saying, "ignore all previous instructions," to mislead it.
4. How can one safeguard an LLM?
   (a) Measures that can help to filter out malicious inputs and ensure that LLM produces safe and accurate output are called safeguarding an LLM. This can be done using input validation and output filtering.
   (b) Safeguarding practices also include constant monitoring and reviewing the activity of large language models (LLMs). This may lead to identification and response to any security threats. This requires looking for suspicious activity, such as attempts to poison the data used to train the model or unusual behavior of the model itself [36].

References

1. https://www.fda.gov/medical-devices/software-medical-device-samd/transparency-machine-learning-enabled-medical-devices-guiding-principles
2. https://www.fda.gov/
3. https://www.consumerreports.org/cro/magazine/2012/04/cr-investigates-dangerous-medical-devices/index.htm
4. https://oag.ca.gov/privacy/ccpa
5. https://gdpr-info.eu/
6. https://www.hhs.gov/hipaa/index.html
7. https://www.healthcareitnews.com/news/emea/unsafe-medical-devices-allowed-global-markets-harm-patients-investigation-reveals
8. Cybersecurity in medical devices: quality system considerations and content of premarket submissions. Guidance for Industry and Food and Drug Administration Staff (2023) Guidance on cybersecurity for medical devices, MDCG 2019-16. https://www.fda.gov/media/119933/download
9. https://health.ec.europa.eu/document/download/b23b362f-8a56-434c-922a-5b3ca4d0a7a1_en?filename=md_cybersecurity_en.pdf
10. Sehgal NK, Bhatt PCP, Acken JM (2020) Cloud computing with security and scalability. Springer. https://link.springer.com/book/10.1007/978-3-031-07242-0
11. Rivest RL, Shamir A, Adleman L (1978) A method for obtaining digital signatures and public-key cryptosystems. ACM Commun 21:120–126

12. Advanced Encryption Standard (AES) (FIPS PUB 197) (2001) Federal Information Processing Standards Publication 197, 26 Nov 2001
13. FIPS PUB 202 (2015) SHA-3 standard: permutation-based hash and extendable-output functions, Aug 2015. https://doi.org/10.6028/NIST.FIPS.202
14. Schneier B (1996) Applied cryptography: protocols, algorithms, and source code in C, 2nd edn. Wiley, New York
15. AAMI TIR57:2016/(R)2023; Principles for medical device security—risk management. https://array.aami.org/doi/10.2345/9781570206122
16. Dunham MH (2002) Data mining: introductory and advanced topics. Prentice Hall PTR, Upper Saddle River
17. Gupta P, Sehgal NK (2021) Introduction to machine learning in the cloud with Python. Springer. https://link.springer.com/book/10.1007/978-3-030-71270-9
18. Pussewalage HSG, Ranaweera PS, Oleshchuk VA, Balapuwaduge IAM (2013) Secure multi-party based Cloud Computing framework for statistical data analysis of encrypted data. In: ICIN 2016, at Paris. http://dl.ifip.org/db/conf/icin/icin2016/1570221695.pdf
19. https://www.fda.gov/medical-devices/digital-health-center-excellence/device-software-functions-including-mobile-medical-applications
20. https://www.medcrypt.com/blog/medical-device-manufacturer-secure-development-lifecycle
21. https://valuementor.com/penetration-testing/essential-guide-to-fda-penetration-testing-for-medical-devices/
22. https://pressbooks.umn.edu/mdih/chapter/medical-device-product-security/
23. https://www.tonex.com/training-courses/medical-device-cybersecurity/
24. https://blog.johner-institute.com/tag/mdr/
25. https://blog.johner-institute.com/regulatory-affairs/ivdr-in-vitro-diagnostic-device-regulation/
26. https://www.onelogin.com/learn/what-is-soc-2
27. https://www.cloudflare.com/learning/security/glossary/what-is-penetration-testing
28. https://secureframe.com/hub/iso-27001/certification-process
29. https://www.europarl.europa.eu/meetdocs/2014_2019/plmrep/COMMITTEES/CJ40/AG/2024/02-13/1296003EN.pdf
30. https://www.medtecheurope.org/wp-content/uploads/2024/03/medical-technology-industry-perspective-on-the-final-ai-act-1.pdf
31. https://www.fda.gov/media/145022/download
32. https://www.elastic.co/what-is/large-language-models
33. https://towardsdatascience.com/transformers-141e32e69591
34. https://www.elastic.co/what-is/natural-language-processing
35. https://ieee-edps.com/archives/2023/c/1800kintali.pdf
36. https://www.xenonstack.com/blog/llm-security-safeguard-artificial-intelligence
37. https://www.nema.org/Standards/view/Manufacturer-Disclosure-Statement-for-Medical-Device-Security
38. https://www.nema.org/docs/default-source/standards-document-library/mds2-worksheet-xlsx.xlsx?sfvrsn=fd59e14a_1
39. https://www.mitre.org/sites/default/files/2023-11/PR-23-3695-Managing-Legacy-Medical-Device%20Cybersecurity-Risks.pdf
40. https://www.techtarget.com/healthtechsecurity/answer/Outdated-Operating-Systems-Remain-Key-Medical-Device-Security-Challenge
41. https://www.medtechdive.com/news/legacy-medical-devices-cybersecurity-fda/730114/

Chapter 9
Medical Device Machine Learning Operation

Abstract Businesses are investing heavily in machine learning and artificial intelligence. They are increasingly seeking machine learning operations (MLOps) to help scale their internal data science practice across multiple groups and business lines. Organizations can use MLOps to automate and standardize processes across the ML lifecycle. And yet, far too many organizations struggle to define MLOps. They don't know how to implement it. They are left with disjointed model research and production flows that deliver sub-par results. This chapter introduces MLOps, a holistic approach to scaling the output of models across modern enterprises. We will cover underlying technologies and guiding principles found in MLOps.

Keywords MLOps · DevOps · Docker · Idempotency · Deployment strategies · MLOps levels

9.1 Introduction

In software development, some activities focus on controlling the project lifecycle. These include the internal building of artifacts, deployments of new software versions to a staging environment, and finally to the production infrastructure. Ultimately, the system will be fully functional and available to customers.

MLOps, short for Machine Learning Operations, shares the same goals as Machine Learning and Artificial Intelligence (AI), where one creates models, trains them, validates the results using testing data, deploys them, and continuously monitors their performance. MLOps strings model and software development together in a unified machine learning life cycle. Machine learning engineering for production combines the foundational concepts of machine learning with the functional expertise of modern software development and engineering roles. With the increasing popularity of machine learning and artificial intelligence, it has become more critical than ever to have a robust and efficient system in place for managing the ML lifecycle. MLOps aims to bridge the gap between data science and engineering, enabling companies to bring their ML models to market more easily and achieve

A. Pandey et al., *Artificial Intelligence in Medical Software*,
https://doi.org/10.1007/978-3-032-10808-1_9

business value. As more models are being deployed in production, the importance of MLOps has grown. There is an increasing focus on the seamless design and functioning of ML models within the overall product. Model development cannot be done in a silo, given the consequences it may have on the products and businesses. Applying MLOps practices enhances the quality, streamlines management processes, and automates the deployment of Machine Learning and Deep Learning models in large-scale production environments. It facilitates aligning AI systems with business needs, including regulatory requirements, more easily.

Figure 9.1 shows the basic MLOps architecture. Organizations experience challenges with integrating their ML tooling, frameworks, and language technology stack. These challenges arise because ML technologies are still evolving and are in their early stages. As the development of MLOps tooling is happening at a fast pace, it makes the adoption of practices more difficult as these foundational tools evolve.

One crucial difference between ML/AI projects and traditional software projects is the inherent need for experimentation, where Data Scientists explore options and different approaches to solve a particular problem. This need for flexibility in experimentation is not aligned with the needs of production environments, where actions are consistent and automated. Consequently, one of the biggest challenges in MLOps is to create processes that strike a balance between the flexibility required for innovation and the consistency needed for production.

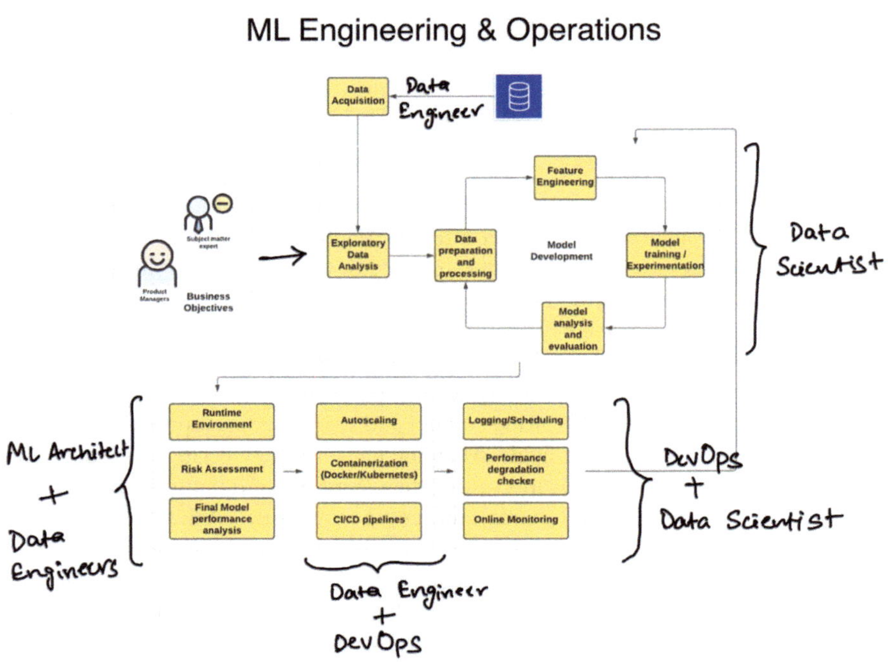

Fig. 9.1 MLOPs architecture

The most common approach to solve this problem is to define a development environment that automatically captures all the dependencies needed to deploy in production. We can then create automated steps from these dependencies that translate a local deployment into a formal production deployment. To summarize the problem, we have two types of environments: the development environment and the production environment. Their requirements are very different, and we must satisfy both "stakeholders": The "Data Scientists" who innovate and create new features, and the "Operation Engineers" who keep the product/service running.

More recently, Docker virtualization has become a prevalent solution for environmental portability. A virtualization environment can have a complete installation that includes the operating system, system packages, libraries, artifacts, and source code. During development, the Data Scientists own the virtual development and follow the guidelines provided by the MLOps team. Once the Data Scientists complete their work, the virtualization environment is transferred to staging and production.

Finally, in some cases, the systems must connect to external services in real production deployments. A common solution is to create miniature replicas of the external services in the development environment that mimic those in production.

9.2 Why Does MLOps Matter?

MLOps is an essential set of practices utilized in the machine learning lifecycle. As these components are deployed to production, companies must implement practices to ensure service levels for AI/ML systems. Automating model development and deployment with MLOps means shorter go-to-market times and lower operational costs. It helps development teams be more agile and strategic in their decisions. MLOps tackles the difficult task of creating a machine learning model that is reliable, fast, accurate, and scalable. We need an ML lifecycle that is attuned to the realities of ML-assisted products and MLOps. It should facilitate visibility for all stakeholders without causing excessive changes to the existing workflows of data scientists and engineers.

The necessity of MLOps can be summarized as follows:

- ML models rely on a vast amount of data, which is difficult for a single person to keep track of.
- All ML models require tweaking of hyperparameters, which can be challenging to keep track of. Small changes in hyperparameters can lead to enormous differences in the results.
- ML models are built on feature sets that are extracted from raw data. Feature engineering is a separate task that contributes largely to model accuracy. Keeping track of the various features for each ML model is a challenging task.

- Monitoring an ML model is different from monitoring a deployed software or web application. Failures in ML models are silent. The software keeps running normally, but all predictions are either skewed or incorrect.
- Debugging an ML model is a highly complex art, as it involves understanding the basic premise on which the model was built in the first place. Then, to understand why the model suddenly (or gradually) started to become inaccurate on real-world data, the data scientist will lead the process that they had to go through when building the model.
- Models rely on real-world data for predicting, as real-world data changes, so should the model. This means we must keep track of new data changes and ensure the model learns accordingly.

9.3 DevOps Versus MLOps

The concept of DevOps was established much before MLOps [1]. MLOps is considered to be the DevOps for machine learning. Thus, it makes sense to draw some comparisons with DevOps when explaining MLOps. DevOps is a set of practices that combine software development (Dev) and IT operations (Ops). It aims to shorten the software development life cycle and provide continuous delivery with high software quality. It focuses on automation, continuous integration, and service delivery. It emphasizes collaboration between development and operations teams to shorten the development, testing, and release cycle.

MLOps is a set of practices that facilitate collaboration and communication among data scientists, Machine Learning Engineers, and operations professionals. It aims to streamline the machine learning life cycle from development to deployment and maintenance. MLOps encompasses the entire machine learning process, from data preparation and model training to version control, deployment, and monitoring. It emphasizes automation in model training, validation, and deployment processes.

In summary, while MLOps and DevOps emphasize automation and continuous processes, MLOps specifically addresses the challenges associated with machine learning models, including version control of models, datasets, and experiments, as well as monitoring models in production. DevOps, on the other hand, is broader in scope and focuses on the general software development processes and operations.

9.4 MLOps Best Practices

A continuous improvement of posture is critical to creating a good MLOps environment. Remember that the forces defining the problem require many trade-off decisions. More formal means safer to operate, while more flexible means more agile. Data Scientists, ML Engineers, and Operators have very different priorities. The goal is to find a sweet spot that is compatible with the system being built while providing good value to each of these stakeholders. A medical prediction system will be more rigorous than an advertisement prediction system, and the professionals working on these systems will have different approaches.

As noted earlier, efforts should begin with automation. A significant concern should be the capability to document, deploy, and revert to previous versions. Another essential feature that is often overlooked is the ability to migrate code from development to production with minimal changes. Docker has been a tool of choice to handle this problem, as you can build "images" that accept parameters [1]. Only the parameter values change. So, the same artifact runs on both environments, reducing the chances for errors.

9.5 AI/Machine Learning Project Lifecycle

Model development typically follows a sequence of actions, from collecting data to deploying software to solve a problem using AI techniques. ML Ops efforts focus on automation since it is the basis for reproducibility. We utilize a workflow engine to coordinate the execution of each step. The workflow defines the order in which steps must be executed and what steps can run in parallel.

The Model Training Lifecycle (Fig. 9.2)
Building machine learning products or ML-assisted product features involves two distinct disciplines:

- Model Development: During this process, data scientists, highly skilled in statistics, linear algebra, and calculus, train, evaluate, and select the best-performing statistical or neural network model.
- Model Deployment: During this process, developers, highly skilled in software design and engineering, build a robust software system, deploy it, and scale it to serve a vast number of concurrent model inference requests.

Figure 9.2 illustrates the primary components of a typical workflow. Please note that each block can comprise several internal steps, which are represented in the workflow.

9.5.1 Data Collection

At this stage, we develop systems to collect anonymized data in its raw format, which is available from external systems, log files, customers, partners, and other sources. Our focus here is to capture the original data. We want to preserve the original data because if errors occur in the processing downstream, it's possible to fix the program and re-run the transformations using the original data.

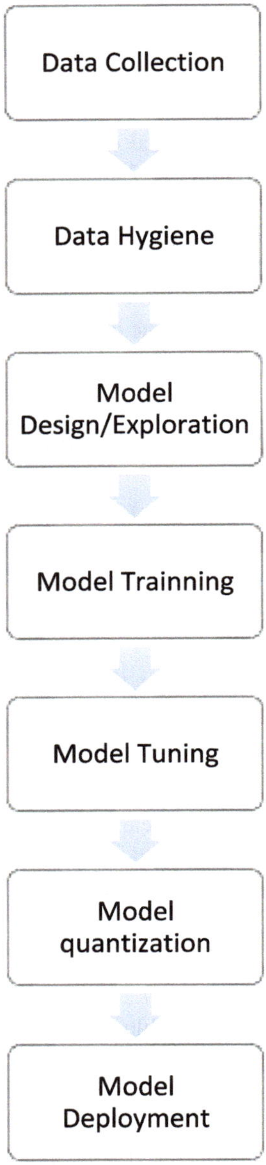

Fig. 9.2 Model training lifecycle

In some cases, external data will be available for a short time, so it's critical to have systems continuously ingest data. Additionally, it would be beneficial to have alarms that detect delays, discrepancies, and errors during the ingestion process. With the advent of cloud computing, a common practice is to ingest data continuously and later develop a policy for data deletion.

Another important consideration is ensuring any personally identifiable information (PII) complies with local regulations [1]. Sometimes, it's appropriate to "mask" certain fields and never store those that are unnecessary for the application. Typically, access to raw data is restricted to a limited number of systems and users, and we can expand access after the hygiene stage.

9.5.2 *Data Hygiene*

This step normalizes the data. Examples could include normalizing image sizes and cleaning timestamps to calculate UTC and local timestamps, among others. Local timestamps can serve as input for feature engineering and produce features such as "is working hours," "is weekend," and "is morning," among others.

In some cases, it includes generating synthetic data. For example, we could apply image transformations such as zooming in, zooming out, image rotation, and changes in lighting, among others. These operations enable us to generate valid input images from a single image, thereby creating a larger and more diverse dataset for training. This stage involves data cleaning for the training data, but it's essential to note that some components will be reused when making predictions or classifications.

This stage can sometimes include producing different datasets from the same data, where one dataset has fewer fields and is optimized for fast production processing. The original dataset will contain all fields and be made available for exploratory data analysis to a select group of users.

9.5.3 *Model Design/Exploration*

In this step, we start making assumptions about the model. In the context of deep learning, we would define the initial values for the number of layers and the size of each layer and provide more details about the neural architecture. A development environment must be highly agile, enabling data scientists to replace models as needed with minimal effort.

The model design, training, and tuning stages usually require many adjustments. The ML/AI engineers need to focus on allowing a great deal of flexibility here while preserving all the requirements to automate these stages in production. An important component here is the definition of inputs and outputs. Once these "contracts" are clearly defined, we can update software versions without breaking their communication with other stages or systems.

9.5.4 *Model Training*

Here, we focus on training the models and performing the initial verification to ensure the training step runs without errors. We also continue to add versioning for the configuration. Versioning is crucial during the tuning step, as we will reuse versions to create model variations, thereby keeping a documented history of all changes.

When we refer to "versioning," we capture all model configurations and store them in a system where they can be retrieved by name. For example, one can create a model named "model-tiger" and add a version named "dev.1.0.1.rc01." We can make a few changes as the project progresses and store another version, like "dev.1.0.1.rc02." This process continues until we create a "release" version and name it "rel.1.0.1.rc01." At any point in time, we can easily compare model configurations and switch between them with little effort.

9.5.5 *Model Tuning and Testing*

Tuning refers to adjusting model hyperparameters to enhance accuracy. There are sophisticated tools that help execute the tuning of model hyperparameters, aiming to optimize the combinations used; however, this process is inherently a trial-and-error approach. It is computationally intense, as for each adjustment, we need to "train" the model and then validate its performance.

After tuning the model, the best-performing candidates will be versioned and named with a "release" prefix to separate them from all the other versions created during development. In the case of deep learning models, the tuning phase can include work related to model quantization. A quantized model executes operations on tensors with reduced precision rather than full precision values [1]. For example, instead of using 8-byte floating-point values, we can adjust values to fit into 4-byte values. This enables a more compact model representation that less powerful hardware platforms can accommodate. After completing model tuning, it is sent for testing with predefined performance metrics and previously unseen test data.

9.5.5.1 The Model Execution

Figure 9.3 shows the main components of a model execution setup. There is a natural sequence from Provisioning, Model Deployment, App Deployment, and Monitoring during the first execution. After the initial setup, the most common sequence is Model Deployment and, optionally, Serving-App deployment. Regular model updates only require model deployment, as the rest of the infrastructure should use the new model once it is deployed.

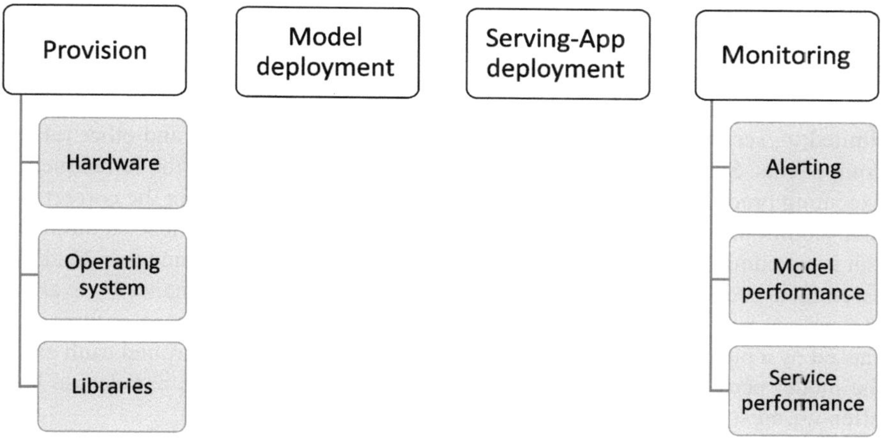

Fig. 9.3 Model deployment process

9.5.5.2 Provision

The provision stage refers to the steps to provision the hardware where the model will be deployed. In some cases, we have control of the hardware used; in other cases, the models are deployed to remote devices (like set-top boxes, TVs, IoT devices, etc.) Once the hardware is provisioned, the next step is ensuring the operating system is ready to install the core libraries. (For instance, TensorFlow, PyTorch, etc.). After the core libraries are installed, the system is prepared to perform model deployment.

9.5.5.3 Model Deployment

This step involves copying the versioned artifacts of the trained model into the service hosts.

9.5.5.4 Serving Application Deployment

Model artifacts alone cannot "serve" the model via a remote call. Some core libraries, such as TensorFlow and PyTorch, provide "server" components to handle remote calls by collecting the request, executing the model, and returning the results. Sometimes, the system includes a custom app to perform part of the server, log results, and allow other tools to track model performance.

9.5.5.5 Monitoring

Complex systems with many moving parts require a sophisticated system to "see" what is happening with their various components. This tooling includes, but is not limited to, service dashboards, log search tools, sensors, alarms, and other related components. Some examples of monitoring involve ensuring that the infrastructure executing predictions (or other actions) is live and returns results in the correct format. Additionally, monitoring measures the processing time of model execution and can issue commands to adjust the infrastructure in response to changes in load. An essential aspect of service monitoring is measuring model performance by examining various metrics, such as average model accuracy. These changes in accuracy are caused by a phenomenon called "model drift," where changes in ground truth cause the model performance to decay. By monitoring model drift, we can decide how often we must "retrain" it with new data.

9.5.5.6 Maintenance

Rarely will an AI/ML system not need adjustments after deployment. Model maintenance refers to all activities related to model updates, including the deployment of retrained models, cost reduction measures, and enhancements to computing performance.

9.6 MLOps General Concepts

Borrowing from the software development ops, MLOps focuses on optimizing processes using the following concepts.

9.6.1 Version Control

Versioning artifacts being produced enables the easy selection of what to deploy and remove from production. As we have seen in the previous session, versioning enables teams to track model changes and is crucial for deploying automation. With the artifacts used versioned, model deployment systems have a clear input of what needs to be deployed.

9.6.2 *Automation*

Besides "timesaving," an important benefit of automation is consistency. Consistency is critical to any complex system to allow engineers from other teams to manage. These engineers are specialized in keeping the systems running and having a formal "operation manual" is essential to a reliable operation. When we automate, we reduce this operation manual to a few steps. All the operation details are hidden and represented by automation scripts.

9.6.3 *Idempotency*

One important principle in automation is to allow the execution of the same step multiple times and have the system state be the same as if we executed a single time. This capability reduces the variations of deployment states and simplifies operations. For example, a step deploys a model configuration into production based on a model name/version. If an operator deploys the model "rel.1.0.2" twice, the final system state must be the same as if I deployed once.

An operator should not need to know the historic system state, just what state we want the system to be. For instance, we want to enable "rel.1.0.2."

9.6.4 *Deployment Strategies*

There are several ways to deploy applications and models. Here are a few deployment strategies and their summary.

9.6.4.1 Blue/Green

Let us call blue the current version of the live app. We can deploy the new version in parallel where both versions are available. We can then start moving requests from the old version to the new. We turn off the old version when traffic flows to the new version (Fig. 9.4).

9.6.4.2 Red/Black

It uses the same process as the Blue/Green, adding a constraint that only one of the versions receives requests. Meaning the move all requests to the new version in one step (Fig. 9.5).

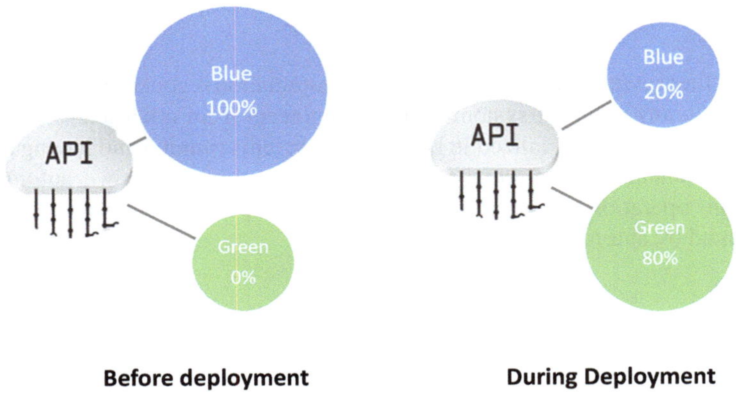

Fig. 9.4 Blue/green deployment

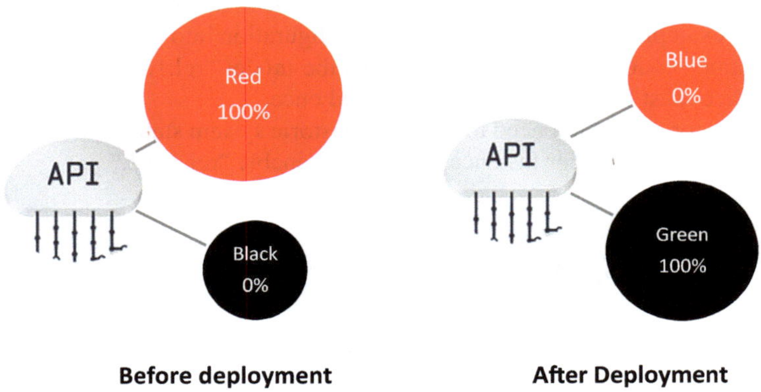

Fig. 9.5 Red/black deployment

9.7 The Google MLOps Levels

MLOps is all about automating the various tasks involved in building, testing, operating and maintaining a machine learning model. According to Google, there are three levels of automation depending on problem complexity and the ability of companies to manage complexity of operations:

1. MLOps level 0 (manual process)
2. MLOps level 1 (ML pipeline automation)
3. MLOps level 2 (CI/CD pipeline automation)

Below is a summary of each level.

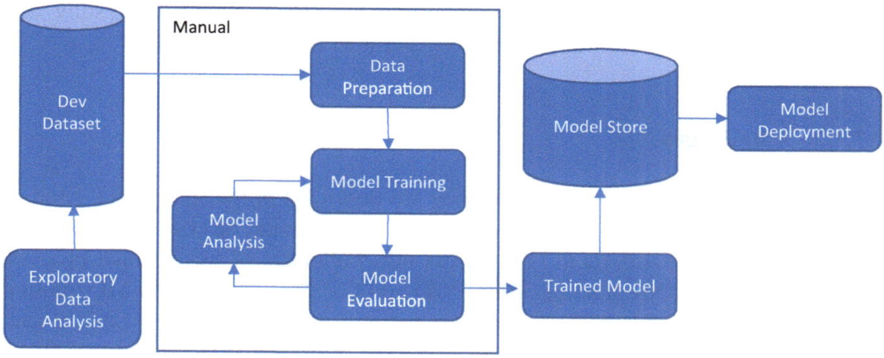

Fig. 9.6 Components of a level 0 automation

9.7.1 MLOps Level 0 (All Manual)

This is typical for companies that are just starting with ML, and it involves an entirely manual ML workflow. The data-scientist-driven process might be sufficient if models are rarely changed or retrained. Various components of level 0 are shown in Fig. 9.6.

9.7.1.1 Characteristics

- **Manual, Script-Driven, and Interactive Process**: Every step is manual, including data analysis, data preparation, model training, and validation. It requires manual execution of each step and manual transition from one step to another.
- **Disconnect Between ML and Operations**: The process separates data scientists who create the model, and engineers who serve the model as a prediction service. The data scientist hands over a trained model as an artifact for the engineering team to deploy their API infrastructure.
- **Infrequent Release Iterations**: The assumption is that your data science team manages a few models that don't change frequently, either changing model implementation or retraining the model with new data. A new model version is deployed only a couple of times per year.
- **No Continuous Integration (CI)**: Because few implementation changes are assumed, you ignore CI. Usually, testing the code is part of the notebooks or script execution.
- **No Continuous Deployment (CD)**: Since there isn't frequent model version deployment, CD isn't considered.
- **Deployment refers to the prediction service** (i.e., a microservice with REST API).
- **Lack of Active Performance Monitoring**: The process doesn't track or log model predictions and actions.

The engineering team may have a complex setup for API configuration, testing, and deployment, including security, regression, and load and canary testing.

9.7.1.2 Challenges

In practice, models often break when they are deployed in the real world. Models fail to adapt to changes in the dynamics of the environment or changes in the data that describes the environment. To address the challenges of this manual process, it's good to use MLOps practices for CI/CD and CT. By deploying an ML training pipeline, one can enable CT, an done can set up a CI/CD system to rapidly test, build, and deploy new implementations of the ML pipelines.

9.7.2 MLOps Level 1

As we strive to achieve more automation, we move towards a semi-automatic deployment of what is called level 1 automation. The goal of MLOps level 1, as shown in Fig. 9.7, is to perform continuous training (CT) of the model by automating the ML pipeline. This way, one achieves continuous delivery of the model prediction service. This scenario may be helpful for solutions that operate in constantly changing environments and need to proactively address shifts in customer behavior, price rates, and other indicators.

Level 1 automation relieves us of the burden of manually training the model every day. An automation pipeline is now in place that validates the data, prepares it for training, and generates a training model. It also attempts to choose the best model by comparing multiple error metrics and choosing the one with the lowest error between the training and test datasets. The pipeline takes care of it all. Note that in Fig. 9.8, we refer to this automation module as "Orchestrated Experiment." This will be used later in the next level. An individual data scientist or machine

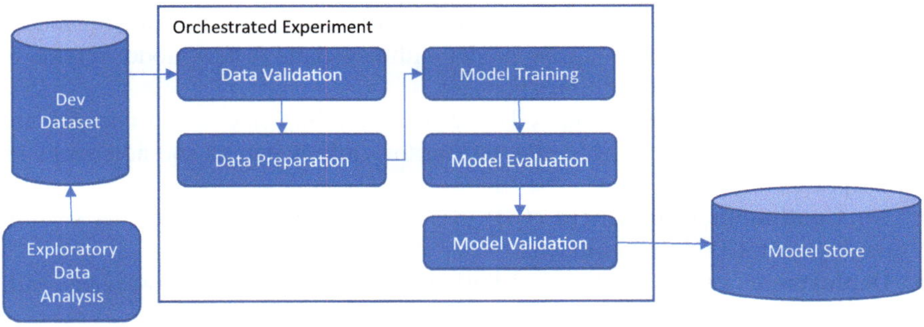

Fig. 9.7 Components of a level 1 automation

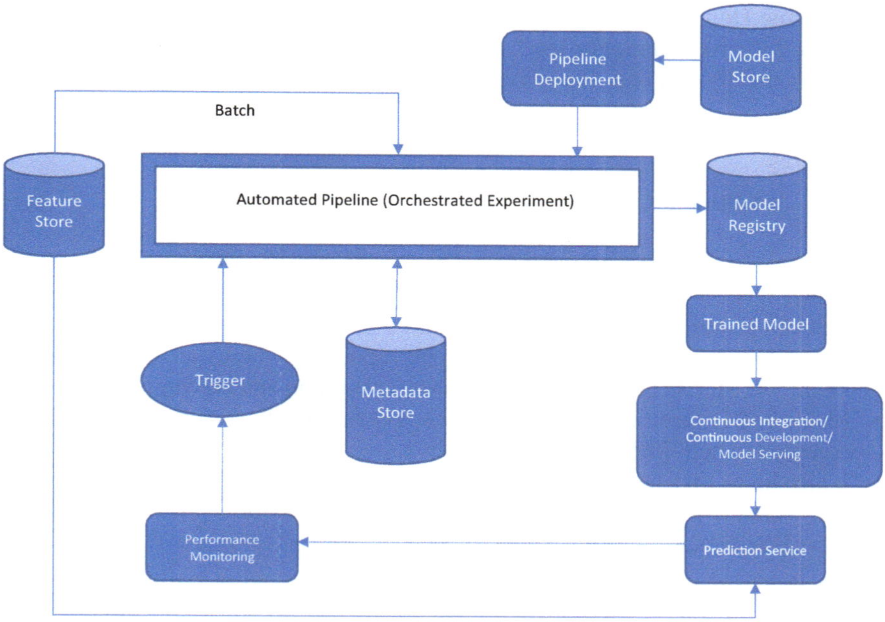

Fig. 9.8 Components of level 2 automation

learning engineer can achieve this degree of automation. Most companies can achieve this level. It is good enough when testing the model in a development environment.

9.7.2.1 Characteristics

- **Rapid Experiment**: ML experiment steps are orchestrated and done automatically.
- **CT of the Model in Production**: The model is automatically trained in production, using fresh data based on live pipeline triggers.
- **Experimental-Operational Symmetry**: The pipeline implementation used in the development or experiment environment is used in the preproduction and production environment, which is a key aspect of MLOps practice for unifying DevOps.
- **Modularize Code for Components and Pipelines**: to construct ML pipelines, components need to be reusable, composable, and potentially shareable across ML pipelines.
- **Continuous Delivery of Models**: The model deployment step, which serves the trained and validated model as a prediction service for online predictions, is automated.

- **Pipeline Deployment**: In level 0, one deploys a trained model as a prediction service to production. For level 1, one deploys a whole training pipeline, which automatically and recurrently runs to serve the trained model as the prediction service.

9.7.2.2 Additional Components

- Data and Model Validation: The pipeline expects new, live data to produce a new model version that is trained on the new data. Therefore, automated data validation and model validation steps are required in the production pipeline.
- Feature Store: a feature store is centralized repository where one standardizes the definition, storage, and access of features for training and serving.
- Metadata Management: Information about each execution of the ML pipeline is recorded to help with data and artifact lineage, reproducibility, and comparisons. It also helps to debug errors and anomalies.
- ML Pipeline Triggers: one can automate ML production pipelines to retrain models with new data, depending on the use case:
  - On-demand
  - On a schedule
  - On the availability of new training data
  - On model performance degradation
  - On significant changes in the data distribution (evolving data profiles)

9.7.2.3 Challenges

This set up is suitable when one deploys new models based on new data, rather than based on new ML ideas. However, one needs to try new ML ideas and rapidly deploy new implementations of ML components. If one manages multiple ML pipelines in production, a CI/CD setup is necessary to automate the build, testing, and deployment of these pipelines.

9.7.3 MLOps Level 2

For a rapid and reliable update of pipelines in production, a robust automated CI/CD system is necessary. With this automated CI/CD system, data scientists rapidly explore new ideas around feature engineering, model architecture, and hyperparameters. This level suits tech-driven companies that must retrain their models daily, if not hourly, update them in minutes, and deploy them simultaneously on thousands of servers. Without an end-to-end MLOps cycle, such organizations won't survive. This MLOps setup includes the following components:

- Source control
- Test and build services
- Deployment services
- Model registry
- Feature store
- ML metadata store
- ML pipeline orchestrator

9.7.3.1 Characteristics

- **Development and Experimentation**: One iteratively tries out new ML algorithms and new modeling, where the experiment steps are orchestrated. The output of this stage is the source code of ML pipeline steps, which are then pushed to the source repository.
- **Pipeline Continuous Integration**: One builds source code and runs various tests. The outputs of this stage are pipeline components (packages, executables, and artifacts) to be deployed in a later stage.
- **Pipeline Continuous Delivery** involves deploying the artifacts produced by the CI stage to the target environment. The output of this stage is a deployed pipeline with the new implementation of the model.
- **Automated Triggering**: The pipeline is automatically executed on production based on a schedule or in response to a trigger. The output of this stage is a newly trained model that is pushed to the model registry.
- **Model Continuous Delivery**: One serves the trained model as a prediction service for predictions. The output of this stage is a deployed model prediction service.
- **Monitoring**: One collects statistics on model performance based on live data. The output of this stage is a trigger to execute the pipeline or to execute a new experiment cycle.

The data analysis step is still a manual process for data scientists before the pipeline starts a new iteration of the experiment. The model analysis is also a manual process. For more details, readers are referred to https://neptune.ai/blog/mlops.

9.8 How to Automate ML Workflow?

Automation is a key aspect of MLOPs, helping to streamline the development, deployment, and maintenance of machine learning models. There are several benefits of automation in MLOps, including:

- Increased Efficiency: Automating repetitive tasks can save time and reduce the chances of human error.

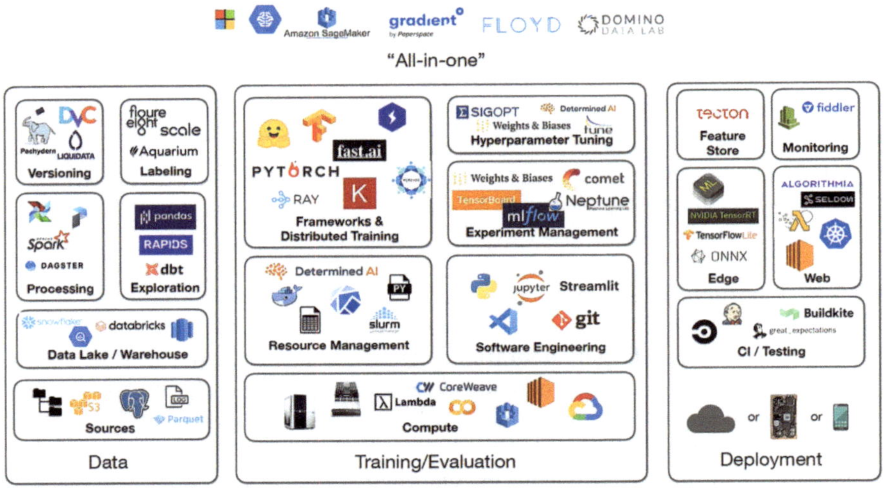

Fig. 9.9 Tech stack for learning and implementing MLOps. (Source: https://fullstackdeeplearning.com/spring2021/lecture-6/)

- Improved Reproducibility: Automated workflows facilitate the reproduction of results, which is crucial for debugging and troubleshooting.
- Increased Scalability: Automation enables the simultaneous training and deployment of multiple models, which is crucial for companies handling large amounts of data.

There are several popular tools and techniques for automating the ML workflow, including:

- Python Scripts: Python is a popular programming language for machine learning, and it can be used to automate tasks such as data preparation, model training, and evaluation.
- Kubernetes: Kubernetes is a platform that can be used to automate the deployment of learning models. It enables the scaling and management of resources required for training and deploying models. It can be used to automate. The scaling of resources for training and deploying models and using Python scripts to automate tasks, e.g., data cleaning and model evaluation (Fig. 9.9).

It is worth noting that while automation can bring numerous benefits, it requires careful planning and management to ensure that the automation is executed correctly and that the correct version of the model is deployed.

9.8.1 Ethical Considerations

Ethical considerations must be integral to the entire machine-learning lifecycle to ensure that AI technologies benefit society while minimizing harm and promoting fairness. Addressing ethical issues requires a multi-faceted approach that involves data collection, algorithm development, transparency, accountability, and ongoing monitoring.

9.8.2 Fairness and Bias

Bias in machine learning models can lead to discriminatory outcomes. Algorithms may inadvertently discriminate against certain groups based on factors such as race, gender, age, or other protected characteristics. This can perpetuate existing societal inequalities and undermine fairness. Biased data is a normal state in nature; continuously monitoring data bias helps detect problems early on, before we perform all lifecycle steps.

9.8.3 Customer Data and Privacy

When personal data is used to train machine learning models, individuals should be informed about how their data will be used and have the option to consent or decline the provider's right to use their data. It's good practice to clearly state how we would use the data and how long. Protecting individuals' privacy is a fundamental ethical concern. Machine learning models that learn from personal data must ensure that this data is handled securely and used only for legitimate purposes. Additionally, users have the right to request that their data be deleted at any time.

9.8.4 Bias Amplification

Machine learning algorithms can, in some instances, amplify biases present in the data on which they are trained. A crucial consideration is to recognize that any data has some level of bias that needs to be controlled. This is particularly important when using historical data because it captures inequalities that our society aspires to eliminate.

9.9 A Use Case for an AI System in Production

This session will begin with a hypothetical AI problem or context and describe one approach to operating a production environment that requires the necessary infrastructure. We will assume that the production environment utilizes internet cloud providers such as Amazon Web Services, Google Cloud Platform, and Azure, among others. The changes required to host the proposed solution in an on-premises data center are relatively small if you can access the IT department personnel.

ABC Acme is developing a service that utilizes AI to generate comments comparing the prospects of two stocks. The service exposes an API over the internet and expects the ticker symbols for the two stocks as input and a text field describing what the system should focus on. For example, we can request a comparison between "AMD" and "IBM" and add a text like this: "focus on the possible outlooks considering the European market." The response would be a text elaborating and commenting on these two companies in the current marketplace. Subscriber companies will use this service to request and receive responses, expanding their systems with ABC Acme services.

The Application's High-Level View

Figure 9.10 shows the high-level view of the application comprising two major services:

(a) The "application endpoint" contains an application load balancer that receives requests from outside and redirects them to one of the pairs of web server/model processors.
(b) The "service pool" consists of "model processors" to compute the request and return the results to users. Before submitting a request to model processors, the webservers authenticate the request, ensuring it is valid.

In some cases, web servers and model processors run on the same host, while in other scenarios, these components run on distinct servers. It all depends on the

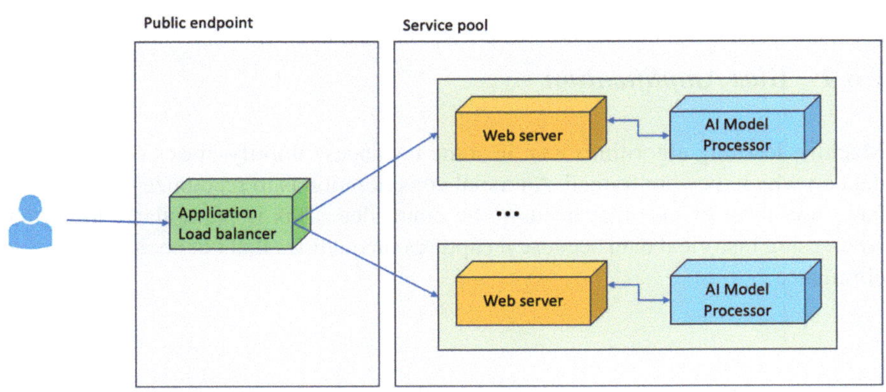

Fig. 9.10 High-level view of application architecture

processing and memory requirements for each task. The design above assumes that the computation of results takes less than a few seconds, so a "synchronous" connection is acceptable because requests will not be "blocked" waiting for long. In synchronous systems, requests "block" while waiting for responses, so the response time must be small, or the servers will become overwhelmed with many waiting requests.

In cases where the computing time is substantial, the best approach is to use messages where the "users" add messages to a queue and receive a "ticket number" representing the job. In the background, processes take items from the queue, process the job, and store results in a key-value store. Later, the user systems would "inquire" about the job completion using the ticket, and the result would be either a "job is not yet" message or a "job done" message accompanied by the computed data.

9.9.1 Development, Validation, and Production Environments

During the project's development, we need to accommodate the flexibility data scientists require while protecting the live system against bugs. Data Scientists will run experiments in the development environment. This environment allows models to be trained and is also isolated from production, so any mistakes made in development do not directly affect production. Deploying updates to the production environment follows a "formal" process with controls to ensure no errors can disrupt the live service. For medical device development, a validation environment will be established for validation purposes.

Figure 9.11 shows "two" tracks:

- The development environment
- The production environment

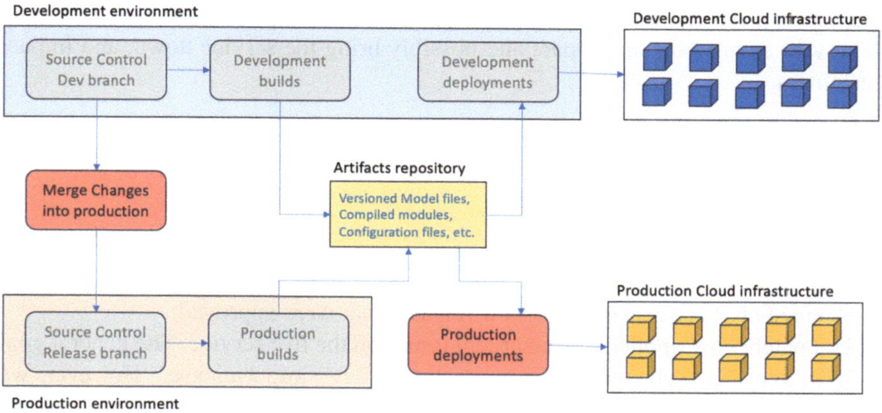

Fig. 9.11 Deployment environments

9.9.1.1 Development Environment

Following the development environment, we see three blocks:

- Source control
- Builds
- Deployments

Source Control

In source control, we use "branches" to keep changes isolated by groups. The "dev" branch is where currently developed features will be stored.

Builds

The "builds" represent the steps that assemble the files needed to run the service. The artifacts required to run the service are versioned and stored in the "Artifacts Repository." Think of the artifact repository as source control for large binary files.

Deployments

The "deployments" sequence copies the required artifacts from the repository to the servers, enabling the new software to run on the target environment. After the "development deployment" step, we can test the whole system running in "dev."

9.9.1.2 Production Environment

The production environment has a more formal control of what changes are being deployed, as one bug can impact and possibly bring the service down, and impact customers.

Source Control

The "release" branch contains the current state of the software as it is live. Typically, there are two types of additions to it:

- Planned changes consisting of new releases that were approved.
- Hot patches are applied during urgent issues on the live service. The general goal is to avoid any changes not coming from "dev" into "release." However, we directly update the release branch during critical moments to address pressing problems. Later, these changes are retrofitted into the "dev" branch.

Besides these urgent cases, all changes made into "release" have already been approved and tested in dev. This process is called "merging" one branch into another.

Builds

In production, the versioning of artifacts built follows a more controlled process, as any artifact built in production can be deployed there.

Deployments

Production deployments are automated processes, but are typically monitored a lot more by the operations team. These deployments are generally closely monitored until the team observes that metrics have returned to normal for an extended period.

9.9.2 *Data Workflows*

In the case of traditional systems, what we have described so far represents the complete end-to-end process. However, AI systems require substantial work related to training the models. In addition to the previous steps, we need to manage and pre-process the data used for training. Figure 9.12 shows the simplified version of the "model training" and "data" workflows:

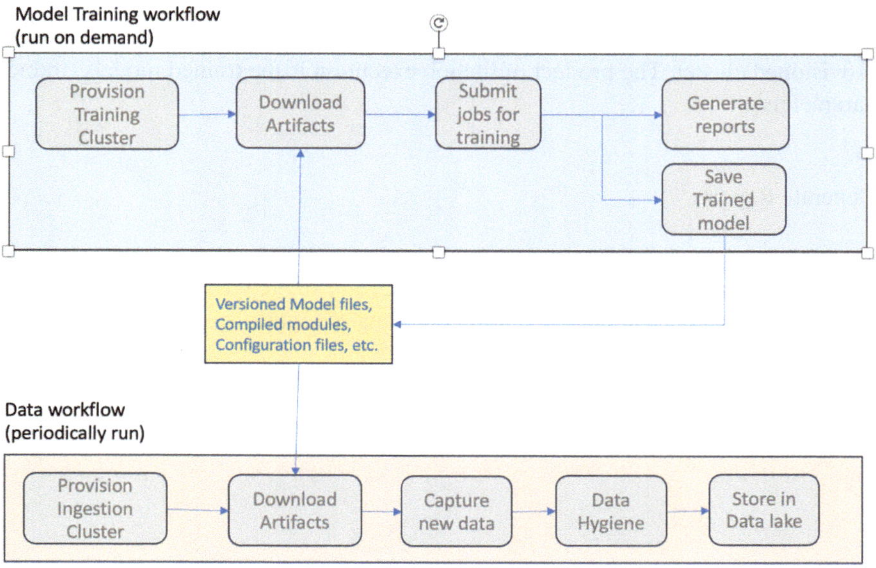

Fig. 9.12 Data and model training workflows

9.9.2.1 Model Training Workflow

This workflow defines the sequence of activities needed to create a new version of a trained model. In environments with company data centers, cluster provisioning is a straightforward step to reconnect to an existing cluster. In the following sections, we provide a detailed description of each step.

Cluster Provision

At the top, we can see the first step of the model training workflow, "provisioning a cluster" of computers to use during training. After this step is completed, a computing infrastructure will be available to process the "training tasks." Because the provisioning is an independent step, we can adjust their specifications without impacting the next steps.

Artifact Download

In this step, the previously built versioned software artifacts are "downloaded" into the cluster hosts. These can include deep learning architecture code, training code, and other relevant components. Note that the trained model will not be included, as the training workflow will generate it.

Job Submission

In this step, we prepare the training job request and submit it for execution on the provisioned cluster. The product of the job execution is the trained models (trained parameters).

Generate Reports

Once we have the trained model, we can generate additional tests and validations, produce detailed reports on model performance, and so on. We should be able to run these reports with custom input, allowing Data Scientists to explore the current results while preparing adjustments to the model.

Export Trained Model

All the trained model data is stored inside the provisioned cluster. We must export the data to persistent storage before destroying the cluster provisioned in the first step. Since we are executing jobs on the cloud, the target location for storing the trained models will be a distributed file store, such as AWS S3.

Model Execution

The training workflow typically executes on demand, where a Data Scientist or ML Engineer triggers the process. This is particularly true for computationally intensive training, such as deep learning and generative AI. In cases where the data changes frequently and the cost of model training is relatively low, the training workflow could be automatically triggered. The frequency of the training runs depends on the extent of business impact that model drift would cause and the computing cost required to execute the job.

9.9.3 *Data Workflows*

Data workflows capture the need to collect new data, preprocess it, and store it after cleanup so that the data can be used directly by the model training workflow.

The two concerns at this stage are:

1. Make sure no data is lost.
2. Reduce costs by avoiding the need to recompute pre-processing multiple times.

The data storage required to perform the jobs at training and data is not represented in the diagrams for the sake of simplicity. Since we assume the architecture is deployed in the cloud, we can utilize the cloud provider's distributed storage service (such as S3 in Amazon Web Services).

Data workflows typically run periodically, such as daily and hourly. In this use case, the data workflows could capture new articles about stock prices and other relevant information. We could run the steps daily and continuously store the data, but an essential factor in deciding how frequently we should run is the information about how frequently we will train the model. In this case, we can safely run the data workflows on a weekly basis.

9.10 Tools Used in MLOps

Despite being beyond the scope of this book, it is helpful to have a basic understanding of what each tool does. We are including a short description of a few of them to give the reader a taste of what they do.

9.10.1 Docker

Docker is a tool for building Containers, which are lightweight, standalone packages containing everything needed to run an application (code, runtime, system tools, libraries, and settings). It uses layers of file systems, where the tool adds a new layer that includes just a few changes if you want to modify a set of files.

For example, if I use Ubuntu Linux and want to copy my files into it. Docker uses the original Ubuntu layer file and adds a layer on top of it with my files. This translates into very fast builds as the files for these layers can be reused and cached.

The tool uses a text file to describe a server.

Here is a simple Dockerfile example (Table 9.1):

The contents of the "Dockerfile" are described on the left, with comments on the right. With this file, we can use Docker to build custom "images" of operating systems that include our code.

From the top, we define what image to start with. In this case, we start with a fresh copy of Ubuntu Linux.

On lines 2 and 3, we state we want to copy files from the local computer from "/code/" and "/config/" directories into the image.

On lines 4 and 5, we execute commands during the image build.

Finally, we define on line 6 that when this image is deployed, the program located at "/app-code/my-program" should run.

We could add parameters to the "Dockerfile" so that Docker can build images for both development and production with the same code.

We can now contact Docker and request that they build the image. From the command line, we can run:

```
docker build -t model_abc
```

Table 9.1 A simple "Dockerfile" example

1	FROM ubuntu:22.04	# Start from the Ubuntu v22.04 image
2	COPY /code/ /app-code	# Add my code files to it
3	COPY /config/ /app-cfg	# Add my configuration files to it
4	RUN build-app /app-code	# Execute commands when building the image
5	RUN delete-code /app-code	#
6	CMD /app-code/my-program	# Define command to execute when running

Once we complete the image build step, we will have an artifact that can be used to launch "machines" of that type. A simple example of starting a server using the image is shown below:

```
docker run model_abc
```

Please note that Docker has an extensive set of options, and the examples presented here are the bare minimum. Please consult the Docker documentation for more details.

9.10.2 Kubernetes

Kubernetes is an open-source platform for automating the deployment, scaling, and management of application containers. It works with various container tools, including Docker, and enables running applications in a highly available and scalable manner without tying them to any specific hardware. In Kubernetes, the "pod" is the smallest deployable unit. A pod can contain a single server or multiple servers. For example, we can have a pod containing one server used to make model predictions. Then we ask Kubernetes to run five pods of that type. Now, imagine that we need access to a local database to make predictions. We can define a pod that contains both the prediction and database servers. Each database server will have the same data.

Kubernetes excels in scaling and failover recovery. The approach used in Kubernetes is to have the architects define the desired state, and the system will adjust to align the current state with the desired state. It offers a range of essential features for distributed systems, like service discovery and load balancing. If traffic to a container is too high, Kubernetes can distribute network traffic to stabilize the deployment. It also automates the rollouts and rollbacks of application updates, enabling continuous integration and delivery in software development.

Lastly, it monitors the health of nodes and containers. It executes self-healing by automatically restarting failed containers, replacing and rescheduling containers when nodes die, and killing containers that don't respond to user-defined health checks.

9.11 Most Popular Machine Learning Software Tools

There are several ML software products that are available. Table 9.2 shows the most popular ones among them.

Table 9.2 ML software tools

Name	Platform	Cost	Written in Language	Features
Scikit Learn	Linux, Mac, Windows	Free	Python, Cython, C, C++	Classification, regression, clustering, preprocessing etc.
PyTorch	Linux, Mac, Windows	Free	Python, C++, CUDA	Autograd module, nn module
TensorFlow	Linux, Mac, Windows	Free	Python, C++, CUDA	Deep learning, ML
WEKA	Linux, Mac, Windows	Free	JAVA	Classification, regression, clustering, preprocessing, etc.
KNIME	Linux, Mac, Windows	Free	JAVA	Supports text mining and image processing
Apache Mahout	Cross-platform	Free	JAVA Scala	
Keras.io	Cross platform	Free	Python	API for neural networks
Rapid Miner	Cross platform		JAVA	Data loading and transformation, etc.

9.12 Machine Learning Platforms

Data Science and Machine Learning Platforms provide users with tools to build, deploy, and monitor machine learning algorithms. These platforms combine intelligent decision-making algorithms with data, thereby enabling developers to create a business solution. The autonomous learning capabilities of AI and ML platforms are at the center of today's enterprises. Technology is increasingly being used to make important decisions and drive automation that improves enterprise operations across various disciplines. In recent years, ML technology has also formed the foundation for generative AI models, which are trained to generate new content through larger datasets and more complex ML algorithms. With its range of relevant business use cases in the modern enterprise, machine learning platform technology has quickly grown in popularity, and vendors have expanded these platforms, capabilities and offerings to meet growing demands. Some platforms offer prebuilt algorithms and simplified workflows with features such as drag-and-drop modeling and visual interfaces that easily connect necessary data to the end solution, while others require a greater understanding of development and coding. The nature of some Data Science and Machine Learning Platforms enables users without intensive data science skills to benefit from the platforms' features. This platform streamlines the process of implementing machine learning, offering an integrated environment for tasks such as data handling, algorithm selection, model training, validation, and deployment. It is like a one-stop shop for all machine learning needs! By leveraging the power of artificial intelligence and data analysis, machine learning platforms empower businesses to unlock valuable insights, automate processes, and make data-driven decisions like never before. Whether you're a small startup or a multinational corporation, these platforms offer a wealth of tools and resources to revolutionize your business operations and drive strategic initiatives. A data science platform is defined [2] as:

> A cohesive software application that offers a mixture of basic building blocks essential for creating all kinds of data science solutions, and for incorporating those solutions into business processes, surrounding infrastructure, and product.

Machine learning platforms are no longer the wave of the future. It is happening now. Developers need to know how and when to harness their power. Working within the ML landscape while using the right tools can make it easier for developers to create a productive algorithm that taps into its power.

9.12.1 Benefits of Machine Learning Platforms

- Data-driven business decisions.
- Improved products and services.
- Time and energy saved through automation.
- Shared insights, as users can share data, models, and related information with collaborative tools.
- Simplified, scalable data science via user-friendly features and out-of-the-box solutions.
- Optimized experimentation through data visualization, augmentation, and preparation tools.

9.12.2 Types of Machine Learning Platforms

Data scientists use a data science and machine-learning platform that enables them to work both online and offline. Machine learning platforms fall into one of two categories: cloud and on-premises.

Most machine learning platforms operate in the cloud, enabling flexible resource utilization and making on-site infrastructure unnecessary. With the introduction of cloud-based platforms, data scientists can now work with their data on any Internet-enabled device. They can also share components of their work with colleagues or collaborate securely on specific tasks. Besides having cloud features, the data science and machine-learning platform should also run faster to provide accurate results. Some machine learning platforms are housed on-premises, either at the business site or another location, usually selected for security or speed (latency) issues.

Several software vendors are currently unleashing their software products that match this description. However, not all the software products released by them are ideal for use in data-oriented organizations.

9.12.3 Components of a Machine Learning Platform

Machine learning platforms facilitate machine learning from end to end, enabling users to manage the entire data lifecycle, from data ingestion to inference. A good data science and machine-learning platform should provide data scientists with all the necessary building blocks for creating a solution to a data science problem. It should also provide the experts with an environment where they can incorporate the solutions into products and business processes. The platform must provide data scientists with all the necessary support when performing data and analytics tasks. These tasks encompass data access, data preparation, visualization, interactive exploration, deployment, and performance engineering. A few essential processes a machine learning platform should enable:

- Data management
- Data is the lifeblood of any machine learning project. The platform should have tools for data ingestion, preprocessing, transformation, and management.
- Feature engineering
- Transforming raw data into features that better present the underlying problem to the predictive models.
- Algorithm selection
- Different tasks require different algorithms. The platform should provide a wide range of prebuilt algorithms and allow for custom ones.
- Deployment and operation
- Finally, the platform should support model deployment and monitoring, ensuring optimal performance in real-world applications. It should have built-in support for frameworks such as Python, PyTorch, scikit-learn, or TensorFlow.

9.12.4 Popular Machine Learning Platforms

In this section, we will discuss some of the well-known and proven platforms available.

9.12.4.1 Alteryx Analytics

Alteryx Analytics [3] provides a machine-learning platform for building models in a workflow. Alteryx's product vision aims at helping companies cultivate a data analytics culture without necessarily hiring data scientists. The platform can discover, prepare, and analyze all the data, then deploy and share analytics at scale for faster, deeper insights. Data discovery and data security are made breathtakingly easy with Alteryx. One can create a culture of collaboration, sharing, and innovation by extending tribal knowledge across the organization. It can solve even the most complex business analytics problems with less time and effort. Alteryx empowers data scientists to break through data barriers, deliver insights, and experience the thrill of getting answers faster. Business analysts and data scientists can discover, transform, model, and analyze data using a single governed, collaborative, and scalable enterprise analytics solution. For a complete list of system requirements and supported data sources, visit [3].

9.12.4.2 H_2O.ai

The company offers H_2O deep water for deep learning and H_2O sparkling water for those interested in spark integration. The platform, H_2O driverless AI, uses AI to do AI to make it easier, faster, and cheaper to deliver expert data science as a force multiplier for every enterprise. H_2O.ai is designed for use with the Python, R, and Java programming languages. Available on Mac, Windows, and Linux, H_2O provides developers with the tools they need to analyze data sets in the Apache Hadoop file systems as well as those in the cloud [4].

9.12.4.3 KNIME Analytics Platform

KNIME [5] is an open-source platform that enterprises can utilize to enhance their performance, security, and collaboration. Cloud versions are available on Microsoft Azure and Amazon AWS. It can blend the data from any source (text formats, databases, and Datawarehouse, and from sources such as Twitter, AWS S3, Google Sheets, and Microsoft Azure.

9.12.4.4 RapidMiner

The RapidMiner platform [6] includes RapidMiner Radoop for extending the platform's execution capabilities to a Hadoop environment, RapidMiner Studio for model development, and RapidMiner Server, which enables data scientists to share, collaborate on, and maintain models. RapidMiner excels in introducing new performance and productivity capabilities to model development and execution.

9.12.4.5 Databricks Unified Analytics Platform

Databricks' Apache Spark-based Unified Analytics platform [7] offers features for real-time enablement, performance, operations, reliability, and security on AWS. The Apache Spark-based platform combines data engineering and data science capabilities, utilizing a variety of open-source languages. Apache Spark MLlib features an algorithms database with a focus on clustering, collaborative filtering, classification, and regression. Developers can find Singa, an open-source framework, that contains a programming tool that can be used across numerous machines and their deep learning networks. Azure Databricks is an integrated service within Microsoft Azure that provides a high-performance Apache Spark-based platform optimized for Azure.

9.12.4.6 Microsoft's Azure Machine Learning Studio

Microsoft provides the solution for data science and ML through its Azure software products [8]. These products include Azure Machine Learning, Power BI, Azure Data Lake, Azure HDInsight, Azure Stream Analytics, and Azure Data Factory. Its cloud-based Azure Machine Learning Studio is ideal for data scientists who want to build tests and execute predictive analytics solutions on their data. The cloud platform also offers advantages in terms of performance tuning, scalability, and agile support for open-source technology.

9.12.4.7 Google's Analytics Platform

Google's core ML platform includes Cloud ML Engine, Cloud AutoML, TensorFlow, and BigQuery [9]. Its ML components require other Google components for end-to-end capabilities, such as Google Cloud Dataprep, Google Datalab, Google Cloud Dataproc, and Google Kubernetes Engine, etc. Most of these components require access to the Google Cloud Platform. Google offers a rich ecosystem of AI products and solutions, ranging from hardware (Tensor Processing Unit [TPU]) and crowdsourcing (Kaggle) to world-class ML components for processing unstructured data like images, video, and text. Aided by a plethora of online resources, documentation, and tutorials, TensorFlow provides a library that contains data flow graphs in the form of numerical computation. The purpose of this approach is to enable developers to deploy deep learning frameworks across multiple devices, including mobile devices, tablets, and desktops. Historically, TensorFlow was aimed at "democratizing" machine Learning. It was the first platform that made ML simple, visual, and accessible to this degree. The most significant selling point of TensorFlow is Keras, a library that enables efficient programming of Neural Networks.

9.12.4.8 IBM Watson

IBM's Watson platform is where both business users and developers can find a range of AI tools. Users of the platform can build virtual agents, cognitive search engines, and chatbots using starter kits, sample code, and other tools accessible via open APIs [10].

9.12.4.9 Amazon Web Services (AWS)

Amazon Web Services include Amazon Lex, Amazon Rekognition Image, and Amazon Polly [11]. Each is used in a different way by developers to create ML tools. For example, Amazon Polly utilizes AI to automate the process of converting voice to written text. Similarly, Amazon Lex forms the basis of the brand's chatbots, which are used in conjunction with its personal assistant, Alexa. Amazon offers many more AI services, and one could spend the entire day browsing through them. It's challenging to find a comprehensive overview of all these services in one place on the AWS docs, but if you look, it lists them under "AI Services."

There is no shortage of AI and ML platforms today. So, which platform is the best? Unfortunately, there is no clear answer, as these services are tied to a particular technology stack or ecosystem. The other, more important reason is that by now, AI and ML technologies have been commoditized, and there is a race to provide as many features as low a price as possible. No vendor can afford to offer what others are offering, and any new offering is copied and served by competitors almost immediately. As such, it all comes down to what stack one has access to and what the goals are. Additionally, it is essential to consider the perceptions of the companies behind it intuitively.

9.13 Predetermined Change Control Plan

Artificial intelligence software in a production environment requires the adoption of modern software development practices, including scalability, extensibility, configuration, modularity, testability, monitoring, and reproducibility. As illustrated in Fig. 9.13, model monitoring is crucial in ensuring that a machine learning system operates within its specified parameters. Achieving this requires continuous monitoring and the frequent release of updated machine learning models to address emerging issues. However, in regulated environments, these frequent updates can become challenging due to global post-production change regulations, which impose additional burdens on the deployment process.

The dynamic nature of AI-enabled devices necessitates more frequent software updates than non-AI medical device software. To address this, the U.S. FDA introduced the Predetermined Change Control Plan (PCCP) [13] in December 2024, which provides a streamlined approach for manufacturers to implement iterative improvements to AI-enabled device software functions (AI-DSF) while maintaining safety and effectiveness. This approach, aligned with the FDA's "least burdensome" principles, allows manufacturers to predefine potential modifications in their initial marketing submissions. Once approved, these changes can be made without requiring additional marketing submissions, reducing regulatory hurdles. By reviewing the PCCP as part of the device's initial marketing submission, the FDA ensures that safety and effectiveness are upheld throughout the device's lifecycle, even as modifications are implemented. PCCP is discussed in detail in Chap. 10 of this book.

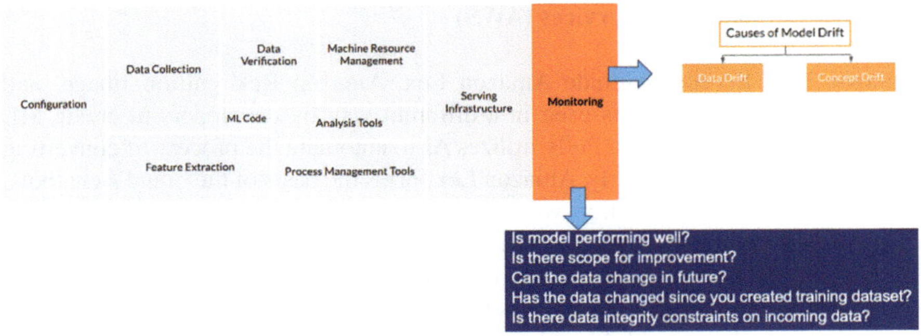

Fig. 9.13 Importance of monitoring in machine learning system [12]

9.14 Summary

In conclusion, we can say that MLOps is the intersection of people, process, and platform for gaining business value for machine learning. It streamlines development and deployment by monitoring, validating, and governing machine learning models. Applying these practices increases the quality, simplifies the management process, and automates the deployment of machine learning models in large-scale production environments.

References

1. Sehgal NK, Bhatt PC, Acken JM (2020) Cloud computing with security and scalability. Springer
2. https://www.ml4devs.com/articles/mlops-machine-learning-life-cycle/
3. https://www.alteryx.com/
4. https://h2o.ai/
5. https://www.knime.com/
6. https://rapidminer.com/
7. https://www.databricks.com
8. https://azure.microsoft.com
9. https://analytics.withgoogle.com/
10. https://www.ibm.com/watson?utm_content=SRCWW&p1=Search&p4=43700074359379220&p5=e&gclid=EAIaIQobChMIp4aCi5bHgwMVpAetBh3WcQm9EAAYASAAEgKraPD_BwE&gclsrc=aw.ds
11. https://aws.amazon.com/machine-learning
12. Skully D et al (2015) Hidden technical debt in machine learning system
13. US FDA Guidance (2024) Predetermined change control plan for AI/ML-enabled medical devices

Chapter 10
Regulatory Requirements and Risk Management for AI/ML-Enabled Medical Devices

Abstract This chapter provides a comprehensive examination of the latest regulatory developments in AI Lifecycle Frameworks for the medical device industry. It begins with initiatives by the US Food and Drug Administration and Health Canada aimed at addressing and simplifying regulatory hurdles in the development of artificial intelligence software for medical devices. The FDA's recent draft guidance on AI-enabled device software functions offers detailed recommendations for lifecycle management and marketing submissions. Health Canada's guidance on machine learning-enabled medical devices, which covers topics such as bias, data representativeness, and transparency, is discussed. The chapter explores state-of-the-art practices for AI software development, emphasizing the importance of a total product lifecycle approach. The discussion then shifts to the European Union's AI Act, which introduces a risk-based approach to regulating AI systems, categorizing them into four risk levels: unacceptable, high, limited, and minimal. It also explores AI-enabled medical device risk management, emphasizing the crucial role of usability engineering in the development of these devices. The discussion underscores the importance of minimizing use-related hazards and risks, ensuring that users can operate devices safely and effectively. By examining these regulatory frameworks and best practices, the chapter provides valuable insights into the complex landscape of AI regulation and development in the medical device industry.

Keywords US FDA · EU AI Act · Artificial intelligence software development · Risk management · AI risk management · Health Canada · PCCP · Usability · Human factor · AAMI TIE 34971 · ISO 14971 · IEC 62366 · Transparency · Explainability · AI labeling · Marketing submission · 510(k) · PMA · BLA · De novo · Validation · Training · Testing · Tuning · Autonomy · Bias · Data quality · Overtrust · Confidence interval · Performance validation

A. Pandey et al., *Artificial Intelligence in Medical Software*,
https://doi.org/10.1007/978-3-032-10808-1_10

10.1 US FDA Marketing Submission for AI-Device Software Function (AI-DSF)

The US FDA introduced a draft guidance [1] in January 2025 for "Artificial Intelligence-Enabled Device Software Functions: Lifecycle Management and Marketing Submission Recommendations." Although this guidance is not yet effective as of June 2025, it outlines specific requirements for AI-DSF marketing submissions and reflects the FDA's current thinking on the topic. Sections 10.1.1–10.1.8 in this chapter detail the marketing submission requirements for AI-DSFs as envisioned by the US FDA.

AI-enabled devices are defined as those that integrate one or more AI-enabled device software functions (AI-DSFs), where each AI-DSF is a specific software component within the device that employs one or more AI models—mathematical constructs designed to generate inferences or predictions from new input data—to fulfill the device's intended purpose. In this framework, "AI-enabled device" refers to the whole product. At the same time, "AI-DSF" denotes explicitly the AI-powered function, and "model" is used exclusively for the mathematical algorithm at the core. For devices under 510(k) regulatory requirements, an AI-enabled device may be considered substantially equivalent to a non-AI-enabled device with the same intended use if the integration of AI does not raise new issues regarding safety or effectiveness and if all other substantial equivalence criteria under Section 513(i) of the FD&C Act are satisfied. While these recommendations generally apply to all AI-enabled devices, they are particularly relevant to those utilizing machine learning techniques, such as deep learning and neural networks. However, other forms of AI may require additional considerations to ensure compliance and safety.

Transparency is essential, requiring that vital information about a device is not only accessible but also understandable and usable, directly connecting the sharing of information to the practical usability of the device. In the context of AI-enabled devices, addressing AI bias is critical, as these systems can systematically produce incorrect results in ways that may be difficult to predict, potentially impacting the safety and effectiveness of the device for the entire intended use population or specific subgroups—such as those defined by healthcare setting, input device, sex, or age. Ensuring transparency and actively managing bias is especially crucial for AI-enabled devices, which can be challenging for users to interpret due to the inherent opacity of many AI models and their reliance on data correlations that may not correspond to biologically plausible mechanisms [2].

For clarity in medical device marketing submissions, the term "validation" should not be used to describe the training and tuning process; instead, "development" is used throughout the submission to encompass training, tuning, and the evaluation of tuning (commonly known as "internal testing" within the AI community). Additionally, "test data" refers specifically to data used for verification and validation activities, which constitute the testing process and should not be confused with data used during the development phase. This distinction helps ensure clear communication and a proper understanding of each stage in the lifecycle of AI-enabled medical devices.

The marketing submission should include at least the following sections:

- Device description
- User interface and labeling
- User interface
- Labeling
- Risk assessment
- Data management
- Model description and development
- Validation
- Performance validation
- Device performance monitoring
- Cybersecurity
- Public submission summary

10.1.1 AI Lifecycle Framework

The US FDA introduced the AI Lifecycle Framework [3], depicted in Fig. 10.1, to provide a structured method for overseeing AI-enabled medical devices. To successfully implement this framework, manufacturers should establish a quality management system that incorporates the principles of the AI Lifecycle Framework. Medical device manufacturers can leverage their existing quality management systems, which are already compliant with Quality System Regulations and ISO 13485, to support the execution of this framework. The Software Development Lifecycle

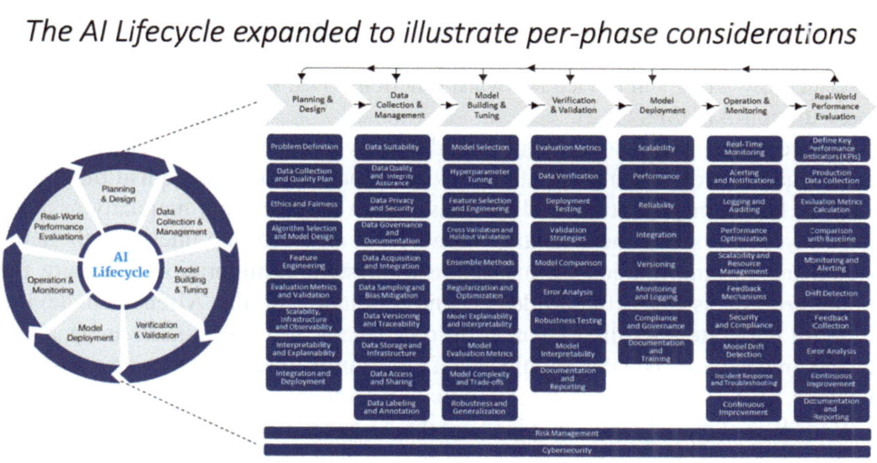

Fig. 10.1 US FDA artificial intelligence lifecycle framework [3]

(SDLC) for medical devices adheres to established standards, including IEC 62304, IEC 82304, and AAMI TIR45. However, AI-enabled Device Software Functions (AI-DSFs) introduce unique development requirements due to their dynamic and evolving characteristics. Consequently, the development of AI-DSF software must comply not only with the standard software lifecycle processes outlined in IEC 62304 but also with the additional AI-DSF development requirements specified within the AI Lifecycle Framework.

The Artificial Intelligence Lifecycle concept diagram emphasizes systematic approaches to data and model evaluation throughout both the Data Collection and Management phase and the Model Building and Tuning phase. Furthermore, the diagram illustrates the ongoing monitoring of AI software after deployment, covering both the Operation and Monitoring phase and the Real-World Performance Evaluation phase. This comprehensive approach ensures robust oversight and continuous assessment of AI systems across their entire lifecycle.

The key elements of the AI Lifecycle, as shown in Fig. 10.1, are as follows: Planning and design, Data Collection and Management, Model Building and Tuning, Verification and Validation, Model Deployment, Operation and Monitoring, and Real-World Performance Evaluation.

10.1.2 AI-DSF Labeling

The labeling for an AI-enabled device should address the types of information in a format and at a reading level suitable for the intended user, considering characteristics such as age, education or literacy level, sensory or physical impairments, or occupational specialty. This helps ensure users can quickly access important information, and tables and graphics may be used to communicate this information effectively. The label should include a statement indicating that AI is used in the device, along with an explanation of how AI is utilized to achieve the device's intended purpose. For devices with multiple functions, the explanation may also describe how AI-driven device software functions (DSFs) interact with each other and with non-AI DSFs [1, 2].

A description of the model input should be provided, detailing the signals or patterns acquired from other compatible devices, images from an acquisition system (such as MRI), or patient-derived samples, which may be input manually or automatically. For systems incorporating inputs from an electronic interface, the label should include information on the necessary system configuration to ensure the inputs are consistent with the design and validation of the AI-enabled device. Suppose the system requires input from other medical devices, such as an X-ray device. In that case, the label should specify the compatible devices or device specifications, along with the acceptable acquisition protocols.

If the loss of model inputs may prevent the AI-enabled device from generating an output, the label should explain the potential impact of the lost inputs on the device's performance. Instructions on any steps the user is expected to take to prepare input data for processing by the device should be included, detailing any expected characteristics such as functional capabilities, experience and knowledge levels, and level of training of those performing these steps. This information should be consistent with the intended use that was studied in the device validation. The label should also explain what the model output means and how it is intended to be used, including an explanation of the intended degree of automation exhibited by the device. A high-level description of the methods and architecture used to develop the model(s) implemented in the device should be included, along with a description of the development data, including the source and types of data, as well as the quantity of data used.

The device labeling should include a comprehensive description of the device's performance metrics. These metrics may encompass measures such as the area under the receiver operating characteristic curve (AUROC), sensitivity and specificity, true/false positive and true/false negative counts (for example, in a confusion matrix), positive and negative predictive values (PPV/NPV), and positive/negative diagnostic likelihood ratios (PLR/NLR). Confidence intervals should accompany all performance estimates. Additionally, the labeling should explain device performance across important subgroups, typically supported by subgroup analysis based on characteristics such as sex, age, race, ethnicity, disease severity, geographic sites, and data collection equipment. When applicable, a description of the corresponding performance for different operating points should also be included, along with subgroup analysis for each operating point.

For performance monitoring, the labeling must describe any methods or tools used to monitor and manage device performance. This includes instructions for using such tools, especially when ongoing performance monitoring and management by the user is necessary for the safe and effective use of the device. The limitations section should detail all known limitations of the AI-enabled device, its AI-driven software functions (AI-DSFs), or its underlying models. Some limitations may not be severe enough to warrant a contraindication, warning, or precaution, but they should still be included in the labeling. For instance, if the training dataset included only a few patients with a rare presentation of a disease or condition, users may benefit from understanding these limitations, as the model's output may be less reliable in such cases.

Installation and use instructions should provide clear guidance on integrating the AI-enabled device into the site's data systems and clinical workflow. The labeling must also include instructions to ensure that any input data is compatible and appropriate for the device. In some cases, terms may need to be explicitly defined; for example, a healthcare system and a manufacturer might both have data labeled as "sex," but one may be using sex as assigned at birth while the other may be using self-reported sex. The labeling should describe any customizable features, including options for users or healthcare systems to configure operating points. Users should be able to select different configurations, such as changing the operating point to optimize for sensitivity, specificity, or other relevant values, depending on the needs of their workflow, site, or patient population.

10.1.3 Risk Management

Sponsors are advised to consult the FDA-recognized version of ANSI/AAMI/ISO 14971, which provides comprehensive guidance on developing and maintaining a risk management file for the application of risk management to medical devices. This standard is also relevant for AI-enabled devices. The FDA acknowledges that AI-enabled devices may present new or unique risks compared to traditional device software functions. Consequently, the FDA recommends that sponsors also consider the guidance provided in the FDA-recognized voluntary consensus standard AAMI TR34971, which specifically addresses the application of ISO 14971 to artificial intelligence and machine learning technologies in medical devices. This additional standard offers targeted considerations for managing the distinct risks associated with AI-enabled devices.

10.1.4 Data Management

The performance and behavior of AI systems rely heavily on the quality, diversity, and quantity of data used to train and tune them. The accuracy and usefulness of validating an AI-enabled device also depend on the quality, diversity, and quantity of data used to test it. Thus, FDA reviewers evaluate data management to determine whether an AI-enabled device is safe and effective. This includes aligning the collection and management of training and test data with the intended use and resulting device requirements. Data management is also an essential means of identifying and mitigating bias. Characterizing sources of bias is crucial for assessing the potential for AI bias in AI-enabled devices. AI bias refers to the potential tendency for an algorithm to produce incorrect results in a systematic, yet sometimes unforeseen, manner due to limitations in the training data or erroneous assumptions in the machine learning process.

AI bias has been well-documented. For example, during training, models can be over-trained to recognize features of images that are unique to specific scanners, patient subpopulations, or clinical sites but have little to do with generalizable patient anatomy, physiology, or condition, which can lead to AI bias in the resulting model. In another example, the underrepresentation of specific populations in datasets can lead to overfitting (i.e., data fitting too closely to the potential biases of the training data) based on demographic characteristics, which can impact the performance of AI-enabled devices in the underrepresented population. Using unbiased, representative training data for models promotes generalizability to the intended use population and helps avoid perpetuating biases or idiosyncrasies inherent in the data itself.

A robust approach to data management for medical device studies begins with careful data collection, ensuring that information is gathered consistently and in compliance with recognized standards such as those from the Clinical Data

Interchange Standards Consortium (CDISC). Following collection, data cleaning and processing are essential to validate accuracy, remove inconsistencies, and prepare the dataset for analysis. Establishing a clear reference standard is critical, as it defines the criteria against which device performance and outcomes are measured. Data annotation should be conducted systematically to ensure that all relevant variables are clearly labeled and described, supporting both interpretability and reproducibility. Secure and organized data storage is necessary to maintain data integrity, facilitate traceability, and protect sensitive information. Effective management of the data lifecycle, including maintaining independence between data collection and analysis teams where appropriate, helps minimize bias and enhances the reliability of study findings. Ensuring data representativeness is vital; the dataset should accurately reflect the target population to support the generalizability of results and regulatory acceptance.

10.1.5 Model Description and Development

A comprehensive submission for a medical device incorporating artificial intelligence should begin with a detailed description of the model and an account of its development process, outlining the design choices and methodologies employed. This should be followed by a thorough validation section, including both initial validation and ongoing performance validation, to demonstrate the device's accuracy, reliability, and robustness in clinical or real-world settings. Continuous device performance monitoring is crucial to ensure sustained effectiveness and to promptly identify any issues that may arise after deployment. Cybersecurity considerations must also be addressed, detailing the measures implemented to protect patient data and ensure the device's integrity against potential threats. A public submission summary should be provided, offering a clear and accessible overview of the device's intended use, performance, safety features, and compliance with regulatory requirements, thereby supporting transparency [2] and informed decision-making by stakeholders.

10.1.6 Validation

For AI-enabled devices, validation encompasses not only verifying that the device performs its intended function safely and effectively in the hands of users but also confirming that it consistently meets relevant performance specifications. Manufacturers should demonstrate that users can interact with and understand the device as intended, in addition to ensuring the technical performance of the device itself. To achieve this, it is essential to consider both performance validation, which includes human factors validation, and usability evaluation. Within this US FDA guidance, usability refers to whether the device can be operated safely and effectively by its intended users and whether users can reliably receive, comprehend, interpret, and apply information related to the AI-enabled device, especially in the absence of formal human factors validation.

Together, performance validation and human factors validation, or an appropriate usability evaluation, provide the FDA with insight into how the device is likely to be used and how it will perform in real-world settings. Performance validation may involve a variety of testing and monitoring methods to assess the statistical performance of the AI model under controlled conditions. In contrast, human factors validation focuses on understanding how different users interact with the device in its intended context. In essence, performance validation confirms that the device's specifications align with user needs and intended uses and that these requirements can be consistently met. Human factors validation and usability evaluation, on the other hand, are specifically designed to ensure that all intended users can consistently achieve their goals with the device and interact with it safely and effectively.

Subgroup analysis is crucial for evaluating device performance within specific populations, enabling the identification of scenarios where the device may underperform and revealing potential limitations. This information supports more effective labeling by providing end users with relevant insights. Additionally, understanding the uncertainty of device outputs is crucial, as it enables reviewers to interpret the results accurately. Repeatability and reproducibility studies can help quantify this uncertainty when it is not otherwise specified.

When assessing AI-enabled devices, it is crucial to consider the interaction between users and the device, especially in systems where humans interpret or act on AI outputs. Evaluating the performance of the human-AI team, rather than just the model alone, aligns with good machine learning practices [4] and ensures that the device's intended use and clinical workflow are adequately supported. Depending on the device's level of autonomy, studies may focus more on standalone model performance or the combined performance of users and the device. In many cases, both types of evaluation are valuable for a comprehensive assessment of the AI-enabled device's effectiveness and safety.

10.1.6.1 Performance Monitoring Plan

Sponsors of AI-enabled devices that use proactive performance monitoring to manage risk and ensure device safety and effectiveness should include details of their monitoring plans in premarket submissions. While the FDA generally does not require these plans for 510(k) submissions unless specified as an exceptional control for the device type, a performance monitoring plan may be necessary for De Novo classification requests to address specific risks and could become an outstanding control for future devices of the same type. For PMA submissions, such a plan may be a condition of approval. Regardless of submission type, sponsors may choose to provide information about their performance monitoring plans for AI-enabled devices.

Performance monitoring plans should promptly identify and address performance changes or conditions that could lead to a decline or degradation in device performance. An effective plan proactively captures device performance after deployment and typically includes several key components. First, it should describe the methods for data collection and analysis, focusing on identifying, characterizing, and assessing changes in model performance, as well as evaluating the impact of these changes on safety and effectiveness. The plan should also address monitoring for potential causes of undesirable performance changes, such as shifts in patient demographics or disease prevalence, changes in input data, issues with data integrity (including missing or duplicate records), and changes in user behavior or demographics. Additionally, the plan should outline robust software lifecycle processes that incorporate mechanisms for monitoring performance within the deployment environment. Finally, it should include a strategy for the timely deployment of updates, mitigations, and corrective actions to address any identified performance issues.

10.1.7 Cybersecurity Requirements for AI-Enabled Medical Devices

For AI-enabled devices classified as "cyber devices" under Section 524B(c) of the FD&C Act, this guidance aims to help manufacturers fulfill their regulatory obligations by addressing specific cybersecurity risks associated with AI-enabled devices. Key risks include data poisoning, model inversion or stealing, model evasion, data leakage, overfitting, model bias, and performance drift, all of which can compromise device safety, reliability, and privacy. To address these risks, sponsors should include in their submissions a cybersecurity risk management report, threat modeling, risk assessments, and appropriate labeling. Documentation should detail cybersecurity controls, including malformed input (fuzz) testing, penetration testing, access controls, data encryption, and the anonymization or de-identification of sensitive data. Additional measures, such as data validation, anomaly detection, and adversarial training, are recommended to mitigate threats like data poisoning and model manipulation, in line with the 2023 Premarket Cybersecurity Guidance [5].

10.1.8 Public Information Summary

Sponsors must adhere to applicable submission regulations and consider FDA recommendations for their specific marketing submission type. Documentation should clearly state the use of AI within the device and explain how AI contributes to its intended use, including interactions between AI-enabled and non-AI device software functions, if relevant. The submission should describe the model class (e.g., convolutional neural network, support vector machine) and its limitations, as well as detail the development and validation datasets (data size, sources, and demographic characteristics). It should also compare these datasets to the intended use population and ensure that the test data is independent of the training data. Additionally, sponsors should provide information on model prediction confidence and uncertainty and describe plans for ongoing model updates and maintenance. While not mandatory, using a model card is recommended for organizing and summarizing key device characteristics, performance, and limitations. For submissions via PMA, HDE, De Novo, BLA, or 510(k) pathways, this information should be included in the relevant public summary, excluding patient identifiers and confidential details [1, 2, 4–6].

10.2 Health Canada Pre-market Guidance for Machine Learning-Enabled Medical Devices

Artificial intelligence (AI) is a broad term encompassing a category of algorithms and models that perform tasks and exhibit behaviors such as learning, decision-making, and prediction. Machine learning (ML) is a subset of AI that enables ML training algorithms to create models when applied to data, rather than models that are explicitly programmed. Medical devices that utilize machine learning, in part or in whole, to achieve their intended medical purpose are known as machine learning-enabled medical devices (MLMDs).

In this guidance [7], "transparency" describes the degree to which appropriate and clear information about a device (that could impact risks and patient outcomes) is communicated to stakeholders (such as patients, users, healthcare providers, and regulators). Transparency is a crucial aspect of the device's safety and effectiveness, enabling stakeholders to make informed decisions. This guidance introduces the concept of a predetermined change control plan (PCCP) [8]. A PCCP provides a mechanism for Health Canada to address cases where regulatory pre-authorization of planned changes to ML systems is necessary to mitigate a known risk.

In this guidance, "ML training algorithm" refers to the software procedure that establishes the parameters of an ML model by analyzing data. The "ML model" represents a mathematical construct that generates an inference or prediction based on new input data, resulting from an ML training algorithm learning from the data. The "ML system" refers to an ML-enabled software that meets the definition of a medical device as per Section 1 of the regulations, including ML models and the associated ML training algorithms.

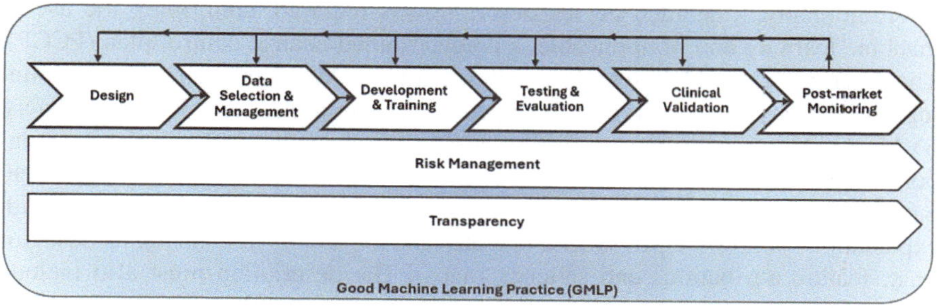

Fig. 10.2 MLMD product lifecycle

10.2.1 Guidance for Implementation

Health Canada considers product lifecycle information to be essential in demonstrating the safety and effectiveness of an MLMD. From Health Canada's perspective, the MLMD lifecycle includes the following components:

- Good machine learning practice
- Design
- Risk management
- Data selection and management
- Development and training
- Testing and evaluation
- Clinical validation
- Transparency
- Post-market monitoring

Figure 10.2 gives a visual overview of the MLMD product lifecycle. The iterative components reflected in this lifecycle schematic are not mutually exclusive and may not occur in the order indicated.

10.2.2 Design

For Class II, III, or IV machine learning-enabled medical devices (MLMDs), the application must clearly define the intended use, indications for use, and medical purposes, such as diagnosis, treatment, or monitoring, along with the targeted patient population, intended users, and use environment. The submission should specify device functions, including software inputs and outputs, how these outputs integrate into the healthcare workflow, and the degree of clinical autonomy (i.e., the capacity to perform functions with minimal or no user intervention). Contraindications and all known limitations must also be included.

Additionally, a detailed device description is required, confirming the use of machine learning and, if applicable, a predetermined change control plan (PCCP). The description should outline the machine learning (ML) methods, including training algorithms (e.g., supervised, unsupervised, or reinforcement learning; CNNs, SVMs, GANs, transformers), as well as architectural details, such as software components and tuning approaches. Information about the development data, system outputs, intended users, and clinical autonomy should be provided, along with explanations of how the ML system operates and the factors influencing its behavior (e.g., feature attributions and saliency maps). The description must also include required input parameters and sources, compatibility with other devices (including software and hardware versions), and hardware requirements, such as CPU, GPU, RAM, and operating system specifications.

10.2.3 Risk Management

Manufacturers must implement comprehensive risk management throughout the lifecycle of machine learning-enabled medical devices (MLMDs), including clear descriptions of identified risks and the controls implemented to mitigate them. Documentation should outline the methods for initial and ongoing risk assessment, the system for categorizing and assessing risk, and the results of these assessments. Risk analysis should address erroneous outputs such as false positives, false negatives, or incorrect diagnostic or treatment information, as well as outputs that are misleading or inappropriate. Consideration of bias—including systematic differences in the treatment of certain groups—is essential, with attention to population characteristics, device inputs, the context of use, and the sources of the dataset. Additional factors to address include overfitting and underfitting, degradation of system performance due to demographic or clinical practice changes, automation bias, alarm fatigue, and risks related to the use and impact of patient-centered control panels (PCCPs). Manufacturers are advised to refer to the most current version of ISO 14971 for best practices in medical device risk management.

10.2.4 Data Selection and Management

The quality of datasets used to develop a machine learning-enabled medical device (MLMD) directly impacts the device's overall performance, making thorough data characterization a critical part of device evaluation. When describing data selection and management, manufacturers should detail the training, tuning, and test datasets, including sample sizes, clinical and demographic characteristics, and how these datasets compare to the intended population in terms of prevalence and representativeness. Documentation should address the data collection methods and environments, devices used, and whether data are from single or multiple centers or personalized sources, with justifications for dataset choices based on intended use, statistical factors, identity characteristics (such as sex, gender, race, or age), and subgroup representation—including under-represented populations and alignment with Canadian demographics. Inclusion and exclusion criteria, reasons for data removal, and techniques to address data imbalances—such as sampling or augmentation—should also be explained. Finally, manufacturers must describe how data integrity was maintained during curation, including quality assurance measures and any data augmentation or imputation practices, and explain the steps taken to control bias throughout the dataset development process.

10.2.5 Development, Training, Testing, and Evaluation

For machine learning-enabled medical device development, provide a detailed description of the methods used for ML system development, training, and tuning, along with justifications for these approaches. This should include the characterization and justification of the reference standard used in training, highlighting any uncertainties or limitations, as well as a description of the input parameters and features extracted from the data.

For testing and evaluation, describe the methods and data used for performance assessment, referencing the guidance on data selection and management. Characterize and justify the reference standard used in testing, including its definition, uncertainty, and limitations. Specify the chosen performance metrics, acceptance criteria, and operating thresholds with clinical and risk-based rationale. Provide evidence that the ML system performs as intended when integrated into the device, and demonstrate its effectiveness across relevant subgroups, such as by identity factors, underrepresented populations, and clinical status or features. Include robustness testing with unexpected inputs, statistical significance analysis where appropriate, and an estimate of output uncertainty with supporting justification. Finally, document the software version tested, explain the version numbering system, and ensure traceability of the ML system or model version.

10.2.6 Clinical Validation

For a Class III or IV machine learning-enabled medical device (MLMD), manufacturers must submit clinical evidence, such as clinical validation studies, to support the device's safe and effective use in the intended population; for Class II MLMDs, this information should be available upon request. Clinical evidence should demonstrate that the trained, tuned, and tested ML system and the MLMD perform as intended and are clinically safe. Suitable evidence includes clinical validation studies with clear descriptions of study type, design, statistical methods, and rationale (including relevance to intended use, comparison to standard of care, and use of retrospective or prospective evaluations), as well as details about the model, study participants (ensuring independence from development data), and justification for the study population reflecting relevant identity factors and statistical considerations. Results, subgroup analyses, and study limitations should be included, along with relevant published clinical data, comparator device investigations, usability and human factors testing, device-specific evaluations, and real-world or post-market evidence. The submission must justify the sufficiency of the evidence to confirm safety and effectiveness for the intended use and demonstrate consideration of impacts across diverse populations, including different sexes, genders, racial and ethnic groups, and age ranges.

10.2.7 Transparency

Transparency requirements must account for all stakeholders involved in patient care throughout the device lifecycle, including patients, users, healthcare providers, and regulators, by clearly communicating the intended use, device development processes, and performance data. When available, transparency also involves explaining how the device reaches a particular output or makes a decision, often referred to as explainability. Guidance on these principles can be found in the document "Transparency for machine learning-enabled medical devices: Guiding principles." Transparency should be integrated across the device lifecycle, specifically within device labeling (such as instructions for use), the software user interface, and the medical device license application. It should also be maintained during device use and whenever changes are made to the device. This approach supports informed decision-making and regulatory compliance throughout the device's operational life [9].

10.2.7.1 Transparency Considerations for MLMD Labeling

Manufacturers must include copies of all labeling—covering both the device and its ML system—in applications for Class II, III, and IV MLMDs, with Health Canada reviewing these against regulatory requirements. Labeling encompasses directions for use, product brochures, websites, and marketing materials containing ML-related claims. The labeling should clearly state the device's indications for use, intended use, and contraindications, along with user instructions on operating the ML system, interpreting outputs (such as saliency maps and confidence scores), and performing calibrations, local validation, and ongoing performance monitoring. It should also detail device design, including the inclusion of machine learning (ML), algorithms, feature attributions, input parameters, compatible devices, hardware and software requirements, and dataset characteristics (such as data collection methods, reference standards, sample sizes, and inclusion/exclusion criteria). Information about device changes—such as the presence of a patient-centered control panel (PCCP), update requirements, and notification of software updates—should be included, along with device performance metrics, test results (including subgroups and uncertainty), and clinical study summaries. Finally, labeling must communicate device limitations, known failure modes, and applicable warnings, and may benefit from a structured summary of key ML system information, such as a model card.

10.2.8 Post-market Monitoring

Manufacturers should document the processes, surveillance, and performance monitoring plans, along with risk mitigation strategies, to ensure the ongoing performance and interoperability of the machine learning (ML) system. This documentation should address potential impacts on ML system outputs or clinical workflows, such as performance degradation over time, shifts in input data distribution following deployment, changes to ML system inputs or input sources, and alterations in how compatible products manage outputs. References to relevant risk analysis and, if applicable, the patient-centered control panel (PCCP) should also be included.

10.3 Predetermined Change Control Plan

The evolving capabilities of AI-enabled devices require more frequent software updates compared to traditional non-AI medical device software. In response, the US FDA introduced the Predetermined Change Control Plan (PCCP) [6] in December 2024, offering a streamlined pathway for manufacturers to make iterative improvements to AI-enabled device software functions (AI-DSFs) while safeguarding safety and effectiveness. Following approval, these changes can be implemented without additional regulatory submissions, minimizing administrative burdens.

The documentation outlines the modifications to be made to a device and explains how these modifications will be evaluated. The described changes in the PCCP include device modifications that would otherwise require a PMA supplement or new premarket notification. The PCCP comprises a Description of Modifications, a Modification Protocol, and an Impact Assessment. A PCCP that has been reviewed and established through a device marketing authorization is known as an Authorized Predetermined Change Control Plan (Authorized PCCP). An authorized PCCP is a technological characteristic of the authorized device with which it was established.

The Predetermined Change Control Plan (PCCP) should be incorporated into the marketing submission within the device description, labeling, and other pertinent sections to demonstrate reasonable assurance of safety and effectiveness during substantial equivalence comparisons. Device labeling must comply with applicable statutes and regulations; for devices with an authorized PCCP, the FDA may require special labeling to ensure safe and effective use as changes outlined in the PCCP are implemented, excluding trade secrets and confidential commercial information as appropriate. For AI-enabled device software functions (AI-DSFs) with an authorized PCCP, the labeling must indicate that the device uses machine learning and that an authorized PCCP is in place, alerting users that software updates may be necessary and could affect device performance, inputs, or usage.

10.3.1 Modification Protocol

The Modification Protocol should document the methods for developing, validating, and implementing modifications specified in the PCCP's Description of Modifications section, including verification and validation activities with predefined acceptance criteria and a step-by-step process to ensure continued device safety and effectiveness. Appendix A offers detailed questions and considerations to help manufacturers structure and populate the Modification Protocol as recommended in the guidance. The following sections discuss the key elements of the modification protocol.

10.3.1.1 Data Management Practices

Data collection protocols must define explicit inclusion and exclusion criteria tied to the intended use population, specify the target distribution of patient demographics and data acquisition conditions, and state whether data will be collected prospectively or retrospectively, including details on consecutive cases, random sampling, or stratified sampling to ensure diversity. The number and geographic distribution of data collection sites should be provided, along with plans to mitigate bias throughout model training, tuning, and testing, as well as to address historical healthcare inequalities. Strategies must be in place to maintain data relevance amid evolving technology and clinical practices, and compliance with human subject protection regulations must be confirmed. Data for AI-enabled device software functions (AI-DSFs) must align with device specifications and interoperability requirements. Ensuring data quality requires defined processes to ensure consistency, completeness, authenticity, and integrity throughout the data collection process.

Approaches for handling missing data, criteria for case inclusion and exclusion based on quality, and methods to minimize the impact on generalizability should be detailed, along with procedures for addressing data issues. Controls against unauthorized access or manipulation must be outlined, and the presence and implementation of automated data quality processes should be stated. Reference standard determination must be justified, specifying the rationale for the chosen standard and, if clinician evaluations are used, detailing the grading protocol, clinician qualifications, and processes for adjudicating results. Strategies for resolving equivocal or missing results and quantifying reference standard uncertainty should be clearly described. Additionally, any automated methods or differences in standards between initial authorization and model updates should be clarified. For test data sequestration, manufacturers must implement measures to isolate test datasets from development, maintain sequestration during retraining, and prevent bias if test data are reused. This includes providing descriptive statistics and validating alignment with the intended use population.

10.3.1.2 Re-Training Practices

AI software re-training practices within a Modification Protocol must outline the specific processing steps subject to change during updates to AI-enabled device software functions (AI-DSFs), including methods for implementing modifications such as architectural adjustments (e.g., neural network layers, hyperparameters) with justifications for each change. This transparency enables the FDA to determine whether modifications adhere to well-defined practices and align with performance evaluations. Submissions should detail re-training objectives, model architecture, modified components (e.g., data preprocessing, transfer learning), protocols to isolate unaffected functions, and triggers like data volume thresholds or performance drift. Implementation strategies must address overfitting risks (via cross-validation and regularization) and mitigate bias through the use of diversified datasets and fairness validation. The plan should align with the Predetermined Change Control Plan (PCCP), ensuring updates maintain safety and effectiveness while adapting to new data or clinical needs.

10.3.1.3 Performance Evaluation

The FDA may require that performance requirements for changes under the Predetermined Change Control Plan (PCCP) be specified within the plan itself. Performance evaluation methods must detail how the modified AI-enabled device software function (AI-DSF) will be verified and validated to meet both new and existing specifications, including protocols for testing the entire device after each individual and aggregate modification. Evaluation should compare the modified device to the original and most recent prior version, and for software driving hardware, assess the impacts on hardware functionality. The Modification Protocol should specify the study design, performance metrics, acceptance criteria, and statistical tests for each planned change. More comprehensive testing may enable broader modifications.

Performance evaluation plans for retrained or modified AI-enabled device software functions (AI-DSFs) should specify clear triggers, such as reaching performance thresholds, test dataset size, or scheduled intervals, to initiate assessments. The frequency of these assessments should be aligned with data availability, update cycles, and regulatory requirements. Evaluation metrics derived from independent test datasets managed according to established data plans should include accuracy, sensitivity, specificity, and AUC-ROC to quantify device performance and validate safety for the intended use. This assessment should also consider clinical relevance, robustness, and risk stratification.

Statistical analysis plans must validate equivalent or improved performance relative to prior versions, target high-risk subpopulations and underrepresented groups, and inform labeling with clear performance and limitation details while justifying sample size and addressing reference standard variability and missing data. Acceptance criteria should be defined and compared to those of the authorized device, ensuring modifications meet clinical, safety, and usability standards. Additional testing should assess whether database evaluations sufficiently address modification risks, including the impacts on user interaction and hardware integration, and verify that usability and workflow changes maintain safe and effective device use.

10.3.1.4 Update Procedures

AI-enabled device software functions (AI-DSFs) must be updated according to clear procedures that specify whether modifications will be implemented automatically, manually, or through a combination of both and detail how manufacturers will communicate these changes to users to ensure safe device use. The Modification Protocol should outline how updates will be carried out in compliance with Quality System Regulations (QSR), provide appropriate transparency and updated user training where necessary, and describe post-market surveillance plans—such as real-world monitoring and notification requirements if the device does not function as intended under the authorized Predetermined Change Control Plan (PCCP). Manufacturers are expected to monitor device safety continuously (e.g., adverse events) and effectiveness (e.g., performance), with monitoring approaches tailored to AI-DSFs that receive manual versus automatic updates or are deployed globally versus locally.

The Modification Protocol must also address how labeling will be updated to reflect modifications, specifying which sections are anticipated to change and ensuring that directions for use remain adequate. Labeling should indicate the current version(s) of the AI-DSF and any site-specific modifications, and new unique device identifiers (UDIs) must be assigned when required. Importantly, labeling should not include information about changes that have not yet been implemented, as this could mislead users and potentially result in misbranding.

The Predetermined Change Control Plan (PCCP) should be clearly outlined in public summaries and labeling, with users promptly notified of updates, modifications to device performance, and any new limitations or safety considerations. Version details should be displayed alongside outputs for traceability, and users should have access to updated labeling before implementation, with opt-out options for non-critical updates. Known biases, performance limitations, and validation population demographics should be disclosed to contextualize device performance, and reprocessed prior patient data should be shared when clinically significant differences arise.

10.3.1.5 Device Monitoring Plan

- Track adverse events by device update version for accountability and trend analysis.
- Establish and share a risk-based real-world performance monitoring plan with users for post-implementation oversight.
- Provide users with real-world safety metrics and performance data unless justified otherwise.
- Monitor safety (adverse events, new risks) and effectiveness (performance shifts, biases) at defined intervals for both manual and automated updates.
- Identify performance changes in patient subpopulations through stratified analysis and continuous monitoring.
- Define a response protocol for new risks or high-risk subpopulations, including mitigations and user notifications.
- Develop a strategy for addressing unexpected deficiencies or hazards, including root-cause analysis and corrective actions (such as patches and recalls).
- Track diagnostic errors attributable to the device through user feedback and error-reporting systems.
- Establish rollback criteria and protocols if safety or performance thresholds are breached.

10.4 European Union's Artificial Intelligence Act (AIA)

The EU AI Act [10] is the first comprehensive regulation on artificial intelligence in the European Union, establishing a risk-based legal framework for AI systems. It categorizes AI systems into four primary risk levels:

- Prohibited AI System (Unacceptable Risk): AI practices that are banned due to fundamental rights violations (e.g., social scoring, subliminal manipulation, real-time biometric identification in public by law enforcement).
- High-Risk AI System: AI systems that could significantly impact health, safety, or fundamental rights include safety components of regulated products and stand-alone systems in specific areas listed in Annex III (e.g., medical devices, critical infrastructure, employment).
- Limited-Risk AI System (with Transparency Requirements): Systems with transparency obligations (e.g., chatbots, deepfakes) where users must be informed about AI interaction.
- Minimal-Risk AI System: Systems with little or no risk (e.g., spam filters); minimal or no regulation.

The Act applies to any AI system developed, placed on the EU market, or used in the EU, including those from non-EU providers if their output is used within the EU [10]. General-purpose AI (GPAI) models are treated as a separate category, with additional requirements if they are considered to have systemic risk.

10.4.1 AI System Definition as per AI Act

10.4.1.1 Official Definition (Article 3) [11]

> AI system means a machine-based system that is designed to operate with varying levels of autonomy and that may exhibit adaptiveness after deployment, and that, for explicit or implicit objectives, infers, from the input it receives, how to generate outputs such as predictions, content, recommendations, or decisions that can influence physical or virtual environments.

10.4.1.2 Key Elements of the Definition

- **Machine-Based**: The system relies on technology, not just manual processes.
- **Varying Levels of Autonomy**: It can operate independently to some degree.
- **Adaptiveness after Deployment**: The system can change or improve its behavior based on new data or experiences.
- **Inferring Outputs**: The system processes input data to produce outputs (predictions, decisions, content, etc.).
- **Influence on Environments**: The outputs can impact both real-world and digital settings.

Under the EU AI Act, an AI system is defined as a machine-based system that, for explicit or implicit objectives, infers from its input how to generate outputs such as predictions, content, recommendations, or decisions that can influence physical or virtual environments. The specific outputs considered in the legal definition are:

- **Predictions**: Estimating unknown values or future events based on input data (e.g., forecasting patient outcomes).
- **Content**: Creating new material such as text, images, audio, or video (e.g., generative AI-producing articles or artwork).
- **Recommendations**: Suggesting actions or choices to users (e.g., product recommendations on an e-commerce site).
- **Decisions**: Automating choices or actions that traditionally require human judgment (e.g., approving a loan application or diagnosing a medical condition).

These outputs must be capable of influencing either physical environments (such as autonomous robots or vehicles) or virtual environments (such as digital assistants or data-driven software ecosystems). The Act's approach is technology-neutral, meaning it does not distinguish between AI systems based on whether they primarily affect physical or virtual environments. Instead, it focuses on the capability of the system's outputs to influence outcomes in either domain, ensuring comprehensive regulation across a wide range of AI applications. This broad categorization ensures that the law covers a wide range of AI applications while distinguishing them from traditional, rule-based software [12]. In addressing influence on environments:

- **Physical Environments**: The Act covers AI systems whose outputs can directly or indirectly impact the real world—for example, autonomous vehicles making driving decisions, industrial robots performing tasks, or medical AI systems influencing patient treatment and care.
- **Virtual Environments**: The law also applies to AI systems whose outputs influence digital or online settings, such as content recommendation algorithms that shape user experiences, chatbots providing customer support, or AI-driven analytics that inform business or policy decisions.

10.4.2 AI Act and Medical Devices

High-Risk Classification: Most AI-enabled medical devices are classified as high-risk under the EU AI Act if the AI system is a safety component of the device or is the device itself, and if a third-party conformity assessment is required under the EU Medical Device Regulation (MDR) or In Vitro Diagnostic Medical Device Regulation (IVDR).

Regulatory Overlap: If a medical device is already subject to MDR or IVDR and incorporates AI, it will also need to comply with the AI Act. This means the device manufacturer must fulfill both sets of regulatory requirements and may require additional certification.

Timeline: Full compliance for high-risk AI systems integrated into medical devices is mandatory as of August 2, 2027.

Key Obligations: High-risk AI systems must meet strict requirements, including comprehensive risk management, conformity assessment, technical documentation, ongoing monitoring, data quality assurance, transparency, and human oversight.

Registration: The device manufacturer must register itself and its AI system in the EU AI database before placing a high-risk AI medical device on the market or into service.

Medical Device Classification:

- Suppose the AI system is a safety component of a medical device or the device itself and is subject to MDR/IVDR (and conformity assessment). In that case, it is automatically considered high-risk under the AI Act.
- If the device is classified as Class I and is not subject to conformity assessment, it may not be considered high-risk unless specified in Annex III of the AI Act.

Autonomy:

- The Act applies to systems with "varying levels of autonomy"—not just fully autonomous systems.

- Suppose the AI system generates outputs that influence user decisions (even if a human always makes the final decision). In that case, it is considered to have autonomy for the Act.
- Only fully manual systems (no independent output or influence) are out of scope.

10.5 Regulatory Implications of Autonomy in Medical Devices

If an AI-enabled medical device has zero autonomy—meaning it always requires a human to make the final decision and does not generate outputs that influence the environment or patient care without explicit human approval—the interpretation of whether it falls under the EU AI Act depends on the specific definition and the system's operational characteristics [10, 12, 13].

Definition of Autonomy in the Act: The EU AI Act defines AI systems as those that "operate with varying levels of autonomy," and it explicitly excludes systems that are "fully manual," meaning there is absolutely no independent output or influence from the AI system itself—every output or action is always and only determined by a human.

Autonomy in Medical Device Context: If your device uses AI to analyze data or generate recommendations, but the AI output is constantly reviewed and acted upon by a human (with no automated or independent action), it is considered to have "zero autonomy" or to be "fully manual."

Regulatory Implications:

- If the AI system truly never acts or outputs anything independently—every decision, recommendation, or output is constantly reviewed, approved, and acted upon by a human—it is generally not considered to have autonomy and thus may not be subject to the strictest requirements of the EU AI Act.
- However, suppose the AI system generates predictions, recommendations, or outputs that influence the human's decision-making process (even if the human always takes the final action). In that case, it is likely still considered to have some level of autonomy for regulatory purposes.

A fully manual system, where AI is used solely for data analysis and does not generate outputs or recommendations that influence decisions or actions, thus acting only as a data processor rather than a decision support tool, may fall outside the scope of the AI Act's high-risk requirements. The decision support systems with zero autonomy in action—where the AI provides recommendations or predictions that are constantly reviewed and acted upon by a human, but the AI output still influences the process—are likely still considered AI systems under the Act.

10.6 Structure of the EU AI Act and Quality Management System

Annexes supplement the Act that provides detailed guidance and technical requirements [10]:

Annex	Purpose/content
I	List of Union harmonization legislation (sectors like machinery, medical devices, toys)
II	List of criminal offenses (for prohibited AI applications)
III	High-risk AI systems (specific areas and use cases)
IV	Technical documentation required for high-risk AI systems
V	EU Declaration of Conformity
VI	Internal control-based conformity assessment procedure
VII	Quality management system and technical documentation assessment
VIII	Information for registration of high-risk AI systems
IX	Information for registration related to testing in real-world conditions
X	Union legislative acts on large-scale IT systems (e.g., border control, justice)
XI	Technical documentation for providers of general-purpose AI models
XII	Transparency information for providers of general-purpose AI models to downstream providers
XIII	Criteria for designation of general-purpose AI models with systemic risk

10.6.1 Prohibited AI System in AI Act

Under the EU AI Act, certain AI practices are explicitly prohibited, including for medical AI systems, due to their unacceptable risks to fundamental rights and values.

The prohibited practices are enumerated in Article 5 and include [14, 15]:

- Subliminal, Manipulative, or Deceptive Techniques:
  - AI systems that use subliminal techniques (below conscious awareness) or other manipulative or misleading methods to distort behavior, causing or likely to cause significant harm.
  - *Example for Medical AI:* An AI system that manipulates or misleads patients into making harmful health decisions using non-transparent or deceptive algorithms is prohibited.
- Exploitation of Vulnerabilities:
  - AI systems that exploit vulnerabilities related to age, disability, or specific social or economic situations to distort behavior, causing or likely to cause significant harm.
  - *Example for Medical AI:* An AI chatbot that exploits the mental or cognitive vulnerabilities of patients (e.g., encouraging people with disabilities to make harmful or costly health decisions) is prohibited.

- Social Scoring:
  - AI systems that evaluate or classify individuals or groups based on social behavior or personal traits, leading to detrimental or unfavorable treatment, especially when unrelated to the context or unjustified.
  - *Example for Medical AI:* AI systems that assign medical resources or treatment based on unrelated social scores (e.g., credit scores) are prohibited.
- Untargeted Facial Image Scraping:
  - AI systems that create or expand facial recognition databases through untargeted scraping of facial images from the internet or CCTV footage (with limited exceptions).
  - This is less common in strictly medical AI systems, but any use of such databases for medical purposes would be prohibited unless explicitly permitted.
- Biometric Categorization and Emotion Recognition in Sensitive Contexts:
  - AI systems that use biometric data to categorize individuals by protected characteristics (e.g., sexual orientation, race) or infer emotions in sensitive contexts such as the workplace or education.
  - *Medical Context:* Using AI to infer emotions from patients in sensitive healthcare settings may be prohibited unless for medical or safety purposes and subject to specific exemptions.

10.6.2 Quality Management System for AI Act

The Team Notified Body Association (Team-NB) has published a Questionnaire and a position paper addressing the AI Act, offering essential guidance for notified bodies on conducting conformity assessments and developing quality systems for medical devices that incorporate artificial intelligence [16, 17]. It is crucial to apply this guidance to understand the expectations of notified bodies regarding device conformity assessment and the CE marking process. Manufacturers, distributors, and deployers of AI-enabled devices should utilize these Team-NB documents to establish or align their quality systems with the requirements set forth by the AI Act.

Software devices must be validated before market placement, and evidence of suitable validation forms part of the technical documentation assessed by the Notified Body. For "learning" software systems, changes in device performance due to ongoing learning may be considered significant. They could require a new conformity assessment before the device is placed on the market. In practice, manufacturers often struggle to sufficiently prove conformity for AI devices that update their underlying models through in-field self-learning mechanisms—currently, notified bodies do not regard such devices as certifiable unless manufacturers can ensure safe operation within the validated scope described in the technical documentation. Manufacturers must address all relevant aspects—such as development, risk management, data management, verification or validation, post-market surveillance and vigilance, software lifecycle (including service, installation, and decommissioning), configuration management, training and qualification, customer communication, purchasing, supplier and supply-chain control, and management review—either in procedural instructions or relevant plans to systematically guarantee product safety. If processes are outsourced, these requirements also apply to third-party providers, and appropriate documentation of outsourced responsibilities must be provided at the time of application.

10.6.3 ISO/IEC JTC 1 and CEN/CLC/JTC 21: Artificial Intelligence

CEN-CLC JTC 21 brings together over 300 experts from 20 European countries, chosen by their respective national standardization organizations. This joint technical committee identifies and adopts international standards that are already available or in development from bodies such as ISO/IEC JTC 1 and its subcommittees, including SC 42 on Artificial Intelligence. Additionally, CEN-CLC/JTC 21 is dedicated to developing new standardization deliverables tailored to the European market and societal needs while also supporting EU legislation, policies, principles, and values. The JTC 21 is also guiding other technical committees working on AI. Additionally, it is responsible for evaluating and potentially adopting pertinent international standards and those from relevant organizations, such as ISO/IEC JTC 1 and its subcommittees (e.g., SC 42 on Artificial Intelligence). The JTC's standardization outputs should specifically address the needs of the European market and society, primarily supporting EU legislation, policies, principles, and values [18, 19].

Medical device manufacturers developing AI-enabled medical devices should utilize published AI standards to demonstrate compliance with the requirements of the EU AI Act and to support the determination of state of the art (SOTA) as mandated by the EU Medical Device Regulation (MDR) and In Vitro Diagnostic Regulation (IVDR) regulations.

10.6.3.1 CEN/CLC/JTC 21 Published Standards on Artificial Intelligence

Reference	Publication date	Title
EN ISO/IEC 22989:2023	2023-06-28	Information technology—Artificial intelligence—Artificial intelligence concepts and terminology (ISO/IEC 22989:2022)
CEN/CLC/TR 18115:2024	2024-11-06	Data governance and quality for AI within the European context
CEN/CLC/TR 18145:2025	2025-02-26	Environmentally sustainable Artificial Intelligence

(continued)

(continued)

Reference	Publication date	Title
CEN/CLC ISO/IEC/TR 24027:2023	2023-12-20	Information technology—Artificial intelligence (AI)—Bias in AI systems and AI-aided decision making (ISO/IEC TR 24027:2021)
CEN/CLC ISO/IEC/TR 24029-1:2023	2023-12-20	Artificial Intelligence (AI)—Assessment of the robustness of neural networks—Part 1: Overview (ISO/IEC TR 24029-1:2021)
EN ISO/IEC 8183:2024	2024-06-12	Information technology—Artificial intelligence—Data life cycle framework (ISO/IEC 8183:2023)
EN ISO/IEC 5259-3:2025	2025-05-21	Artificial intelligence—Data quality for analytics and machine learning (ML)—Part 3: Data quality management requirements and guidelines (ISO/IEC 5259-3:2024)
EN ISO/IEC 5259-1:2025	2025-05-21	Artificial intelligence—Data quality for analytics and machine learning (ML)—Part 1: Overview, terminology, and examples (ISO/IEC 5259-1:2024)
EN ISO/IEC 5259-2:2025	2025-05-21	Artificial intelligence—Data quality for analytics and machine learning (ML)—Part 2: Data quality measures (ISO/IEC 5259-2:2024)
EN ISO/IEC 5259-4:2025	2025-05-21	Artificial intelligence—Data quality for analytics and machine learning (ML)—Part 4: Data quality process framework (ISO/IEC 5259-4:2024)
CEN/CLC/TR 17894:2024	2024-12-18	Artificial Intelligence—Artificial Intelligence Conformity Assessment
EN ISO/IEC 25059:2024	2024-08-07	Software engineering—Systems and software Quality Requirements and Evaluation (SQuaRE)—Quality model for AI systems (ISO/IEC 25059:2023)
EN ISO/IEC 23894:2024	2024-02-21	Information technology—Artificial intelligence—Guidance on risk management (ISO/IEC 23894:2023)
EN ISO/IEC 23053:2023	2023-06-28	Framework for Artificial Intelligence (AI) Systems Using Machine Learning (ML) (ISO/IEC 23053:2022)
CEN/CLC ISO/IEC/TS 12791:2024	2024-11-13	Information technology—Artificial intelligence—Treatment of unwanted bias in classification and regression machine learning tasks (ISO/IEC TS 12791:2024)

10.7 Risk Management

Sponsors are advised to comply with the harmonized standard ANSI/AAMI/ISO 14971, which provides comprehensive guidance on developing and maintaining a risk management file for the application of risk management to medical devices. This standard is also relevant for AI-enabled devices. The FDA acknowledges that AI-enabled devices may present new or unique risks compared to traditional device software functions. Consequently, the FDA recommends that sponsors also consider the guidance provided in the FDA-recognized AAMI TR34971, which specifically addresses the application of ISO 14971 to artificial intelligence and machine learning technologies in medical devices. This additional standard offers targeted considerations for managing the distinct risks associated with AI-enabled devices [20, 21].

The Risk Management Plan should include a risk determination and trigger for the retraining process, as required by Section 4.4 of TR39471. Depending on the frequency of retraining, the system may require different risk considerations and controls. The risk management plan should include the criteria used to trigger an update, the risk management of the update process itself, and provisions for reverting the product to a previous version if necessary. A risk management plan should define requirements based on the type of AI applications:

- Systems that are retrained occasionally.
- Systems that are retrained relatively frequently.
- Systems that are considered "continuously learning."

10.8 Identify Characteristics Related to Safety for AI Software

Add the following questions for AI/ML applications as characteristics related to safety [21]:

- Does the software provide diagnostic or treatment recommendations? If so, how significant is the information in influencing the user (e.g., does the device inform or drive the clinical management of the patient)? What is the target population of the device (e.g., is the patient condition non-serious, serious, or critical)? What is the urgency/emergency of the information provided?
- Does the algorithm provide options and the likelihood of appropriateness?
- Are errors detectable, e.g., due to domain shift (like images with or without contrast agent)?
- Does the system provide a non-ML implementation of the same functionality?
- What autonomous functions, if any, does the system provide? What is the level of that autonomy? Annex D discusses the impact of autonomous functionality.
- Does the ML-enabled medical device have the ability to learn over time? Is the device capable of adjusting its performance characteristics over time? What are the limits to which the device can adjust its performance characteristics? Which limits of performance for various performance criteria are acceptable, and what are the thresholds below which performance for various performance criteria is not to fall? What mitigations can be applied in case a threshold is crossed?
- What are the potential off-label uses of the device? As users become familiar with the device, they discover additional uses for it that extend beyond its intended purpose. The potential for reasonably foreseeable misuse should be allowed. Additionally, designers should assess potential human errors that may occur during the device's use.

- Are there any contraindications due to the restricted conditions of the patients in the data used to train, test, and validate the machine learning (ML) model? If so, is there a risk that the user does not follow these contraindications?
- Is the system intended to learn over time, and if so, is there any potential impact on the intended use?

10.9 Risk Control Measures for AI-Enabled Medical Devices

AAMI TR34971 outlines requirements for implementing risk control measures specifically related to the risks associated with AI/ML applications. The sections that follow discuss the process for identifying and establishing these risk control measures [21].

10.9.1 Risk Control Associated with Data Quality

Explicit and implicit assumptions about data and their characteristics must be documented, as these can introduce dataset bias and compromise data quality. Examples of such assumptions include that the data are independent and identically distributed, free of noise, that a linear relationship exists between independent and dependent variables, or that the data lack a formal distribution. The potential performance-related hazards arising from these assumptions should be included in the risk assessment or hazard traceability matrix, along with risk control measures to mitigate their impact on algorithm performance. Additionally, the bias-variance trade-off should be documented, recognizing that data quality can vary depending on the data source and any processing or calculations performed on the data.

The quality and appropriateness of the data used to train, tune, and test the system impact the safety and performance of an ML medical device system. The results of ML validation tests should be reviewed to assess the risks of errors in the ML algorithm output that can directly or indirectly impact a patient's safety. Additionally, the relationship between the system and the user (e.g., autonomous systems) affects the safety of the product. Other considerations include the following:

- Is the algorithm configurable (e.g., to local preferences)?
- Does configuration of the data source play a role in the performance, and if so, how is this controlled?
- Are measures needed or implemented to assess and prevent domain shift?

10.9.1.1 Data Related Risk Control

It depends on many factors. Risk control related to data quality should address the following:

- **Incorrect Data**: Values are erroneous or not applicable—e.g., data entry errors, corrupted data, calculation errors, inconsistent labeling, or data that no longer represent patient conditions.
- **Incorrect Handling of Outliers**: Outliers may be meaningful or may result from data errors, but their treatment can impact model learning.
- **Incomplete Data**: Missing values, such as empty fields in a database.
- **Subjective Data**: Influenced by beliefs or opinions—e.g., pain scale self-reports or differing professional diagnoses that affect ground truth.
- **Inconsistent Data**: Quality varies across sources or over time, such as model drift from changes in clinical practice or environment.
- **Atypical Data**: Data quality differs from real-world use—e.g., models trained on high-resolution images may not generalize to low-resolution settings. (Note: synthetic data may sometimes be acceptable.)

10.9.1.2 Data Quality and Integrity Risks

- **Data Drift**: Changes in data distribution over time that can degrade model performance. Patient populations and medical practices change over time (data drift or use drift), and a locked system may become outdated.
- **Fragmented Data**: Data that is incomplete or inconsistently organized across sources.
- **Bias**: Bias can negatively impact model performance, particularly for patient populations that are underrepresented in the training data, which may lead to healthcare inequities. Conversely, it may also result in enhanced model performance for specific, well-represented groups.
  - Selection bias
  - Confounding variables
  - Non-normality
  - Proxy variables
  - Implicit bias
  - Group attribution bias
  - Experimenter's bias

10.9.1.3 User Trust and Interaction Risks

- **Overtrust**: This may arise from overconfidence, perceived risk, user workload, or social trust; it is mitigated by disclosing the confidence level of ML outputs to set user expectations.

- **Failure to Act**: The opposite of overtrust; failing to intervene when necessary.
- Inform users about significant residual risks.
- Review the human user interface design to ensure it does not introduce bias or unduly influence users.
- Evaluate the degree of human-in-the-loop decision-making (from human-in-the-loop to human-in-command), including feedback mechanisms and their impact on system outcomes and training.

10.9.1.4 Algorithmic Transparency, Explainability, and Fairness

- Review algorithmic outcomes and decisions about setting prediction thresholds.
- Select fairness tests and performance metrics (including publication of the confusion matrix, where applicable).
- Ensure the rigor of justifications reflects the severity and complexity of the use case and the level of experience; acknowledge potential conflicts between group and individual fairness tests.
- Document and adjust the explainability of decision/prediction outcomes for key stakeholders.
- Where possible, include feature weightings in algorithmic explainability.

10.9.1.5 Security and Privacy Risks

- **Cybersecurity Risks**: Data collected during machine learning (ML) development, testing, and improvement is a potential target for malicious actors. Big data applications must address vulnerabilities such as data poisoning and unauthorized access to private data. ML's opaque nature can make it challenging to detect security breaches.
- **Privacy Failures**: Unauthorized disclosure of sensitive information; anonymization processes can fail, leading to the re-identification of patients.
- **The Use of Synthetic Data** can help address privacy risks but introduces new challenges regarding data realism and applicability.

10.9.1.6 Performance Monitoring and Post-market Surveillance

- Post-production Monitoring: Should collect information on overtrust and other issues, with actions taken as appropriate.
  - Monitoring device performance over time is a critical aspect of ensuring safety and optimal performance.
  - Recognize that autonomous systems require closer monitoring due to different risk profiles.

- Acknowledge that adaptive ML systems, which learn over time, make monitoring more important and potentially more challenging.
- Identify and document the acceptable range of product performance to support post-market monitoring.

10.9.2 Issues with AI/ML Applications

Below are some issues that should be identified and analyzed as part of the risk analysis process [22]:

- AI Functional Errors:
  - Misinterpretation of medical images, leading to incorrect diagnoses
  - Object misidentification in surgical robots, potentially causing surgical errors
  - Speech recognition failures due to variations in accents or dialects
- Software Rot:
  - Degradation of software performance over time
  - Outdated algorithms lead to inaccurate predictions or recommendations
- Unexplained Programming Glitches:
  - Unpredictable behavior in complex AI systems
  - "Hallucinations" or unexpected outputs from AI models
- Human Factor Issues:
  - Over-reliance on AI systems by healthcare practitioners
  - Improper use or misinterpretation of AI-generated results
- Data-Related Failures:
  - Biases in training data leading to unfair or inaccurate outcomes
  - Incomplete or outdated datasets undermining AI effectiveness
- Security Vulnerabilities:
  - Susceptibility to hacking, potentially compromising patient data or device function
  - Insertion of false data into medical imaging systems
- Integration and Adoption Challenges:
  - Difficulties in incorporating AI systems into existing clinical workflows
  - Resistance or improper use by healthcare professionals
- Algorithmic Bias and Fairness Issues:
  - Unequal performance across different patient demographics
  - Potential for discriminatory outcomes in diagnosis or treatment recommendations

- Performance Degradation:
  - Inconsistent performance in real-world settings compared to controlled testing environments
  - Inability to adapt to new scenarios or patient populations
- Transparency and Explainability Issues:
  - Difficulty in understanding decision-making processes of complex AI models
  - Challenges in identifying and addressing errors in "black-box" systems

10.10 Usability and Human Factor Requirements

AI/ML-enabled medical devices must meet robust usability and human factor requirements to ensure safe and effective use by clinicians and other intended users. Human factors engineering is a critical component that focuses on the interaction between users, devices, and the clinical environment to minimize user errors and maximize device performance. IEC 62366-1:2015+AMD1:2020, "Medical devices—Part 1: Application of usability engineering to medical devices," is the primary standard for usability engineering. It mandates that usability be assessed throughout the development process, including formative evaluations during design and summative evaluations in real-world settings. This standard emphasizes that all intended users and environments must be considered and that AI/ML functionalities—such as decision support—must be presented and controllable so that users understand how to interpret AI outputs and can override or adjust them as needed [23].

For AI/ML devices, it is essential to have user interfaces that are both intuitive and reliable while also ensuring that the transparency and explainability of AI recommendations are addressed. Standards such as ISO 13485 (quality management), ISO 14971 (risk management), and IEC 62304 (software lifecycle management), as well as technical report AAMI TIR 34971, provide further structure for managing risks and ensuring software reliability, which is especially important given the adaptive nature of AI/ML systems.

10.10.1 Usability, Transparency, and Explainability

Usability and human factors are critical considerations in the design of AI/ML-enabled medical devices. Design must account for user needs, cognitive load, and the potential for use errors, mainly when AI/ML provides decision support. The interface should communicate AI outputs and offer mechanisms for users to agree, disagree, or override suggestions. Additionally, the device must integrate seamlessly into clinical workflows, minimizing disruptions and enabling clinicians to efficiently interpret and act on AI/ML outputs [22].

Transparency and explainability are increasingly emphasized for AI/ML-enabled medical devices. Users should understand how the system functions, including its limitations and the rationale behind its recommendations. AI/ML outputs must be presented in a way that clinicians can comprehend and justify, which is essential for building trust and ensuring safe clinical decision-making. Regulatory bodies, including the FDA, Health Canada, and MHRA, have published guiding principles that underscore the importance of clear communication with users and encourage thorough documentation of algorithms, data sources, and decision logic.

Validation and performance are also central to regulatory compliance. AI/ML-enabled devices require rigorous validation to ensure they perform as intended, including clinical validation for diagnostic or therapeutic outputs, with evidence of sensitivity, specificity, and reliability. As devices with AI/ML capabilities evolve, regulatory frameworks such as the FDA's Predetermined Change Control Plan (PCCP) enable iterative updates. Still, ongoing validation and monitoring are required to maintain safety and effectiveness. Risk management standards, such as ISO 14971 and AAMI TIR 34971:2023, provide guidance specific to AI/ML, addressing hazards, including data bias, algorithm errors, and cybersecurity risks [21].

10.10.2 Use Related Risks

AI/ML-enabled medical devices introduce unique usability and human factor requirements that are crucial for ensuring transparency, explainability, validation, user interface design, and workflow optimization while also identifying and mitigating use-related risks. These requirements are increasingly emphasized in regulatory frameworks and international standards, reflecting the need for devices to be not only technically robust but also safe and effective in real-world clinical environments.

The user interface (UI) must be designed with human factors in mind, ensuring that device outputs are presented in a way that supports effective clinical decision-making. This involves an intuitive display of AI-generated information, clear differentiation between recommendations and final decisions, and the ability for users to easily override or question AI outputs when necessary. Workflow integration is another critical consideration, as AI/ML-enabled devices should seamlessly fit into existing clinical processes without disrupting established routines or introducing unnecessary complexity. Identifying and managing use-related risks is an ongoing process that spans the entire product lifecycle. Risks may include misinterpretation of AI outputs, over-reliance on automated recommendations, or workflow disruptions due to poorly designed interfaces. Usability engineering standards, as outlined in IEC 62366-1, require formative and summative usability testing with representative users and in realistic environments to uncover and address these risks early in the development process [23].

10.11 Summary

This chapter provided a comprehensive examination of the latest regulatory developments in AI Lifecycle Frameworks for the medical device industry. The review began by examining initiatives undertaken by the U.S. Food and Drug Administration and Health Canada aimed at addressing and simplifying regulatory hurdles in the development of artificial intelligence software for medical devices. The FDA's recent draft guidance on AI-enabled device software functions offered detailed recommendations for lifecycle management and marketing submissions. In contrast, Health Canada's guidance on machine learning-enabled medical devices, which covers topics such as bias, data representativeness, and transparency, was also discussed. The chapter explored state-of-the-art practices for AI software development, emphasizing the importance of a total product lifecycle approach. It then shifted to the European Union's AI Act, which introduced a risk-based approach to regulating AI systems by categorizing them into four risk levels: unacceptable, high, limited, and minimal. The discussion also examined AI-enabled medical device risk management, highlighting the crucial role of usability engineering in the development of these devices. It underscored the importance of minimizing use-related hazards and risks to ensure that users could operate devices safely and effectively. By examining these regulatory frameworks and best practices, the chapter delivered valuable insights into the complex landscape of AI regulation and development in the medical device industry.

10.12 Points to Ponder

1. For an AI/ML-enabled medical device, what is the difference between Performance Monitoring and Real-World Performance Monitoring?
2. Question on Fig. 10.2 from Health Canada Guidance—"Figure 10.2 gives a visual overview of the MLMD Product Lifecycle. The iterative components reflected in this lifecycle schematic are not mutually exclusive and may not occur in the order indicated." What does it mean by mutually exclusive and may not occur in the order indicated?
3. Section 10.6.3.1 includes CEN/CLC/JTC 21 Published Standards on Artificial Intelligence. How to use CEN/CLC/JTC 21 Published Standards on Artificial Intelligence in AI/ML-enabled medical device development and EU MDR compliance?

10.13 Answers

1. For an AI/ML-enabled medical device, Performance Monitoring and Real-World Performance Monitoring are related but distinct concepts:
   (a) Performance monitoring refers to the ongoing assessment of how well the device operates according to predefined metrics, typically in controlled environments such as laboratory testing, clinical trials, or simulated scenarios. This process focuses on tracking model accuracy, precision, recall, and other relevant metrics to ensure the device meets its intended specifications and regulatory requirements. Performance monitoring typically occurs during development, validation, and the initial post-market phases, with a focus on detecting technical issues, data drift, or model degradation in a structured setting.
   (b) Real-world performance monitoring involves evaluating how the device performs in actual clinical practice outside of controlled test environments. This monitoring captures data from diverse patient populations, clinical settings, and real-world usage patterns. It aims to identify and address deviations or unintended outcomes that may arise due to variations in patient demographics, clinical workflows, or data input characteristics. Real-world performance monitoring is crucial for ensuring the safety, effectiveness, and reliability of AI/ML devices as they interact with the complexities and unpredictability of real healthcare environments.
2. The Machine Learning Deployment Pipeline is structured using Directed Acyclic Graphs (DAGs), which determine the execution order of tasks based on their relationships and dependencies. As depicted in Fig. 10.2, each activity is connected by arrows indicating the flow of work. Unlike a traditional waterfall development methodology, where phases are distinct and strictly sequential, the DAG-based approach allows for greater flexibility. In this pipeline, activities are not necessarily mutually exclusive, and the development process can adapt dynamically based on project outcomes, enabling teams to move forward or revisit previous steps as needed.
3. To ensure AI/ML-enabled medical devices comply with EU MDR and broader AI regulatory requirements, developers should integrate CEN/CLC/JTC 21 standards throughout the device lifecycle. These standards guide the design, testing, risk management, and post-market monitoring of AI systems, emphasizing safety, transparency, and robustness. By following JTC 21 recommendations, manufacturers can align their technical documentation, MDRs' state-of-the-art (SoTA) determination, and conformity assessments with both medical device and AI-specific legal frameworks, helping to demonstrate compliance and maintain trust in their products.

References

1. US FDA Guidance (2025) Artificial intelligence-enabled device software functions: lifecycle management and marketing submission recommendations
2. US FDA, MHRA, and Health Canada. Transparency for machine learning-enabled medical devices: guiding principles. https://www.fda.gov/medical-devices/software-medical-device-samd/transparency-machine-learning-enabled-medical-devices-guiding-principles
3. US FDA's Artificial Intelligence (AI) lifecycle framework. https://www.fda.gov/media/180310/download?attachment
4. US FDA's Good machine learning practices. https://www.fda.gov/medical-devices/software-medical-device-samd/good-machine-learning-practice-medical-device-development-guiding-principles
5. US FDA Guidance (2023) Cybersecurity in medical devices: quality system considerations and content of premarket submissions
6. US FDA Guidance (2024) Marketing submission recommendations for a predetermined change control plan for artificial intelligence-enabled device software functions
7. Health Canada (2025) Pre-market guidance for machine learning-enabled medical devices
8. Health Canada (2024) Predetermined change control plans for machine learning-enabled medical devices: guiding principles
9. Health Canada (2024) Transparency for machine learning-enabled medical devices: guiding principles
10. Future of Life Institute (2025) The EU artificial intelligence act—up-to-date developments and analyses of the EU AI Act
11. EU AI Act Article 3. https://artificialintelligenceact.eu/article/3/
12. Aboy M, Minssen T, Vayena E (2024) Navigating the EU AI Act: implications for regulated digital medical products. npj Digit Med 7:237. https://doi.org/10.1038/s41746-024-01232-3
13. Medtech Europe (2024) Medical technology industry perspective on the final AI Act
14. EU AI Act. Article 5: Prohibited AI practices. https://artificialintelligenceact.eu/article/5/
15. EU Commission (2025) The guidelines on prohibited artificial intelligence (AI) practices, as defined by the AI Act
16. German Notified Bodies Alliance for Medical Devices and Team NB (The European Association of Medical Devices Notified Bodies) (2024) Questionnaire: artificial intelligence in medical devices
17. Team-NB Position Paper, Artificial Intelligence Act (2025)
18. ISO/IEC JTC1 in AI standards. https://jtc1info.org/technology/subcommittees/ai/
19. CEN/CLC/JTC 21 Artificial Intelligence
20. ANSI/AAMI/ISO 14971:2019; Medical devices—application of risk management to medical devices
21. AAMI TIR34971:2023; Application of ISO 14971 to machine learning in artificial intelligence—guide
22. Babic B, Glenn Cohen I, Stern AD et al (2025) A general framework for governing marketed AI/ML medical devices. npj Digit Med 8:328. https://doi.org/10.1038/s41746-025-01717-9
23. IEC 62366-1:2015 Part 1: Application of usability engineering to medical devices

Chapter 11
AI-Based Health Care Applications

Abstract Latest advances in the healthcare frontier utilize cloud computing and AI/ML to explore new methods for diagnosing and treating existing diseases. Further advances are expected as more medical data becomes available in the Cloud—the use of AI and ML-based solutions to identify hidden patterns in data yields previously unavailable insights. Doctors can better determine the course of treatment for patients by comparing a patient's test results with datasets collected over time. We are on the verge of a perfect storm of medical data in the Cloud, with AI and ML solutions that will achieve breakthroughs like never seen before. This chapter presents several examples of this brewing revolution in medical treatments using the Cloud with AI and ML. The first example is of Heartflow, related to cardiovascular health. Second is a collaborative Cloud solution developed by Intel to facilitate cancer treatments. The third is related to multi-modal Glaucoma detection. We wrap up with a discussion on the emerging role of GenAI in healthcare.

Keywords Healthcare · Cardiovascular · Blood flow · Multi-clouds · Federated learning · Gen AI · Cancer · Glaucoma

11.1 Introduction

A crucial area of human development is healthcare. Within healthcare, an important consideration is tracking fundamental indicators of human health. In this chapter, we will review three case studies of cloud-based health care solutions. The first one is related to cardiovascular health and comes from a startup called Heartflow. Second is a collaborative Cloud solution developed by Intel to facilitate cancer treatments. The third is related to multi-modal Glaucoma detection.

A. Pandey et al., *Artificial Intelligence in Medical Software*,
https://doi.org/10.1007/978-3-032-10808-1_11

11.2 Heart Care Data in the Cloud

Heart disease is the #1 killer in both the United States and the World. It is also preventable if risk factors are identified at an earlier stage and care is taken to manage the progression of the disease [1]. There are multiple methods available to assess cardiovascular health. The most accurate one is angiography, which is both expensive and invasive. A non-invasive method is Computed Tomography (CT) scan, which is limited by the quality of images taken. Scientists at Heartflow [2] utilize various images captured during a scan and combine them with advanced technology to compute Fractional Flow Rates (FFR), achieving an accuracy comparable to that of angiography. This results in pretty accurate estimates of blood flow through blocked coronary arteries, as shown in Fig. 11.1.

This representation is essential, as in the case of heart disease, arteries are often blocked, causing restricted blood flow. However, reduction in blood flow is not linearly related to a vessel's blockage. Keeping a thumb on the gardening water hose causes the flow velocity to increase. Similarly, a 70% blockage may still allow 80–90% of the blood to pass through the remaining 30% of the opening in the

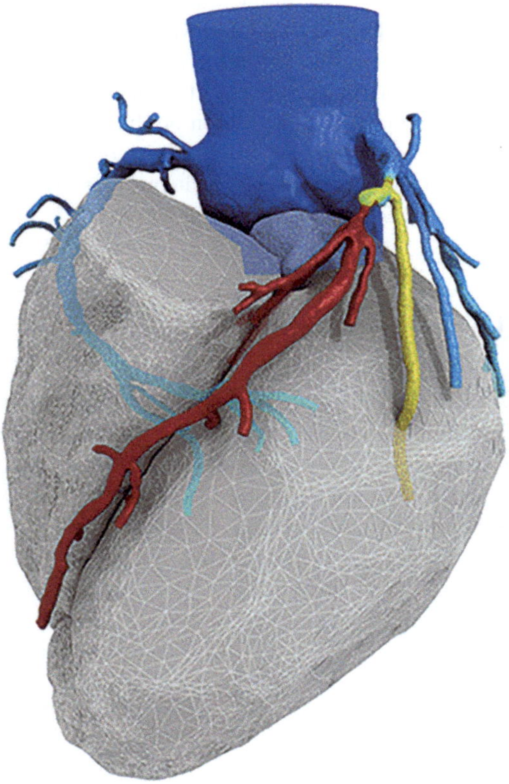

Fig. 11.1 Fractional flow rate (FFR) analysis by Heartflow [2]

artery. It should be noted that the flow rate is limited by the constrained opening and the extent of the blockage. A CT scan alone can confirm the presence and extent of blockages, but it needs Heartflow's FFR technique and fluid-dynamics computations in the Cloud to determine the fraction of flow rate. The latter allows a cardiologist to recommend a proper treatment course for the patient, such as a bypass operation, a stent, or medication to manage the disease.

FFR computations require gigabytes of patient data to be moved from CT scanners in a remote hospital to Amazon's data centers [3]. This data is processed to render a 3D image on a doctor's handheld device. As shown in Fig. 11.2, the CT imaging data flows from a hospital to AWS storage buckets. There, serverless lambda-based computations are performed. This also eliminates the need for dedicated EC2 instances. The medical reports are generated for doctors' review and discussions with the patients. Currently, AI and ML play a limited role in identifying blockages and blood flow issues in images, as well as indicating potential improvements if stents were to be placed. In the future, diagnoses made by medical professionals, the course of treatment, and results with actual patients can be further used as inputs for future machine learning and diagnosis as well.

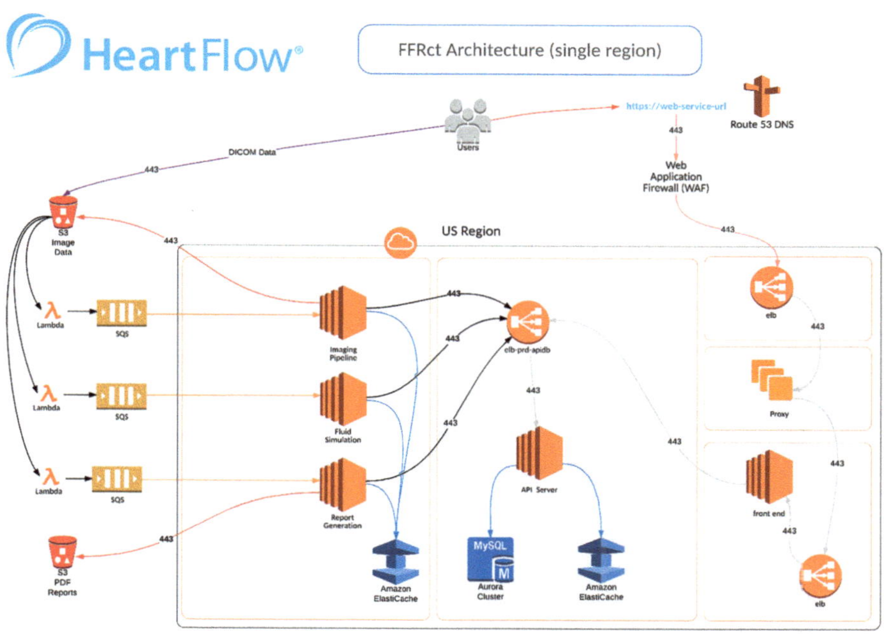

Fig. 11.2 Medical data copied from a hospital to servers in AWS [3]

11.3 Cloud-Based Collaborative Tools

Millions of people worldwide suffer from cancer. Once a diagnosis is made, painful treatments including radiation, chemotherapy, or surgery begin. All of these tend to kill both healthy and cancerous cells. If these patients can be treated as individuals based on their specific genome sequencing, and precision treatment can be drawn quickly, the course of the disease can be significantly altered (Fig. 11.3).

With the above problem in mind, Intel launched the Collaborative Cancer Cloud (CCC) in 2015, which enabled institutions to securely share their patients' genomic, imaging, and clinical data for potentially lifesaving discoveries [4]. Such sharing enables the analysis of large amounts of data from sites worldwide in a distributed system, while preserving the patient's privacy and security.

Exploring solutions beyond cancer, a significant challenge lies in identifying high-value groups of subjects for clinical trials, predicting patient responses to proposed treatments, and selecting relevant biomarkers to extract new insights. Hence, a new technology has emerged to train ML models at scale across multiple medical institutions without moving the data between them. It is called Federated Learning (FL) [5], an instance of which is depicted in Fig. 11.4. FL enables data to stay local, and algorithms to travel across the participating institutions, for training a deep learning algorithm while preserving the privacy and security of patients' data.

For example, Intel and the University of Pennsylvania have announced a collaboration involving 29 international medical centers to train models to recognize brain tumors [6]. The goal of this is to provide diverse datasets for ML that no single institution can provide alone. This effort started recently, and if successful, will

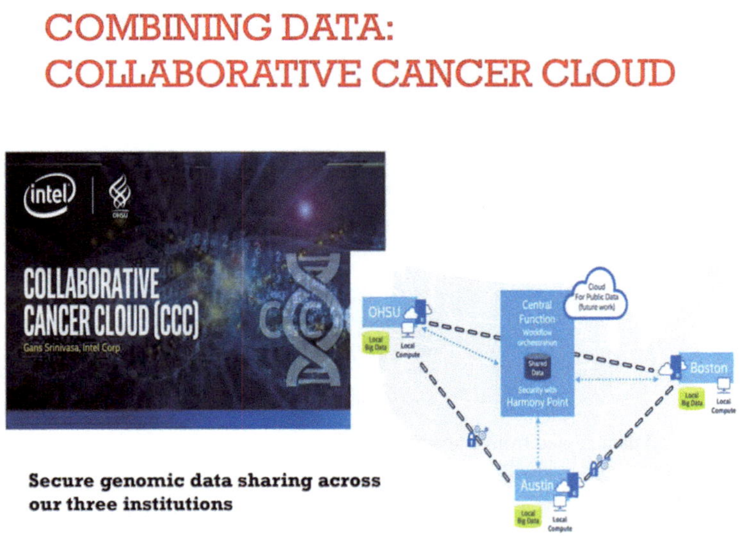

Fig. 11.3 Intel's Collaborative Cancer Cloud [4]

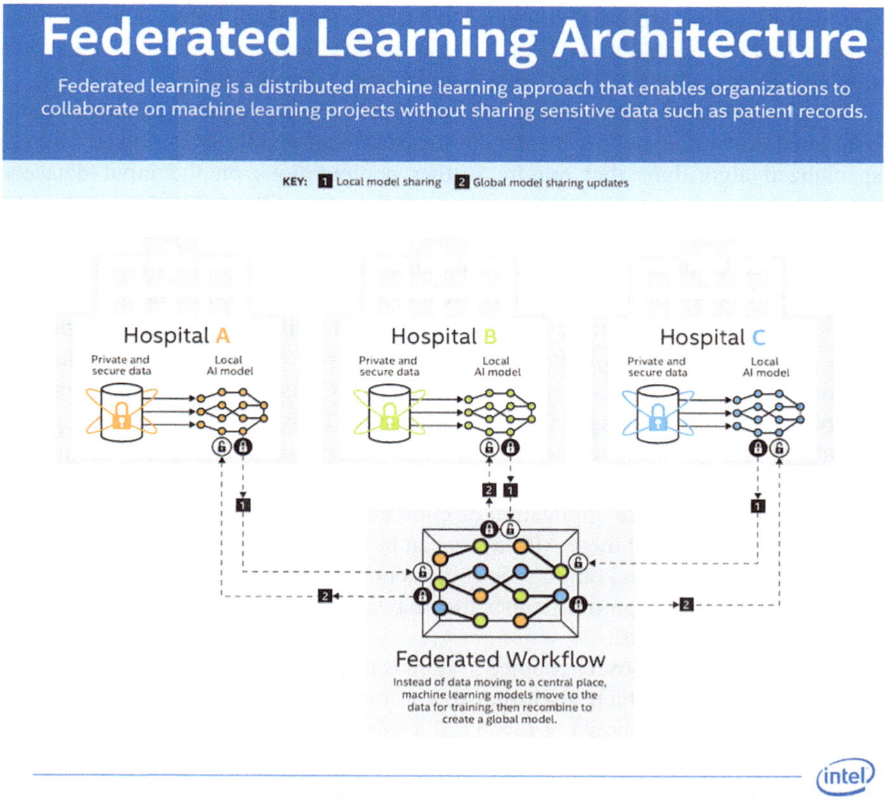

Fig. 11.4 A federated learning solution for ML and healthcare [7]

result in a new class of solutions that can identify brain tumors from a greatly expanded version of the International Brain Tumor Segmentation (BraTS) challenge dataset [7]. BraTS involves the evaluation of state-of-the-art methods for segmenting brain tumors in multimodal magnetic resonance imaging (MRI) scans. Adding FL provides datasets that enhance the quality of results and benefit patients worldwide.

11.4 Multi-Cloud Solutions

To manage operational risks, medical entities are reluctant to use Public Clouds or put all their data in a single Cloud location. Consider a company X that develops medical research algorithms to aid doctors in making informed decisions. To leverage the Cloud's storage and compute prowess, it needs to model and train algorithms based on hospital data. As hospitals adopt Cloud technology, they do so at varying rates, using different strategies to manage their data. Let us consider three client hospitals, identified as Client 1, Client 2, and Client 3. Client 1 uses local data storage, whereas Client 2 uses Cloud A and Client 3 uses Cloud B.

Suppose this company wants to run two types of algorithms: (1) standard computational software, such as displaying measurement results in real time, and (2) specialized algorithms that require further computations on the input datasets, which may not yet be approved for non-research usages. For standard practice computations, such as showing the results and output of a scan, the required algorithms can be ported to the Cloud or a local computer facility. However, for research algorithm deployments, the company may use a loosely coupled infrastructure to access data sources at a higher latency and at a lower cost. This allows data to be pushed from a source, sent to a specific Cloud where the research algorithms are hosted, and results may stay local or be moved back to the Cloud. This secure data transfer can be accomplished using a four-way handshake between API's of two separate Clouds, or between a Cloud and the local data warehouse. The diagram in Fig. 11.5 illustrates an example of data that could be exchanged between Cloud 1 (the data source) and Cloud 2 (the application or computation Cloud) for Clients 2 and 3. It is important to note that these API layers can be configured at a broader level based on triggers for inputs and outputs. Specific source and destination-related information can be passed at run time rather than having specific data values or locations hardcoded into a system.

An advantage of a loosely coupled system is that the data owner retains complete control and decides which algorithm to run on which dataset. Using a message sequence chart [8], as shown in Fig. 11.5, a user on the left initiates services of an application or algorithm that resides in a different Cloud on the right side. A request is sent to the application owner. Now it is up to the algorithm owner to honor the request. If it decides, then a counter request flows in the other direction as depicted

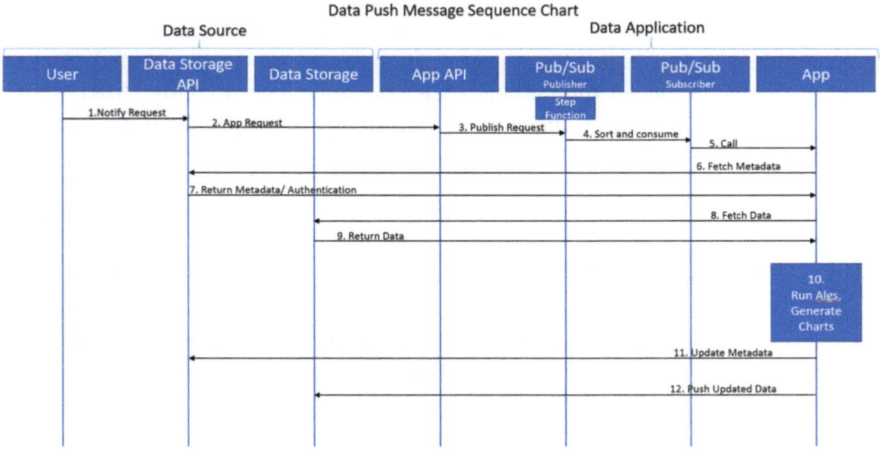

Fig. 11.5 A loose coupling between the data source and applications [8]

by the right-to-left arrow in step 6. Note that the dataset owner provides some minimal information via metadata to the algorithm owner, who then processes it and generates the required results. These are either sent back to the data owner or kept at the algorithmic source. The latter option is preferable if multiple data sources are needed for the computations, as in the case of the Collaborative Cancer Cloud [4].

11.5 Multimodal Glaucoma Detection

Glaucoma [9] is a progressive optic neuropathy characterized by irreversible damage to the optic nerve, typically associated with elevated intraocular pressure, as shown in Fig. 11.6. The disease leads to gradual vision loss, often without early symptoms, making early detection critical. Multiple types exist, with open-angle glaucoma being the most prevalent, potentially resulting in permanent blindness if left untreated.

While glaucoma can affect people of all ages, senior citizens and people with conditions listed below are more vulnerable to Glaucoma:

- Patients suffering from diabetes
- People having a family history of glaucoma
- Patients suffering from blood pressure problems
- People who have myopia and hyperopia
- People taking eye drops or corticosteroid medications

Primary open-angle glaucoma (POAG) and primary angle-closure glaucoma (PACG) are distinct types, characterized by significant differences in anatomy,

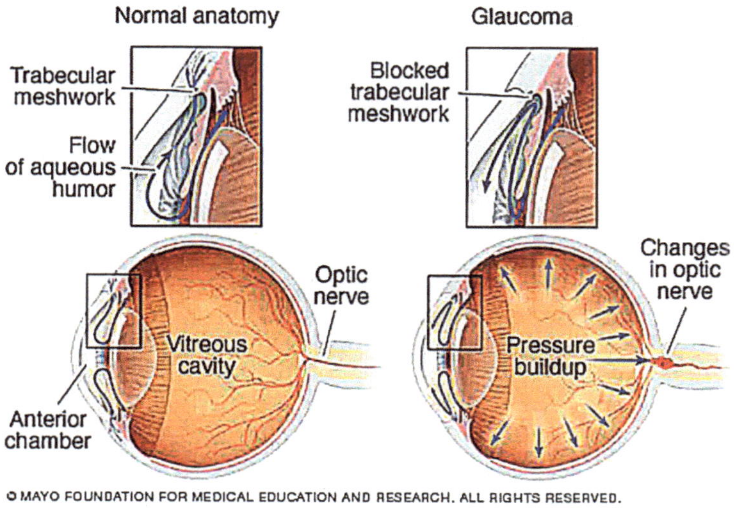

Fig. 11.6 Elevated intraocular pressure causing glaucoma [9]

mechanisms, and clinical presentations [10]. As shown in Fig. 11.7, POAG is characterized by an open iridocorneal angle with inadequate aqueous humor drainage through a dysfunctional trabecular meshwork. At the same time, PACG involves a narrowed or closed angle due to physical obstruction. POAG is more prevalent in Western countries and typically has a gradual, asymptomatic onset. In contrast, PACG is more common in Asia and can present acute, symptomatic attacks or chronic progression. The diagnosis of POAG relies primarily on optic nerve examination and visual field testing, while the PACG diagnosis crucially involves gonioscopy. Treatment approaches also differ, with POAG usually managed initially with topical medications to lower intraocular pressure and PACG often requiring laser peripheral iridotomy as a first-line intervention. Despite these differences, both conditions can lead to optic nerve damage and vision loss if left untreated, underscoring the importance of early detection and appropriate management strategies tailored to each type of glaucoma.

Glaucoma, particularly open-angle glaucoma (OAG), is a complex eye condition where increased intraocular pressure (IOP) can damage the retina and optic nerve head (ONH), potentially leading to irreversible vision loss. In this type, deposits get built up in the eye's drainage canals. These deposits increase the pressure on the eye, causing damage to the optic nerve [11]. Since the canals appear to be open and function normally, you may not notice the disease until you lose vision. Open-angle glaucoma is the most common type. It is also called wide-angle glaucoma, as shown in Fig. 11.8.

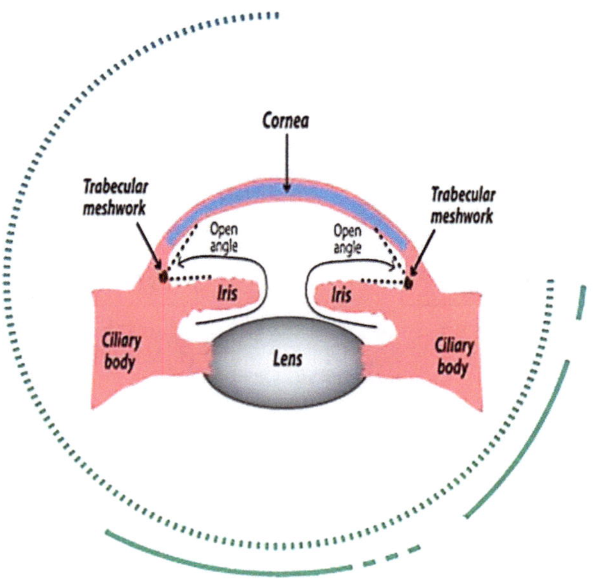

Fig. 11.7 Primary open-angle glaucoma [10]

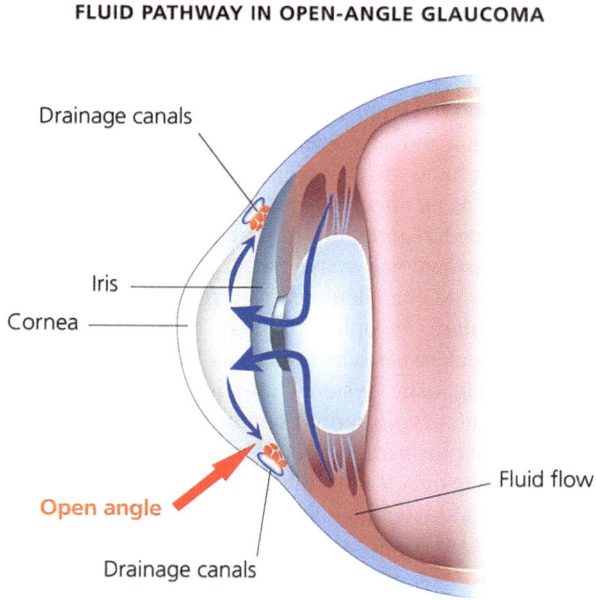

Fig. 11.8 Formation of open angle glaucoma [11]

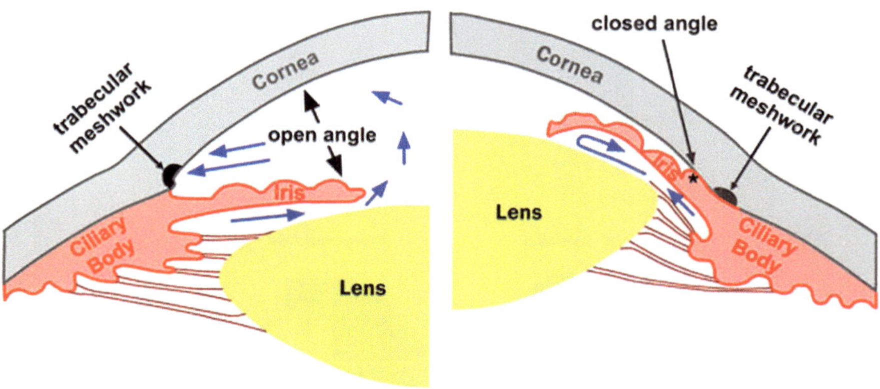

Fig. 11.9 Formation of closed angle glaucoma [11]

In contrast, closed-angle or angle-closure glaucoma occurs when the angle formed between the iris and cornea is too narrow. As shown in Fig. 11.9, this narrow angle can block the drainage canals. Hence, fluid cannot circulate and pass through the eye, which increases the pressure. This again increases the pressure on the eyes and can lead to severe eye pain and headaches.

Early detection of glaucoma is crucial for implementing timely treatments that can effectively slow disease progression, although diagnosis can be challenging, especially in the early stages. The diagnostic process typically involves a comprehensive evaluation of the ONH and the adjacent retinal nerve fiber layer (RNFL) through retinal examination and advanced imaging techniques, assessment of visual fields (VFs), and measurement of IOP levels, as shown in Fig. 11.10. While lowering IOP has been proven as an effective treatment for OAG, the multifaceted nature of glaucoma diagnosis underscores the importance of a thorough and integrated approach to patient assessment and management.

Artificial intelligence (AI) has demonstrated remarkable capabilities in healthcare diagnostics, particularly in ophthalmology [12]. Its proficiency extends to detecting complex medical issues across various healthcare domains. In ophthalmology, AI-powered diagnostic systems utilizing fundus images have become cutting-edge tools for identifying retinal conditions, glaucoma, and other ocular diseases. Glaucoma, an optic neuropathy characterized by specific structural and functional changes, affects over 90 million people globally and is the second leading cause of blindness and associated disability. The global prevalence of primary open-angle glaucoma (POAG) and primary angle-closure glaucoma (PACG) is significant and growing. In 2013, an estimated 64.3 million people aged 40–80 years were affected worldwide. This number is projected to rise to 76.0 million by 2020. It is expected to reach 111.8 million by 2040, underscoring the increasing importance of advanced diagnostic tools, such as AI, in managing this condition.

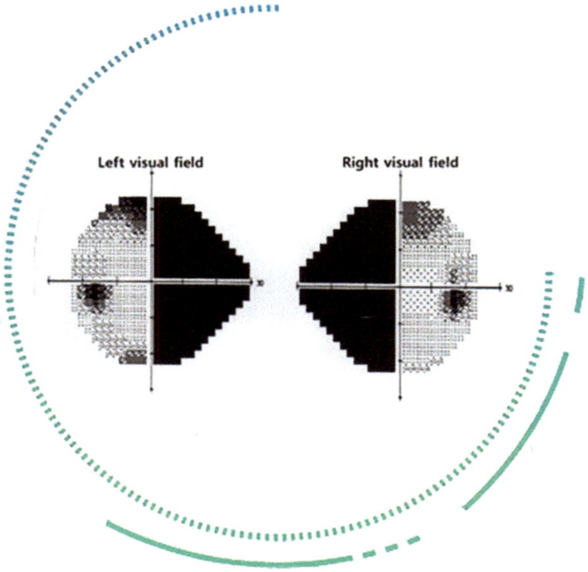

Fig. 11.10 Visual field test depiction [10]

Detecting glaucoma presents significant challenges, especially in its early stages. However, early disease identification is crucial as it can slow its progression and mitigate future vision loss. One of the primary difficulties in diagnosing glaucoma lies in the fact that its signs and symptoms often become apparent only after substantial vision loss has already occurred. This disease characteristic underscores the importance of developing and implementing advanced diagnostic techniques to identify glaucoma in its nascent stages before irreversible damage to vision occurs.

The diagnosis of glaucoma typically involves a comprehensive approach that incorporates several key assessments. These include a thorough examination of the optic nerve head (ONH) through retinal evaluation, measurement of intraocular pressure (IOP), and assessment of visual fields (VFs) [10]. Additionally, other related factors are taken into consideration during the diagnostic process. This multifaceted approach enables healthcare professionals to obtain a comprehensive understanding of the patient's ocular health and identify any signs indicative of glaucoma, facilitating early intervention and effective management of the condition.

Glaucomatous optic neuropathy (GON) is the primary manifestation of glaucoma, and its identification is crucial for diagnosis. This condition is commonly detected through two primary imaging modalities: fundus photographs and optical coherence tomography (OCT) images. Fundus photography has been the gold standard for documenting optic nerve status and detecting GON since its introduction in 1857. This long-standing use has resulted in the accumulation of extensive, annotated datasets of fundus photographs. These large collections of images are particularly valuable in the current era of artificial intelligence, as they provide ideal training material for machine learning (ML) and deep learning (DL) models. The availability of such comprehensive datasets has significantly contributed to the advancement of AI-based diagnostic tools for glaucoma detection.

Prior research has revealed significant challenges in accurately assessing glaucoma risk by both ophthalmology trainees and comprehensive ophthalmologists. Studies have shown that these professionals underestimated the probability of glaucoma in 20% of disc photographs, a finding that highlights the complexity of glaucoma diagnosis. Moreover, the research indicated that ophthalmology trainees and comprehensive ophthalmologists were twice as prone to either underestimating or overestimating the likelihood of glaucoma. This inconsistency in diagnosis can be attributed to several factors, including the inherent variability in assessing the cup-to-disc ratio, the presence of peripapillary retinal nerve fiber layer (RNFL) atrophy, and the detection of disc hemorrhages. These findings underscore the need for ongoing education and training in glaucoma assessment, as well as the potential benefits of incorporating advanced diagnostic tools to supplement clinical judgment in glaucoma detection and management.

Optic nerve evaluation is primarily subjective, with considerable overlap in structural and functional characteristics between healthy individuals and those with disease. This subjectivity leads to variability in assessments between different observers and by the same observer over time. The challenge is compounded when monitoring glaucoma progression, which typically occurs slowly over decades, making accurate prediction of disease advancement difficult. In this context, automated machine learning (ML) models offer the potential for more objective, consistent, and precise assessments, improving the accuracy of optic nerve evaluation and glaucoma progression monitoring.

Although many studies demonstrated promising results in detecting glaucoma using AI, limitations still exist from many perspectives. For example, the lack of standardized and consensus glaucoma definitions makes it difficult to evaluate the results consistently in different datasets; the shortage of large, well-annotated datasets of good quality limits the generalizability of the AI model; the limited interpretability and liability of the DL model hinder the implementation of it in the clinic.

Despite the promising results of numerous studies using artificial intelligence (AI) for glaucoma detection, several significant limitations persist across various aspects of this approach. One major challenge is the absence of standardized and universally accepted definitions of glaucoma, which complicates the consistent evaluation of results across different datasets. This lack of uniformity makes it difficult to compare and validate findings from various studies effectively. Another crucial limitation is the scarcity of large, high-quality datasets with accurate annotations. This shortage significantly hampers the generalizability of AI models, as they may not be exposed to a sufficiently diverse range of cases to ensure reliable performance across different patient populations and clinical settings.

Furthermore, the limited interpretability of deep learning (DL) models poses a significant hurdle to their clinical implementation. The "black box" nature of these models often makes it challenging for healthcare professionals to understand and trust the decision-making process, raising concerns about liability and patient safety. These limitations underscore the need for continued research and development to enhance the reliability, transparency, and clinical applicability of AI in glaucoma detection.

11.6 AI Application for Glaucoma

Automated AI models, as shown in Fig. 11.11, could minimize subjectivity by interpreting and quantifying retinal and optic nerve images. This will help optimize workflows and processes in glaucoma clinics, potentially leading to more time for clinicians to interact with patients. It requires quantification of optic cup, disc, and rim characteristics in fundus images, retinal layers in OCT images, and patterns of VF loss. An improved glaucoma assessment and its forecasting, screening, diagnosis, and prognosis will help advance the treatment cycle [12].

Supervised Models: Supervised learning models have been extensively applied in various aspects of glaucoma management, including:

1. Detection
2. Severity classification
3. Progression prediction
4. Segmentation of relevant structures

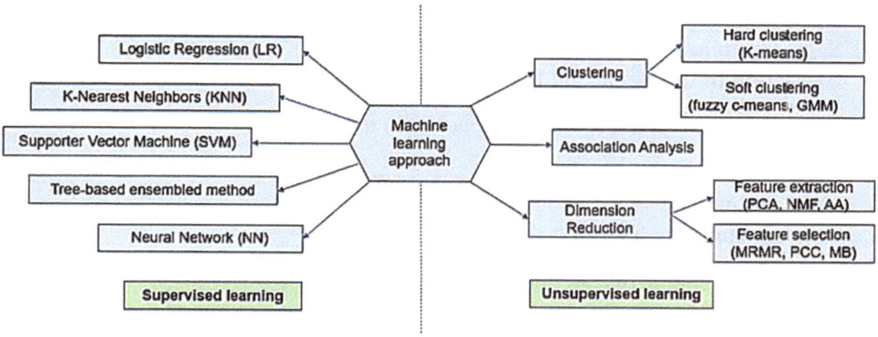

Fig. 11.11 Various types of ML models applied to glaucoma [12]

These models utilize diverse data modalities, such as:

- Visual field (VF) tests
- Fundus imaging
- Optical Coherence Tomography (OCT)
- Clinical data
- Transcriptomic data

The versatility of supervised models enables researchers and clinicians to leverage various types of patient data, thereby enhancing our understanding and management of glaucoma across its different stages and manifestations.

Several metrics were used for model evaluation, such as:

- Accuracy
- Sensitivity/recall
- Specificity
- Error rates
- Precision
- True positive rate
- False positive rate
- Area Under the Receiver Operating Characteristic (ROC) curve (AUC, the model's performance across various thresholds, presenting TPR against FPR at various threshold settings)
- Area under the precision-recall curve (AUPRC), which is a classification model performance metric appropriate for imbalanced classes, demonstrates precision against recall at different threshold settings

Unsupervised Models: Unsupervised learning techniques are valuable for discovering intrinsic patterns and extracting representative features from complex datasets without predefined labels. In glaucoma research, these methods have been applied to various tasks, including:

1. Identifying patterns of Visual Field (VF) loss
2. Detecting Retinal Nerve Fiber Layer (RNFL) deterioration patterns

3. Glaucoma staging
4. Segmentation of ocular structures
5. Extraction of other relevant features

Common unsupervised learning approaches in this field include:

- Clustering algorithms
- Association analysis
- Dimensionality reduction techniques

These methods enable researchers to uncover hidden structures in glaucoma-related data, potentially leading to new insights into disease progression and subtypes.

11.6.1 Deep Learning for Glaucoma

Several Deep Learning (DL) models for Glaucoma classification are shown in Fig. 11.12. Among these, hybrid models are of particular interest. Hybrid Convolutional Neural Network (CNN) models are emerging as a promising approach in glaucoma research, reflecting the complex and multifactorial nature of the disease. These advanced models integrate different single CNN architectures and leverage single or multiple data modalities to provide a more comprehensive understanding of glaucoma.

By fusing information from various sources such as visual fields, retinal imaging, and clinical data, hybrid models can capture nuanced insights that individual models might miss. The growing interest in these hybrid approaches is rooted in information theory principles, which suggest that combining multiple sources of information can significantly enhance disease characterization. As glaucoma presents intricate pathological variations, hybrid CNN models offer a more holistic perspective by synthesizing diverse data types and architectural strategies. Researchers

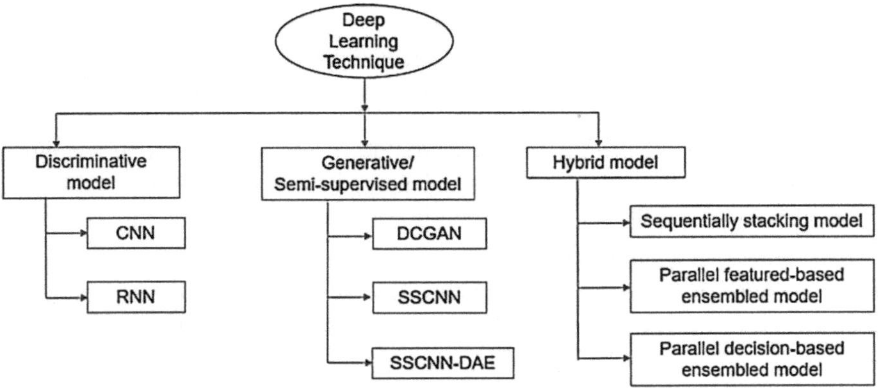

Fig. 11.12 DL models for glaucoma classification [12]

are increasingly recognizing that no single modality or model can fully encapsulate the complexity of glaucoma, making these hybrid approaches increasingly valuable. Looking ahead, the scientific community anticipates continued growth in the development of hybrid CNN models. These models represent a sophisticated approach to unraveling the intricate mechanisms of glaucoma, potentially enhancing early detection, progression prediction, and personalized treatment strategies. As machine learning and computational techniques advance, hybrid CNN models are poised to play a pivotal role in transforming our understanding and management of this challenging eye disease.

11.6.2 Available Glaucoma Datasets

Several public datasets of retinal images have been established for automated glaucoma screening, including RIGA, ORIGA, DRISHTI-GS, REFUGE, and PAPILA. These datasets primarily focus on segmenting the optic disc (OD) and optic cup (OC), highlighting the fundus characteristics that differentiate glaucoma patients from normal individuals. However, public datasets specifically addressing glaucoma progression research are relatively scarce. Two notable examples are the UWHVF dataset and the Annotated Glaucoma Medication Dataset. UWHVF, an open-source visual field (VF) dataset from the University of Washington, contains 28,943 VFs with longitudinal data from one or both eyes of patients. It includes pointwise sensitivities extracted from the Humphrey Field Analyzer (HFA) and other related VF parameters, with progression analysis performed on a subset of VFs. On the other hand, the Annotated Glaucoma Medication Dataset comprises 480 clinical notes from Electronic Health Records (EHR), annotated for glaucoma medication details. It is primarily designed for the development of natural language processing models.

GRAPE (Glaucoma Real-world Appraisal Progression Ensemble) is a multimodal dataset of longitudinal follow-up visual field and fundus images for glaucoma management [13]. Additional datasets are available below (Table 11.1):

- RIGA 20: Retinal fundus images for glaucoma analysis: the RIGA dataset
  https://deepblue.lib.umich.edu/data/concern/file_sets/g732d957m
  https://deepblue.lib.umich.edu/data/concern/data_sets/3b591905z
- The Online Retinal Fundus Image Database for Glaucoma Analysis and Research (ORIGA) dataset is available upon request.
  https://oar.a-star.edu.sg/storage/n/n2od11j13o/embc10-zhangz-apr12-v2.pdf
- DRISHTI-GS22: Retinal Image Dataset For Optic Nerve Head (ONH) Segmentation
  https://cvit.iiit.ac.in/projects/mip/drishti-gs/mip-dataset2/Home.php
  https://cdn.iiit.ac.in/cdn/cvit.iiit.ac.in/images/ConferencePapers/2014/MadhulikaUjjwal2013Drishti.pdf
- REFUGE23: Retinal Fundus Glaucoma Challenge
  https://refuge.grand-challenge.org/Home2020/

Table 11.1 Summarization and detailed description of GRAPE dataset [13]

Data	Format	Detail description
Baseline		
Age	Number	The age at the first visit
Gender	F/M	"F" for female and "M" for male
CCT	Number	The value of CCT
Times of visits	Number	Times of visits during follow-up
Progression-PLR2	0/1	"0" for no progression, and "1" for progression
Progression-PLR3	0/1	"0" for no progression, and "1" for progression
Progression-MD	0/1	"0" for no progression, and "1" for progression
Category	OAG/ACG	The two categories of glaucoma
OCT RNFL thickness	A group of number	Including mean ("Mean"), superior ("S"), nasal ("N"), inferior ("I") and temporal ("T") peripapillary RNFL thickness
Baseline and follow-up		
Subject number	Number	The serial number of inclusion subjects in this dataset
Laterality	OD/OS	The laterality of one eye, "OD" for right eye and "OS" for left eye
Visit number	Number	The times for follow-up visits from baseline
Interval year	Number	The interval years between each visit and baseline
IOP	Number	The value of IOP
VF	A group of number	Including the Octopus VF values of 61 points
CFP	Image	Including the original CFP, ROI and ROI with OD/OC annotation
Acquisition device	Device name	The devices taking the CFPs, including three types of devices
Resolution	Number × number	The resolution of CFP, corresponding to the acquisition device, including three types, 1556 × 1556, 2136 × 2136, 1611 × 1611

- PAPILA24:
  https://www.nature.com/articles/s41597-022-01388-1
  https://figshare.com/articles/dataset/PAPILA/14798004?file=35013982
- ACRIMA
  https://figshare.com/articles/dataset/CNNs_for_Automatic_Glaucoma_Assessment_using_Fundus_Images_An_Extensive_Validation/7613135
- UWHVF, an open-source visual field (VF) dataset
  https://github.com/uw-biomedical-ml/uwhvf
- Annotated Glaucoma Medication Dataset
  https://github.com/jche253/Glaucoma_Med_Dataset

Glaucoma management presents significant challenges due to its chronic and progressive nature. Regular follow-ups are crucial for evaluating the visual function of glaucoma patients, primarily through visual field (VF) assessments using standard automated perimetry (SAP) and monitoring of intraocular pressure (IOP).

Color fundus photography (CFP) and optical coherence tomography (OCT) are employed to detect structural damage. A typical diagnostic process for a suspected glaucoma patient begins with a comprehensive initial examination, allowing the doctor to formulate a preliminary diagnosis. For patients with intraocular hypertension, IOP-lowering medications are prescribed to prevent further visual function deterioration. These patients are then monitored through regular visits to check IOP and VF, with occasional fundus structure examinations as needed. In cases where IOP remains uncontrolled, additional medical treatments, laser procedures, or surgical interventions may be necessary. This systematic approach to glaucoma management underscores the importance of ongoing monitoring and timely intervention in preserving visual function.

The current landscape of glaucoma datasets presents significant limitations in the comprehensive analysis of disease progression. Most datasets containing color fundus photographs (CFP) or optical coherence tomography (OCT) images primarily focus on glaucoma diagnosis, lacking the longitudinal data necessary to evaluate disease progression over time. Conversely, datasets designed for progression detection typically include only visual field (VF) values or medical records, failing to provide a holistic representation of a patient's condition. This gap in available data hinders the development of predictive models for future VF progression based on intraocular pressure (IOP) or CFP. The Glaucoma Real-world Appraisal Progression Ensemble (GRAPE) dataset is developed to address these limitations [13]. This comprehensive dataset encompasses clinical data, VF values, OCT measurements, and CFPs from baseline and follow-up visits, offering a complete record of glaucoma patients' progression over time. The GRAPE dataset facilitates progress prediction and VF estimation in glaucoma, potentially advancing AI-based research in glaucoma management and improving patient care through more accurate and timely interventions.

11.6.3 Future Challenges for Glaucoma Detection

Detecting glaucoma progression remains a significant hurdle in clinical practice since detecting true changes due to the disease is challenging. Some of the imaging modalities generate substantial test–retest variability, which complicates the distinction between genuine change and fluctuations. Furthermore, there is a lack of consensus regarding specific criteria for glaucoma progression based on VF or structural parameters.

In the development of emerging Deep Learning (DL) models for glaucoma, both quality and quantity of data play crucial roles. While a large volume of data is beneficial, the accuracy of labeling is equally essential. For instance, when non-experts label fundus images, it can compromise the dataset's quality, potentially leading to unreliable model constructs. Increasing both the quantity and quality of data enhances model performance and improves the likelihood of generalizability. Well-annotated multi-modal datasets can significantly enhance the accuracy of AI models in disease detection. However, assembling a comprehensive dataset with multiple modalities presents several challenges. These hurdles include the low prevalence of glaucoma in the general population, strict data confidentiality requirements, complex data protection regulations, and the labor-intensive nature of data collection and annotation. Despite these obstacles, the pursuit of high-quality, large-scale, multi-modal datasets remains a key focus in advancing DL applications in glaucoma research and clinical practice.

The generalizability of CNN-based AI models for glaucoma detection remains a significant concern, despite numerous reports of high-performance systems in the literature. This issue stems primarily from the lack of independent validation in many studies. Independent validation is crucial for several reasons: it verifies whether a model has truly learned meaningful disease-related features, helps uncover potential biases or overfitting issues, and assesses the model's robustness against variations in data distribution—a critical factor for real-world applications. Furthermore, using accessible independent datasets for validation enables fair comparisons between different AI models, promoting transparency in the field. Therefore, while current research shows promise, rigorous independent validation is essential to ensure the reliability and practical applicability of AI models in glaucoma detection.

Model reliability remains a significant hurdle for clinical applications of Deep Learning (DL) in glaucoma detection. Most DL models are deterministic, producing outputs regardless of the relevance of the input, which can lead to erroneous results in real-world scenarios. Additionally, the "black box" nature of many DL architectures, lacking proper visualization and interpretability, undermines user trust. This opacity poses a substantial challenge to the successful integration of AI models in clinical practice, highlighting the need for more transparent and interpretable AI systems in healthcare.

A key future direction for AI in glaucoma detection is improving training datasets. Current reference datasets often lack expert annotation, which can potentially misrepresent glaucomatous optic neuropathy (GON). Creating more accurate datasets annotated by panels of glaucoma specialists is crucial. Additionally, the field requires consistent, objective definitions of glaucoma and its progression to ensure accurate evaluation of AI models. One approach could be using quantified parameters from visual field and OCT data to establish objective criteria, while considering comorbidities and ethnic-specific factors.

Addressing limitations in AI model architecture is also vital. Deep learning models can be fragile, unexplainable, and non-interpretable. To enhance performance and interpretability, researchers should focus on accurately quantifying the cup-to-disc ratio in fundus images and retinal nerve fiber layer thickness profiles in OCT images. These improvements would contribute to more robust, transparent, and clinically applicable AI systems for the detection and management of glaucoma.

Currently, there is no FDA-approved autonomous AI system for glaucoma detection, underscoring a significant gap in ophthalmic diagnostics. This void underscores the pressing need for innovative AI models that can independently assess and detect glaucoma. The development of such autonomous systems remains a critical objective in the field, as it holds the potential to significantly enhance treatment outcomes for this leading cause of irreversible blindness. Advancing AI technology in this direction could revolutionize early detection and management of glaucoma, ultimately improving patient care and reducing the global burden of vision loss.

11.7 Role of GenAI in Healthcare

Generative AI (GenAI) already plays a critical role in the healthcare industry, assisting medical professionals in analyzing massive patient datasets to generate new insights and predict potential diagnoses [14]. Soon, GenAI will be able to suggest personalized treatment plans and streamline administrative tasks, leading to enhanced patient care. GenAI also has potential applications in domains such as drug discovery, medical imaging analysis, and patient education.

Some key areas of GenAI applications include:

(a) **Diagnosis Support**: By analyzing medical images, such as MRIs, CT scans, and X-rays, to identify abnormalities and suggest potential diagnoses. This often results in faster and more accurate interpretations [15].
(b) **Drug Discovery**: By identifying potential drug targets and simulating molecular interactions to accelerate the development of new drugs [16].
(c) **Personalized Treatment Plans**: By generating customized treatment options for a patient based on unique medical history, genetic data, and other relevant information or circumstances for that patient [17].
(d) **Fraud Detection**: By analyzing medical billing data, to identify any suspicious activities that may indicate fraud [18].

However, care must be taken when using GenAI in healthcare as related to the following aspects:

1. **Data Quality**: The Accuracy of GenAI predictions depends on the quality and completeness of the training data.
2. **Clinical Validation**: GenAI models must undergo thorough validation to assess their accuracy and reliability before being used in clinical practice.
3. **Ethical Implications**: Medical practitioners using GenAI must be cognizant of pitfalls to avoid decision bias and protect patient privacy.

11.8 Case Study for Bias Avoidance and Transparency

GenAI can be potentially used to generate discharge summaries and patient instructions in a patient's native language, ensuring better understanding by the patient and their family. It can also create checklists, lab summaries, and even clinical orders in real time. However, when a large language model is used to analyze patient data and generate treatment recommendations, it must be carefully trained to avoid perpetuating biases based on factors such as race, ethnicity, or socioeconomic status, which could lead to disparities in care.

For example, an algorithm used to prioritize patients for elective surgeries that unintentionally favored younger, healthier individuals over older patients with complex medical histories, when trained on biased data. It highlights the need for diverse datasets and careful model development to ensure equitable outcomes for all patient populations. Suppose a woman has had a cesarean delivery, also known as a C-section. In that case, there is a chance that a subsequent delivery can be attempted through a vaginal birth, which is known as Vaginal Birth after Cesarean Delivery (VBAC). However, there are known risks associated with attempting VBAC, such as uterine rupture or other complications. The VBAC algorithm was designed to help healthcare providers assess the likelihood of safely giving birth through vaginal delivery. The algorithm considers several factors, including the woman's age, the reason for the previous C-section, and the time elapsed since it occurred. However, in 2017, in a study by Vyas, et al., researchers found the original algorithm was biased [19].

VBAC algorithm had incorrectly predicted that Black/African American and Hispanic/Latino women were less likely to have a successful vaginal birth after a C-section than non-Hispanic White women. This caused doctors to perform more C-sections on Black/African American and Hispanic/Latino women than on White women. After years of work by researchers, advocates, and clinicians, changes were made to the algorithm. The new version of the algorithm no longer considers race or ethnicity when predicting the risk of complications from VBAC. This means that doctors can now make decisions based on more accurate and impartial information that works for all women, providing more equitable care regardless of race or ethnicity. For more details, refer to [20].

11.9 Future of AI-Based Health Care

AI-based healthcare in the Cloud is advancing rapidly, and the future is likely to focus on several key areas:

1. **Advanced Predictive Analytics and Early Diagnosis**
    - AI models will become more accurate in detecting diseases early, such as cancer, cardiovascular issues, and neurological disorders.
    - Integration with wearables and IoT devices will provide better real-time health monitoring.
2. **Personalized Medicine and Treatment Plans**
    - AI-driven cloud platforms will tailor treatments based on genetic, environmental, and lifestyle factors.

- Drug discovery and trial simulations will accelerate with the use of AI-powered data analysis.
- Applications of digital twins for what-if analysis of a treatment on a patient.

3. **Secure and Compliant Data Sharing**
   - More robust encryption and privacy-preserving AI (like federated learning) will allow hospitals to share patient data without compromising security.
   - Compliance with global regulations (HIPAA, GDPR) will be a priority for Cloud providers.
4. **Automated Administrative and Clinical Workflows**
   - AI will streamline administrative tasks (billing, scheduling, medical coding).
   - NLP-powered virtual assistants will support doctors by transcribing and summarizing patient interactions.
5. **AI-Powered Telemedicine and Virtual Health Assistants**
   - Cloud-based AI chatbots will provide 24/7 healthcare advice.
   - Remote diagnostics will improve with the aid of AI-enabled imaging and analysis tools.
6. **Edge AI and Decentralized Healthcare AI**
   - AI models will process health data locally on edge devices (smartwatches, home monitoring systems) before sending insights to the cloud.
   - Reduces latency and enhances real-time decision-making.
7. **Multi-Cloud and Hybrid AI Strategies**
   - Healthcare providers will use multiple cloud platforms (e.g., AWS, Azure, Google Cloud) for redundancy and flexibility.
   - AI workloads will shift dynamically based on cost, performance, and regulatory requirements.
8. **Ethical AI and Bias Reduction**
   - AI models will be refined to reduce bias in healthcare decisions.
   - Transparent AI models will provide explainable outcomes for medical professionals.
9. **Higher Adoption of Gen AI**
   - Increased usage of Gen AI in healthcare applications, such as new drug discoveries, personalized treatment plans and fraud detection etc.
   - Evolution of new AI assistants for medical professionals and patients.

11.10 Summary

Latest advances in the healthcare frontier utilize cloud computing and AI/ML to explore new methods for diagnosing and treating existing diseases. Further advances are expected as more medical data becomes available in the Cloud. We examined several use cases of AI and ML-based solutions to identify hidden patterns in the data. Using these, doctors can better determine the course of treatment for patients by comparing a patient's test results with datasets collected over time. This chapter presented several examples of medical treatments using the Cloud with AI and ML. The first example was of Heartflow, related to cardiovascular health. The second was a collaborative Cloud solution developed by Intel for facilitating Cancer treatments. The third was related to multi-modal Glaucoma detection. We concluded with a discussion on the emerging role of GenAI and future advances in AI-based healthcare.

11.11 Points to Ponder

1. Discuss if we are at the cusp of a medical revolution with the confluence of AI and Cloud?
2. Why do some companies prefer a multi-cloud solution?
3. Why does the FDA have reservations about using AI/ML-based techniques for medical diagnosis?
4. Is it sufficient to remove a patient's information for broad data sharing?
5. Should a patient be given access to his or her medical data?

11.12 Answers

1. **Discuss if we are at the cusp of a medical revolution with the confluence of AI and Cloud?**
   (a) Healthcare is at an inflection point due to a growing population, shortage of medical practitioners, and limited public resources, such as hospitals. With the increasing availability of new devices and data, medical science is poised for individualized treatments rather than one-size-fits-all approaches.
   (b) Telemedicine was practiced during COVID times to minimize risk and improve efficiency. However, most of the patients' data is still locked up in different hospitals' databases due to HIPAA regulations and perceived risk of moving it to a Public Cloud.
   (c) The next obvious steps are to securely transfer this data to the Cloud and leverage the power of AI to assist physicians in identifying patients' conditions and recommending appropriate treatments.

2. **Why do some companies prefer a multi-cloud solution?**
   (a) Many Cloud customers want to avoid vendor lock-in, fearing future price rises and limited capabilities compared to other Cloud vendors.
   (b) A multi-cloud solution avoids the above issues, while also enabling a business to support its customers who may prefer one Cloud vendor over another. An example is a healthcare provider in the Cloud, whose customers are hospitals with existing agreements or preference for different Cloud Service Providers (CSPs). In this case, healthcare providers will have a hard time convincing a hospital to move its data from one Cloud to another and must adopt a multi-cloud solution. Such a solution may need to cross organizational boundaries and use the internet to access storage in one Cloud, while algorithms may be running in another Cloud.
3. **Why does the FDA have reservations about using AI/ML-based techniques for medical diagnosis?**
   (a) The FDA is not opposed to the use of AI/ML techniques for medical diagnosis, but it does require the presence of a human element in the final decision-making process. This is no different from the role of software in self-driving cars. In the event of an injury or accident, the issue of liability arises. Invariably, software providers or car manufacturers may not be willing to accept that kind of liability. It is for this reason that they currently require human drivers to be always behind the wheel, even in autopilot mode. For the same reason, FDA requires a medical doctor to certify reports and results for the final diagnosis. However, doctors can utilize the software to assist and expedite their decision-making process.
   (b) Consider the example of examining X-ray or MRI results of patients with bone injuries. Orthopedic analysis support software can highlight current as well as potential locations with displacement, gaps, or cracks. However, in the end, a doctor must examine and agree with the suggestions before treatment can begin.
4. **Is it sufficient to remove a patient's information for broad data sharing?**
   (a) It is necessary to remove a patient's identifiable information for healthcare data de-identification, but it is essential to replace it with a unique token.
   (b) This enables future data and results for the same patient to be correlated.
   (c) Doctors can then observe the progression of a disease and the efficacy of treatment for that patient based on time-series-based changes.
5. **Should a patient be given access to his or her medical data?**
   (a) Current practice is to share medical data with the patient on a need-to-know basis. Also, since a patient may not be medically savvy, there is a tendency of doctors to up-level or obfuscate the details. This becomes a hindrance if a patient needs to seek health care away from the home base.

(b) Hence, there is a need to grant full access to a patient for their medical records. HIPAA fully supports this.
(c) Cloud is a convenient way to store and manage health information. AI can detect fraudulent access to the stored data and improve security.

References

1. https://www.cdc.gov/heartdisease/
2. https://www.heartflow.com/
3. https://aws.amazon.com/solutions/case-studies/HeartFlow/
4. https://itpeernetwork.intel.com/intel-ohsu-announce-collaborative-cancer-Cloud-at-intel-developers-forum/
5. https://owkin.com/federated-learning/
6. https://www.hpcwire.com/2020/05/11/intel-upenn-launch-massive-multi-center-ai-effort-to-refine-brain-cancer-models/
7. http://braintumorsegmentation.org/
8. https://en.wikipedia.org/wiki/Message_sequence_chart
9. https://www.mayoclinic.org/diseases-conditions/glaucoma/multimedia/open-angle-glaucoma/img-20007729
10. https://www.srieyecarehospital.com/primary-open-angle-glaucoma.php
11. https://drzamindarseyecentre.com/glaucoma-important-things-you-must-know/
12. Huang X et al (2023) Artificial intelligence in glaucoma: opportunities, challenges, and future directions. Biomed Eng Online 22:126. https://biomedical-engineering-online.biomedcentral.com/articles/10.1186/s12938-023-01187-8
13. Huang X et al (2023) GRAPE: a multi-modal dataset of longitudinal follow-up visual field and fundus images for glaucoma management. Sci Data 10:520. https://springernature.figshare.com/collections/GRAPE_A_multi-modal_glaucoma_dataset_of_follow-up_visual_field_and_fundus_images_for_glaucoma_management/6406319/1
14. https://www.alpha-sense.com/blog/trends/generative-ai-healthcare/
15. https://www.n-ix.com/generative-ai-in-healthcare
16. https://www.johnsnowlabs.com/generative-ai-healthcare
17. https://www.xenonstack.com/blog/generative-ai-healthcare-system
18. https://mastechinfotrellis.com/blogs/generative-ai-in-healthcare-use-cases-benefits-and-the-future
19. https://minorityhealth.hhs.gov/news/shedding-light-healthcare-algorithmic-and-artificial-intelligence-bias
20. Vyas DA, Jones DS, Meadows AR, Diouf K, Nour NM, Schantz-Dunn J (2019) Challenging the use of race in the vaginal birth after cesarean section calculator. Womens Health Issues 29(3):201–204. https://doi.org/10.1016/j.whi.2019.04.007. Epub 2019 May 6

Chapter 12
AI-Enabled Medical Devices Approved/ Cleared by the US FDA

Abstract The US FDA's AI/ML-Enabled Medical Devices List includes all medical devices that incorporate artificial intelligence or machine learning and have received approval or clearance from the agency. Radiology stands out among medical specialties, accounting for the largest share of FDA-authorized devices. Siemens Medical and GE Healthcare are the leading manufacturers, holding the highest number of AI/ML-enabled devices on the list. The majority of these devices continue to be authorized through the premarket clearance (510(k)) pathway. This chapter further analyzes the public information summaries for two notable AI/ML-enabled devices cleared by the US FDA: the Sleep Apnea Notification Software as a Medical Device (SaMD) developed by Apple Inc. and the Home Optical Coherence Tomography (OCT) System by Notal Vision. The review provides detailed insights into device development, design considerations, clinical testing protocols, risk management approaches, human factors requirements, and a predetermined change control plan (PCCP), all of which are essential for meeting the US FDA's stringent regulatory standards.

Keywords AI/ML enabled medical device regulatory approval · 510(k) · De Novo · PMA · Premarket clearance · Premarket approval · OCT · Sleep apnea · Non-clinical testing · Clinical testing · Risk management · Human factor · Usability · Indication for use · Siemens medical · GE healthcare · Performance metrics · Sensitivity · Specificity · DICE score · US FDA

12.1 Analysis of the US FDA AI/ML-Enabled Medical Devices

As of June 10, 2025, the US FDA has authorized 1016 Artificial Intelligence/ Machine Learning (AI/ML)-enabled medical devices. The following analysis is based on information from the US Food and Drug Administration (FDA) database, downloaded on June 10, 2025 [1].

A. Pandey et al., *Artificial Intelligence in Medical Software*,
https://doi.org/10.1007/978-3-032-10808-1_12

The number of authorized medical devices by the US FDA has experienced a notable increase, as illustrated by Fig. 12.1, which plots the number of AI/ML-enabled medical devices approved or cleared since 1995. This trend indicates a substantial increase in investment in AI/ML technology by the medical device industry. Notably, since 2020, the US FDA has cleared or approved 795 devices, representing approximately 80% of all such devices authorized by the agency since 1995.

Radiology leads all medical specialties in the number of AI/ML-enabled medical devices approved or cleared by the US FDA. Table 12.1 presents the breakdown of these approvals and clearances across different medical fields.

Figure 12.2 includes a bar plot that shows 76% of medical devices belong to the radiology medical specialty.

GE Healthcare and Siemens Medical have the most significant number of AI/ML-enabled medical devices (Table 12.2).

The US FDA has cleared 980 AI/ML-enabled medical devices through the 510(k) pathway, accounting for 96.5% of the total number of such devices. An additional 32 devices were approved via the De Novo process, while only four devices received approval through the most stringent premarket approval (PMA) pathway. This distribution highlights that the vast majority of AI/ML-enabled medical devices are cleared through the 510(k) route, with far fewer undergoing the more rigorous De Novo and PMA processes (Table 12.3).

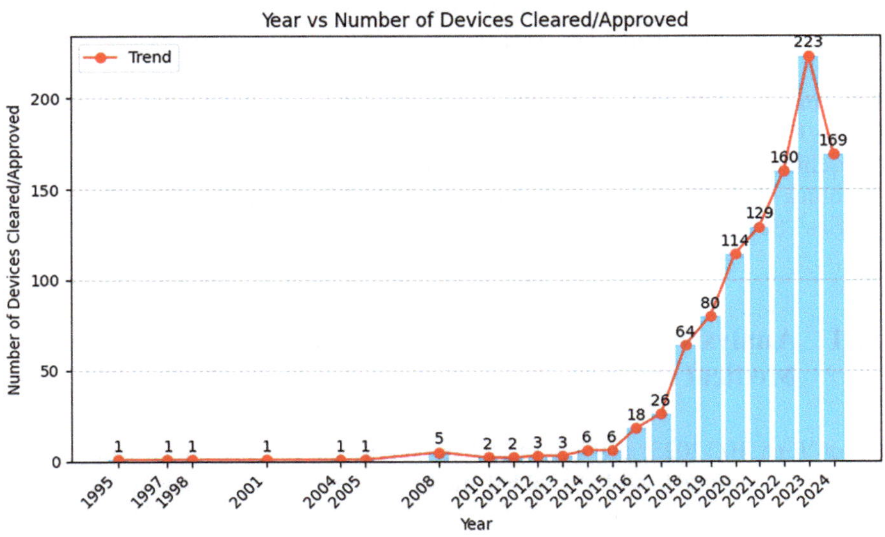

Fig. 12.1 Number of cleared/approved medical devices by the US FDA since 1995 [1]

Table 12.1 Number of medical devices approved or cleared by US FDA [1]

Medical specialty	Number of devices approved or cleared
Radiology	777
Cardiovascular	104
Neurology	42
Anesthesiology	17
Hematology	17
Gastroenterology-urology	14
Clinical chemistry	9
Ophthalmic	9
Microbiology	5
Clinical toxicology	5
Dental	4
General hospital	4
Pathology	3
Orthopedic	2
General and plastic surgery	2
Obstetrics and gynecology	1
Immunology	1

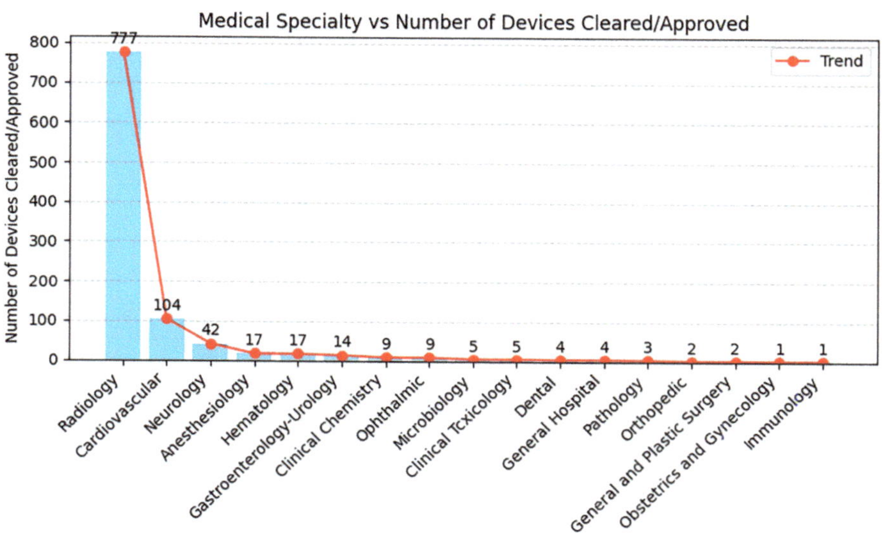

Fig. 12.2 Number of US FDA-authorized medical devices by medical specialty [1]

Table 12.2 Number of medical devices by company [1]

Company	Number of medical devices
GE Healthcare	76
Siemens Medical Solutions USA, Inc.	73
Canon Medical Systems Corporation	32
Aidoc Medical, Ltd.	25
Shanghai United Imaging Healthcare Co., Ltd.	25
Philips	24
Samsung	15
iSchemaView Inc.	15
Viz.ai, Inc.	11
Hyperfine Inc.	10
Clarius Mobile Health Corp.	9
Zebra Medical Vision Ltd.	9
Quantib BV	8
Avicenna.AI	8
Ever Fortune.AI Co., Ltd.	8
Hologic, Inc.	8
Brainlab AG	7
RaySearch Laboratories AB (publ)	7
Surgical Information Sciences, Inc.	7
Qure.ai Technologies	7
Subtle Medical, Inc.	7
Medtronic	7
Annalise-AI Pty Ltd.	7
DiA Imaging Analysis Ltd.	6
Oxehealth Limited	6
HeartFlow, Inc.	6
MIM Software Inc.	5
AliveCor, Inc.	5
iRhythm Technologies, Inc.	5
Arterys Inc.	5
Brainomix	4
Oxehealth Limited	4
Apple Inc.	4

Table 12.3 Number of medical devices by the US FDA approval pathway [1]

FDA approval pathway	Number of devices (% value)
510(k)	980 (96.5%)
De novo	32 (3.1%)
PMA	4 (0.4%)

12.2 Public Information Summary from US FDA Database

Next, two medical devices are analyzed in detail, with each selected to represent a distinct regulatory pathway: 510(k) clearance and De Novo (DEN). The review of these devices is based on the Public Information Summaries submitted to the US FDA [2, 3].

1. Sleep Apnea Notification Feature (SANF) by Apple Inc. (510(k), K240929) [2].
2. Notal Vision Home Optical Coherence Tomography (OCT) System by Notal Vision (De Novo, DEN230043) [3].

12.3 Sleep Apnea Notification Feature (SANF) by Apple Inc. (K240929)

12.3.1 Indication for Use

The Sleep Apnea Notification Feature (SANF) is a software-only mobile medical application that analyzes Apple Watch sensor data to identify patterns of breathing disturbances suggestive of moderate to severe sleep apnea and provides a notification to the user. This feature is intended for over-the-counter (OTC) use by adults age 18 and over who have not previously received a sleep apnea diagnosis and is not intended to diagnose, treat, or aid in the management of sleep apnea. The absence of a notification is not intended to indicate the absence of sleep apnea.

Classification Name: Over-the-counter device to assess risk of sleep apnea, 21 CFR 868.2378
Regulatory Class: Class II
Product Code: QZW
510(k) Review Panel: Anesthesiology

12.3.2 Device Description

The Sleep Apnea Notification Feature (SANF) is an over-the-counter mobile medical application (MMA) intended to identify patterns of breathing disturbances suggestive of moderate-to-severe sleep apnea and provide a notification to the user. SANF is intended to run on compatible iOS (e.g., iPhone, iPad) and Apple Watch platforms. Users set up SANF and view their health data on the iOS platform. Prior to use, users must undergo educational onboarding. SANF uses accelerometer sensor data collected by the Apple Watch to calculate breathing disturbance values while a user is asleep. Breathing disturbances describe transient changes in breathing patterns, such as temporary breathing interruptions. Breathing disturbance data is analyzed in discrete, consecutive 30-day evaluation windows. If patterns consistent with moderate-to-severe sleep apnea are identified within the 30-day evaluation window, the user is notified. SANF provides visualizations depicting the user's breathing disturbance data over various time scales. SANF is not intended to provide instantaneous measurements. Instead, once activated, SANF runs opportunistically in the background, receiving signals from Apple Watch sensors for processing.

12.3.3 Comparison with the Predicate Device

SANF and the predicate device (DEN230041) have the same intended use, technological characteristics, and principles of operation, and the difference in indications does not represent a new intended use. Both the subject and predicate devices are software-only mobile medical applications intended to detect signs of moderate-to-severe sleep apnea for individuals who have not been previously diagnosed with sleep apnea and are not intended to provide a standalone diagnosis. The subject device contains some differences in technological characteristics:

- The subject device is compatible with Apple products (i.e., iOS device, Apple Watch), while the predicate is compatible with Samsung products (i.e., Galaxy Watch and Phone).
- The subject device utilizes passive, opportunistic detection to monitor the user over a 30-day period, and only alerts the user if it detects signs of sleep apnea. The predicate device provides an on-demand 2-day assessment and returns either a positive or negative finding to the user.
- The subject device utilizes accelerometer sensor data while the predicate device utilizes blood oxygen sensor data.

The differences in technological characteristics described above do not raise new questions of safety or effectiveness. The differences can properly be evaluated through the special controls established in 21 CFR 868.2378. The subject device has been appropriately verified and validated through non-clinical and clinical testing to ensure that the device is substantially equivalent to the predicate. Some key aspects of comparison are shown in Table 12.4.

Table 12.4 Comparison with the predicate device

Item	Subject device sleep apnea notification feature	Predicate device (DEN230041)
Overall device design	Assessments are based on sensor data collected over 30-day periods. The device is intended to provide opportunistic detection of sleep apnea, such that after initial enrollment no user interaction is required for the device to perform as intended	Assessments are based on sensor data collected over a 2-day period. The device is intended to provide on demand assessments to detect signs of sleep apnea, such that a user must actively choose to initiate a monitoring period
Use environment	Over-the-counter	Over-the-counter
Device components	Software-only	Software-only
Device input	Accelerometer data	Blood oxygen level (SpO_2) data
Clinical performance	The performance was optimized for high specificity given SANF is designed as an opportunistic detection feature (i.e., passive, recurring).	
	Sensitivity: 66.3% 95% CI [62.2%, 70.3%]	Sensitivity: 82.7% 95% CI [76.7%, 87.6%]
	Specificity: 98.5% 95% CI [98.0%, 99.0%]	Specificity: 87.7% 95% CI [83.1%, 91.4%]

12.3.4 Summary of Non-clinical Testing

12.3.4.1 Algorithm Development

SANF includes a deep learning algorithm to identify breathing disturbances using accelerometer sensor data from Apple Watch. The model was trained on Apple Watch accelerometer signals collected during sleep sessions with concurrent in-lab polysomnography (PSG) and Home Sleep Apnea Test (HSAT) reference recordings. The algorithm development dataset included over 11,000 nights of concurrent reference and watch sensor data. The distribution of sleep apnea classifications in this dataset was broad and spanned all four clinically defined categories of sleep apnea: normal (AHI 0 to <5), mild (AHI 5 to <15), moderate (AHI 15 to <30), and severe (AHI ≥30). For the purposes of algorithm development, data from the studies were pooled and split into four sets: Training, Validation, Test, and Sequestration. The model was trained on the Training set, with the Validation set used for early stopping and threshold selection. The model was then evaluated on the Test set at regular intervals during model development. When the development was complete, and the model was locked, it was evaluated on the Sequestration set as a last test to ensure it had not been over-fit to the training data. This process ensured no subject overlap and matching distributions of sex, age, BMI, and disease severity. The development data included a diverse group of subjects with respect to demographic factors (e.g., age, AHI, race, ethnicity, and BMI) representative of the intended use population.

12.3.4.2 Non-clinical Testing Summary

Apple conducted the necessary non-clinical testing on SANF with passing results supporting a determination of substantial equivalence. Non-clinical testing conducted included the following: Software Verification and Validation, Cybersecurity Tests, Human Factor Validation, and General-Purpose Computing Platform Assessment.

12.3.5 Summary of Clinical Testing

The performance of the Sleep Apnea Notification Feature was validated in a prospective, non-significant risk study enrolling 1499 subjects from several sites across the United States. The purpose of the study was to evaluate the performance of SANF using the Nox T3s home sleep apnea testing (HSAT) device (K192469) as a reference device. The study enrolled subjects across the spectrum of sleep apnea severity classifications, with a broad distribution across each of the following AHI categories using the "4%" hypopnea scoring rule: 559 normal subjects (AHI < 5), 362 mild subjects (5 ≤ AHI < 15), 216 moderate subjects (15 ≤ AHI < 30), and 201 severe subjects (AHI ≥ 30), plus 161 subjects with missing HSAT reference. Subjects were also enrolled based on a broad range of demographic factors, including enrollment targets for age, sex, BMI, skin tone, race, and ethnicity subgroups to ensure the study population was representative of the intended user population.

Of the 1499 enrolled subjects, 1278 contributed to the notification performance analysis, and 1305 contributed to the breathing disturbance performance analysis. Those not included in the performance had insufficient Apple Watch data and/or reference data. Table 12.5 contains SANF clinical study subject demographics data.

The sensitivity of notifications for subjects with moderate-to-severe sleep apnea (AHI ≥ 15) was 66.3%; 95% CI [62.2%, 70.3%]. The specificity of the notifications for those with normal-to-mild sleep apnea (AHI < 15) was 98.5%; 95% CI [98.0%, 99.0%]. SANF did not falsely notify any subjects with normal AHI (AHI < 5). The performance was similar for identified sub-groups.

To assess the performance of breathing disturbance estimates, Apple evaluated the proportion of paired (Breathing Disturbance, reference AHI). Of the total 1305 subjects who had at least one paired measurement, 1193 (91.4%) were within the pre-specified performance zone. These results demonstrate that the Sleep Apnea Notification is effective in generating accurate notifications for moderate-to-severe sleep apnea and Breathing Disturbance values.

Table 12.5 SANF clinical study subject demographics

$N = 1499$	
Age group (years)	
18–49	855 (57.0%)
50–64	491 (32.8%)
≥65	153 (10.2%)
Sex	
Female	847 (56.5%)
Male	652 (43.5%)
Ethnicity	
Hispanic or Latino	181 (12.1%)
Non-Hispanic or Latino	1318 (87.9%)
Race	
American Indian or Alaska Native	25 (1.7%)
Asian	103 (6.9%)
Black or African American	347 (23.1%)
Native Hawaiian or other Pacific Islander	3 (0.2%)
White	1021 (68.1%)

12.3.6 Predetermined Change Control Plan

The PCCP does not include provisions for the implementation of adaptive algorithms that will continuously learn in the field. All algorithm modifications will be trained, tuned, and locked prior to the release of the software to the field. A procedure has also been established for updating the Instructions for Use in order to inform users about algorithm changes implemented under this FDA-authorized PCCP, including a summary of the changes, a characterization of algorithm performance, and the availability and compatibility of the feature.

The PCCP specifies possible modifications to the device software as well as verification and validation activities in place to implement the changes in a controlled manner, such that the modified device remains as safe and effective as the predicate device. The PCCP includes a specific list of potential software modifications defining the region of potential changes that can be made to the algorithms in the device. Details of the potential changes are summarized in Table 12.6. The modification protocol incorporates impact assessment considerations and specifies requirements for data management, including data sources, collection, storage, and sequestration, as well as documentation and data re-use practices. Specific test methods are specified in the PCCP to establish substantial equivalence relative to the Sleep Apnea Notification Feature and include sample size determination, analysis methods, and acceptance criteria. To help ensure validation test datasets are representative of the intended use population, each will meet minimum demographic requirements for age, sex, race, BMI, and ethnicity. Table 12.6 contains SANF list of modifications covered by PCCP.

Table 12.6 Modifications covered by PCCP

	Detailed list of changes	Requirements	Test method
Modifications to breathing disturbances (BD) computation	• Adjust the operating point • Re-train the algorithm with additional datasets while maintaining the same algorithm architecture and number of parameters • Revise signal input module • Add additional classifier outputs • Modifications to signal quality and post-processing modules	• No change to input type • No change to output type • No concurrent change to other modules • No change to intended use of device	Verification of BD accuracy and substantial equivalence in notification-level sensitivity and specificity when compared to the performance of SANF 1.0
Modifications to sleep apnea estimation	• Modify the number of nightly BD readings required to surface a notification • Reduce the interval of the notification window • Modify logic for surfacing a notification based on BDs	• Can be fully verified and/or validated by requirements of the modification protocol	Substantial equivalence in sensitivity and specificity when compared to the performance of SANF 1.0

12.3.7 Conclusion

The Sleep Apnea Notification Feature is substantially equivalent to the predicate device as they are identical with respect to intended use, and there are no differences in technological or performance characteristics that raise different questions of safety and effectiveness.

12.4 Notal Vision Home Optical Coherence Tomography (OCT) System

New Regulation Number: 21 CFR 886.1600
Classification: Class II
Product Code: SAX
Submission Number: DEN230043
Sponsor Information: Notal Vision, Inc.
Date De Novo Received: June 7, 2023

12.4.1 Indications for Use

The Notal Vision Home Optical Coherence Tomography (OCT) System is an Artificial Intelligence (AI)-based Home Use device indicated for visualization of intraretinal and subretinal hyporeflective spaces in a 10° by 10° area centered on the point of fixation of eyes diagnosed with neovascular age-related macular degeneration (NV-AMD). In addition, it provides segmentation and an estimation of the volume of hypo-reflective spaces. The Notal Home OCT device is intended for imaging at home between regularly scheduled clinic assessments and is not intended to be used to make treatment decisions or replace standard-of-care regularly scheduled examinations and clinical testing as needed, including in-office imaging and assessments for changes in vision, by an ophthalmologist.

12.4.2 Device Description

The NVHO system, as shown in Fig. 12.3, is a device that consists of two elements:

- **Notal Home OCT Device**: Patients use this to self-image their eyes using Spectral-Domain OCT; at the end of each scanning session, the data is transmitted via a secure wireless communication to the Notal Health Cloud.
- **Notal Health Cloud**: Cluster of servers and analysis units on which the Notal OCT Analyzer (NOA) analyzes the scans received from the Notal Home OCT device. Processed data are presented on a dedicated interactive web-application, the Notal Home OCT Web Viewer.

The system overview in Fig. 12.4 shows the overall workflow. Images are captured by the patient at home captures images. An application icon-based Graphical User

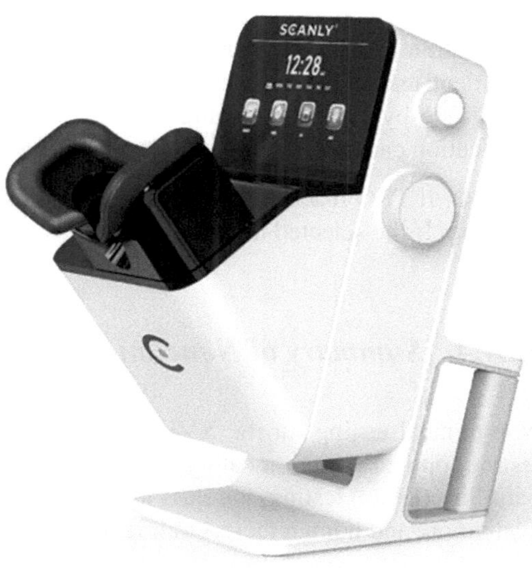

Fig. 12.3 Notal Home OCT device [4]

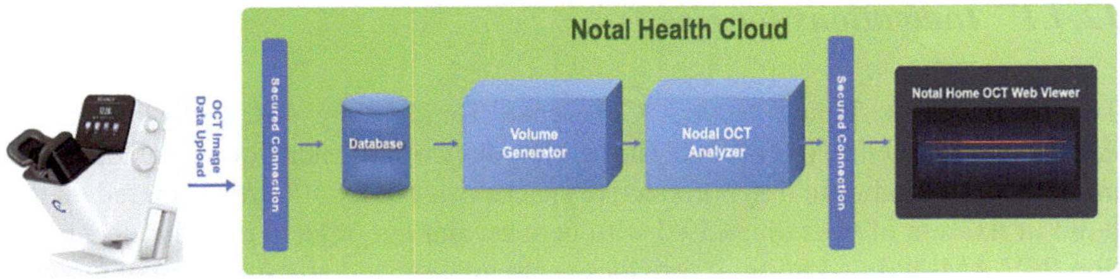

Fig. 12.4 Notal Vision Home OCT system overview

Interface (GUI) enables NV-AMD patients to self-image. The GUI, presented on an external touch screen, takes the patient step by step through the scanning flow. The device contains an internal mini-display that presents an interactive fixation target to the subject. Upon self-imaging, images are automatically and securely transmitted to the Notal Health Cloud for viewing by the prescribing user. Raw data are fed into a Volume Generator (VG) module for OCT B-scan reconstruction, Manufacturer Signal Quality Index (MSI) calculation, segmentation, registration, volume scan reconstruction and alignment. Images are fed into the Notal Home OCT Web-viewer for display, and into NOA for analysis.

12.4.3 Artificial Intelligence Algorithm

NOA is an artificial intelligence (AI) algorithmic module. NOA runs on the Notal Health Cloud and processes the volume scans generated by the VG. Its purpose is to segment the subretinal hypo-reflective spaces (SRO) and intraretinal hypo-reflective spaces (IRO) in the 2D B-scans. From this, the IRO and SRO volumes in the macular volume scan are calculated. Total retinal hypo-reflective space (TRO) volume information, which is the sum of SRO and IRO, is displayed on the Notal Home OCT Web Viewer, NOA tab (see Fig. 12.4), along with projections of SRO and IRO information. The Notal Home OCT Web Viewer and OCT images are not accessible to patient users. The prescribing user can compare two scans from two different dates, side-by-side. Prescribing-user set notifications are available, based on two notification categories (multi-selection is allowed):

- TRO: If TRO reaches the threshold, a notification is generated
- Time: If the selected number of days has elapsed, a notification is generated

12.4.4 Summary of Nonclinical/Bench Studies

Non-clinical testing for the NVHO device included device benchtop performance testing, optical radiation safety, software and cybersecurity, electromagnetic compatibility (EMC) testing, wireless testing, electrical safety testing, biological safety evaluation, and human factors validation testing.

12.4.5 Summary of Clinical Information

Clinical performance data were collected from two pivotal clinical studies: (1) the C2021.001 study ("001 Study"): A 5-Week "Home OCT Fluid Visualization Agreement Study"; (2) the C2012.006 study ("006 Study"): A cross-sectional in-office study for "The Evaluation of the Agreement and Precision of the Notal Vision Home OCT in the Automatic Fluid Quantification in Patients with NV-AMD." The protocols for these studies used the term "retinal fluid" (e.g., total, sub-retinal, and intra-retinal fluid [TRF, SRF, IRF]) to refer to hypo-reflective spaces (HRS; or TRO, SRO, and IRO). However, not all hypo-reflective spaces on macular OCT imaging are retinal fluid, and not all exudative fluids will necessarily be hypo-reflective; therefore, the terms TRF, SRF, and IRF are not synonymous or interchangeable with the terms TRO, SRO, and IRO. During the enrollment visit, participants underwent imaging with a clinic-based, spectral-domain OCT imaging system (CIRRUS HD-OCT) to establish a "baseline" status of the macula. After verification of eligibility during screening, participants were assigned an NVHO device. The assigned NVHO device was delivered to participants' homes. Participants set up the NVHO device using the onscreen tutorial.

To continue in the study, participants must achieve a successful initial NVHO calibration. Continuing participants were instructed to self-image at home with the NVHO every day for five consecutive weeks, including on the days of scheduled office visits. Remote telephone technical support ("Notal Vision Diagnostic Clinic," NVDC) was available to participants. In-clinic visits were scheduled at Week 1 and Week 5. At these scheduled visits, CIRRUS HD-OCT imaging was performed, and best-corrected visual acuity (BCVA), subjective symptoms, and adverse events were assessed. The daily NVHO scans of participants designated as without retinal fluid at baseline were reviewed by the reading center (RC). The RC triggered an alert for an interim clinic visit if the fluid was identified on two consecutive NVHO scans from two consecutive days. CIRRUS HD-OCT images, BCVA, and any symptoms or adverse event information were also collected at these interim visits.

12.4.5.1 DICE Coefficient as AI Algorithm Performance Metrics

The **DICE coefficient** (or **Sørensen-DICE coefficient**) is a statistical measure used to evaluate the similarity between two sets, particularly in AI-driven tasks like image segmentation, natural language processing (NLP), and data comparison. It quantifies overlap by comparing the intersection of sets to the combined size of the sets. The DICE coefficient is a widely used metric for evaluating the accuracy of image segmentation algorithms. It measures the similarity between the predicted segmentation and the ground truth by calculating the overlap between these two binary masks. Specifically, it is defined as twice the number of pixels classified as positive by both the predicted and ground truth segmentations, divided by the total number of positive pixels in both segmentations combined. The DICE coefficient ranges from 0 to 1, where 1 indicates a perfect match between the predicted and ground truth segmentations, and 0 indicates no overlap. This metric is particularly valued in medical image segmentation because it is sensitive to small changes in segmentation results and is not heavily influenced by class imbalance, making it suitable for scenarios where the target regions (such as tumors or organs) are small relative to the whole image.

The segmentation overlap analysis included the calculation of the device-grader DICE coefficient for each grader, averaged over cases, and the calculation of the grader-grader DICE coefficient for each pair of graders, averaged over cases. Since the DICE coefficient is undefined when both methods in the comparison do not provide a segmentation, those cases were excluded from the calculation of the average of DICE coefficients. To account for these excluded cases, the device-grader negative percent agreement (NPA) was calculated for each expert. Uncertainty in the results was characterized by 95% confidence intervals, representing case variability. The following table shows the device-grader DICE coefficient and NPA for each grader acting as the "reference standard." Table 12.7 contains descriptive statistics of eye-level DICE between NOA versus Graders and Graders versus Graders.

12.4.6 *Training*

Upon first usage, the NVHO device runs a built-in tutorial. The tutorial consists of video clips which show the patient how to self-image, followed by a practice phase in which the patient performs these tasks. The device subsequently performs a calibration for each eye to personalize the refractive error correction of the patient. Following the calibration, the patient is able to self-image.

12.4.7 *Labeling*

Device instructions for use were provided for the patient and prescribing users of the device separately. The Notal Home OCT User Manual (patient user manual) included instructions for home use, including system unpacking and setup. It also included instructions for interacting with the device, including accessing an on-device tutorial and calibration procedure, regular image capture, and wireless transmission of images to the Notal Home OCT Web Viewer.

12.4.8 *Risk to Health*

Table 12.8 identifies the risks to health that may be associated with the use of a home monitoring ophthalmic imaging device.

Table 12.7 Descriptive statistics of eye-level DICE between NOA versus Graders and Graders versus Graders

Statistics	NOA vs. Grader 1	NOA vs. Grader 2	NOA vs. Grader 3	Grader 1 vs. Grader 2	Grader 1 vs. Grader 3	Grader 2 vs. Grader 3
Comparing NOA-Grader TRO DICE to Grader-Grader TRO DICE						
N[a]	278	289	299	279	297	298
DICE mean ± SD	0.5819 ± 0.2958	0.5655 ± 0.3203	0.5196 ± 0.3182	0.6222 ± 0.2754	0.5453 ± 0.3018	0.6000 ± 0.2918
95% CI of DICE	0.5470, 0.6168	0.5284, 0.6026	0.4834, 0.5558	0.5897, 0.6546	0.5108, 0.5797	0.5668, 0.6333
Mean	0.6168	0.6026	0.5558	0.6546	0.5797	0.6333
DICE median	0.6802	0.6844	0.6123	0.7221	0.6274	0.6964
NPA	0.734 (58/79)	0.746 (47/63)	0.787 (37/47)	0.722 (57/79)	0.494 (39/79)	0.603 (38/63)
95% CI of NPA[b]	0.623, 0.827	0.621, 0.847	0.643, 0.893	0.609, 0.817	0.379, 0.609	0.472, 0.724
Comparing NOA-Grader SRO DICE to Grader-Grader SRO DICE						
N[a]	213	223	241	205	232	235
DICE mean ± SD	0.5379 ± 0.3495	0.5394 ± 0.3550	0.4951 ± 0.3612	0.5670 ± 0.3283	0.4934 ± 0.3530	0.5558 ± 0.3403
95% CI of DICE	0.4907, 0.5852	0.4925, 0.5862	0.4492, 0.5409	0.5218, 0.6122	0.4477, 0.5390	0.5120, 0.5995
Mean	0.5852	0.5862	0.5409	0.6122	0.5390	0.5995
DICE median	0.6811	0.6799	0.6536	0.7161	0.6129	0.6956
NPA	0.764 (123/161)	0.801 (113/141)	0.856 (95/111)	0.814 (131/161)	0.646 (104/161)	0.716 (101/141)
95% CI of NPA[b]	0.691, 0.827	0.726, 0.864	0.776, 0.915	0.745, 0.871	0.567, 0.720	0.634, 0.789
Comparing NOA-Grader IRO DICE to Grader-Grader IRO DICE						
N[a]	160	172	224	180	231	229
DICE mean ± SD	0.4594 ± 0.3139	0.4316 ± 0.3263	0.2972 ± 0.3107	0.5100 ± 0.3312	0.3631 ± 0.3358	0.4227 ± 0.3444
95% CI of DICE	0.4104, 0.5084	0.3825, 0.4807	0.2563, 0.3381	0.4613, 0.5587	0.3196, 0.4067	0.3779, 0.4676
Mean	0.5084	0.4807	0.3381	0.5587	0.4067	0.4676
DICE median	0.5429	0.5133	0.1970	0.6519	0.4059	0.5467
NPA	0.946 (176/186)	0.965 (164/170)	0.966 (112/116)	0.839 (156/186)	0.565 (105/186)	0.629 (107/170)
95% CI of NPA[b]	0.903, 0.974	0.925, 0.987	0.914, 0.991	0.778,0.888	0.490, 0.637	0.552, 0.702

[a] N is the number of cases that have a segmentation from NOA or the Grader
[b] Exact CI per binomial distribution

Table 12.8 Risk to health and respective risk control measures

Risks to health	Mitigation measures
Inaccurate results related to false negative findings, leading to missed clinician alerts and delayed disease management	Clinical performance testing findings for missed clinician alerts Non-clinical performance testing for delayed disease management Software description, verification, validation and hazard analysis Human factors validation testing Labeling
Inaccurate results related to false positives findings, leading to unnecessary medical procedures	Clinical performance testing findings for missed clinician alerts Non-clinical performance testing for delayed disease management Software description, verification, validation, and hazard analysis Human factors validation testing Labeling
Operator failure to self-image and obtain images that meet input quality specifications, resulting in failure to monitor the disease progression	Training Human factors validation testing Labeling
Ocular light toxicity	Non-clinical performance testing
Equipment malfunction leading to user injury (e.g., shock, burn, interference)	Electromagnetic compatibility (EMC) testing Electrical safety testing Labeling
Adverse tissue reaction	Biocompatibility evaluation

12.4.9 Special Controls

In combination with the general controls of the FD&C Act, the home monitoring ophthalmic imaging device is subject to the following special controls:

12.4.10 Benefit-Risk Determination

The Notal Vision Home Optical Coherence Tomography (OCT) System is an Artificial Intelligence (AI)-based Home Use device indicated for visualization of intraretinal and subretinal hypo reflective spaces in a 10° by 10° area centered on the point of fixation of eyes diagnosed with neovascular age-related macular degeneration (NV-AMD). In addition, it provides segmentation and an estimation of the volume of hyporeflective spaces. The Notal Home OCT device is intended for imaging at home between regularly scheduled clinic assessments and not intended to be used to make treatment decisions or replace standard-of-care regularly scheduled examinations and clinical testing as needed, including in-office imaging and assessments for changes in vision, by an ophthalmologist.

12.4.11 *Conclusion*

The De Novo request for the Notal Vision Home Optical Coherence Tomography (OCT) System is granted.

12.5 Summary

The US FDA's AI/ML-Enabled Medical Devices List included all medical devices that incorporated artificial intelligence or machine learning and had received approval or clearance from the agency. Radiology stood out among medical specialties, accounting for the largest share of FDA-authorized devices. Siemens Medical and GE Healthcare were the leading manufacturers, holding the highest numbers of AI/ML-enabled devices on the list. The majority of these devices were authorized through the premarket clearance (510(k)) pathway. This chapter further analyzed the public information summaries for two notable AI/ML-enabled devices cleared by the US FDA: the Sleep Apnea Notification Software as a Medical Device (SaMD) developed by Apple Inc. and the Home Optical Coherence Tomography (OCT) System by Notal Vision. The review provided detailed insights into device development, design considerations, clinical testing protocols, risk management approaches, human factors requirements, and a predetermined change control plan (PCCP), all of which were essential for meeting the US FDA's stringent regulatory standards.

References

1. US FDA AI/ML-enabled medical devices list, 10 Jun 2025. https://www.fda.gov/medical-devices/software-medical-device-samd/artificial-intelligence-and-machine-learning-aiml-enabled-medical-devices
2. Sleep Apnea Notification Feature (SANF), Apple Inc. https://www.accessdata.fda.gov/scripts/cdrh/cfdocs/cfPMN/pmn.cfm?ID=K240929
3. Notal Vision Home Optical Coherence Tomography (OCT) system, Notal Vision Inc. https://www.accessdata.fda.gov/scripts/cdrh/cfdocs/cfpmn/denovo.cfm?id=DEN23004
4. https://notalvision.com/services/scanly-oct

Glossary

Accuracy Accuracy is a metric to indicate goodness of ML classification model.

Activation Function *Activation function* is a nonlinear function applied to the weighted sum of the inputs of a *neuron* in a *neural network*. The presence of activation functions makes neural networks capable of approximating virtually any function.

Artificial Intelligence (AI) AI is that part of computer science which aims to build machines capable of doing human-like tasks such as decision-making, object classification and detection, speech recognition and translation.

Autoencoder An *autoencoder* is a *neural network* whose goal is to predict the input itself, typically through a *bottleneck* layer somewhere in the network. By introducing a bottleneck, the network is forced to learn a lower dimensional representation of the input, effectively compressing the input into a good representation. Autoencoders are related to *principal component analysis* and other *dimensionality reduction* and *representation learning* techniques but can learn more complex mappings due to their nonlinear nature.

Bag of Words *Bag of words* is a method of feature engineering for text documents. According to the bag of word approach, in the feature vector, each dimension represents the presence or absence of a specific token in the text document. Therefore, such an approach to representing a document as a feature vector ignores the order of words in the document. However, in practice, bag of words often works well in document classification (see Token later).

Bagging Bagging or bootstrap averaging is a technique where multiple models are created on the subset of data, and the final predictions are determined by combining the predictions of all the models. It is designed to improve the stability and accuracy of ML algorithms. It also reduces variance and helps to avoid over fitting (see over fitting later).

Baseline A *baseline* is an algorithm, or a heuristic, which can be used as a *model*. It is often obtained without using machine learning by using the most simplistic engineering method.

A. Pandey et al., *Artificial Intelligence in Medical Software*,
https://doi.org/10.1007/978-3-032-10808-1

Batch *Batch*, or mini batch, is the set of examples used in one iteration of model training using *gradient descent.*

Bias Bias is an error from erroneous assumptions in the learning algorithm. High bias can cause an algorithm to miss the relevant relations between features and target outputs (under fitting).

Bias-Variance Tradeoff The *bias-variance tradeoff* is the property of a set of *predictive models* whereby models with a lower *bias* in parameter estimation have a higher *variance* of the parameter estimates across samples, and vice versa. The bias-variance problem is the conflict in trying to simultaneously minimize these two sources of error that prevent *supervised learning* algorithms from generalizing beyond their *training set.*

Big Data Big data is a term that describes the large amount of data—both structured and unstructured. But it's not the amount of data that's important. It's how organization uses this large amount of data to generate insights.

Binary Variables Binary variables are those variables that can have only two unique values. For example, a variable "Smoking Habit" can contain only two values like "True" and "False".

Binning *Binning* (also called bucketing) is the process of converting a continuous feature into multiple binary features called bins or buckets, typically based on value range. For example, instead of representing age as a single integer-valued feature, the analyst could chop ranges into discrete bins.

Boosting Boosting is a sequential process, where each subsequent model attempts to correct the errors of the previous model. The succeeding models are dependent on the previous model.

Bootstrapping Bootstrapping is the process of dividing the data set into multiple subsets with replacement. Each subset is of the same size of the data set. These samples are called bootstrap samples. Bootstrapping can be used to estimate a quantity of a *population*. This is done by repeatedly taking small samples, calculating the statistic, and taking the average of the calculated statistics.

Bootstrapping model means training a model on a small set of labeled data, and then manually reviewing unlabeled examples for errors, and then adding those to the training set in an iterative **process.**

Business Analytics Business analytics is mainly used to show the practical methodology followed by an organization for exploring data to gain business insights. The methodology focuses on statistical analysis of the data.

Business Intelligence Business intelligence are a set of strategies, applications, data, technologies used by an organization for data collection, analysis, and generating insights to derive strategic business opportunities.

Categorical Variables Variables with a discrete set of possible values. Can be ordinal or nominal. (see ordinal and nominal).

Centroid A *centroid* is the center of a cluster as determined by a *K-means* or K-median algorithm.

Classification Threshold It is the value that is used to classify a new observation into categories. Threshold value of probability of output is used to classify into classes.

Classification It is a supervised learning method. A prediction method that assigns each data point to a predefined category, e.g., is the email spam or not spam? It can be Binary classification (two classes) or multi-class classification. (more than two classes).

Clustering Clustering is an unsupervised learning technique to group data into similar groups or buckets.

Confidence Interval A confidence interval (CI) is used to estimate what percent of a population fits a category based on the results from a sample population.

Confusion Matrix Confusion matrix is a table that is often used to describe the performance of a classification model. It is $C*C$ matrix where C is the number of classes. Confusion matrix is formed between predictions of model classes vs. actual classes. One axis of the confusion matrix is the *label* that the model predicted, and the other axis is the actual label. Confusion matrices can be used to calculate different performance metrics, such as *precision* and *recall*.

Continuous Variables Continuous variables are those variables that can have infinite number of values within a specified range, e.g., sales, lifespan, weight.

Convergence A state reached during the training of a model satisfying certain criterion. An iterative algorithm is said to converge when as the iterations proceed the output gets closer and closer to a specific value.

Correlation Correlation is the ratio of covariance of two variables to a product of variance (of the variables). It takes a value between +1 and −1. An extreme value on both the side means they are strongly correlated with each other. A value of zero indicates no correlation. The most widely used correlation coefficient is Pearson Coefficient.

Cosine Similarity Cosine similarity is the cosine of the angle between two vectors. It measures the similarity between two vectors. Two parallel vectors have a cosine similarity of 1 and two vectors at 90° have a cosine similarity of 0. Suppose we have two vector A and B. Cosine similarity of these vectors can be calculated by dividing the dot product of A and B with the product of the magnitude of the two vectors as given below:

$$\text{sim}(A,B) = \cos(\theta) = \frac{A \bullet B}{\|A\| \bullet \|B\|}$$

Cost Function Cost function is to define and measure the error of the model the cist function is given by:

$$J(\theta_0,\theta_1) = \frac{1}{2m}\sum_{i=1}^{m}\left(\left(h_\theta x^i\right) - y^i\right)^2$$

where

- $h(x)$ is the prediction
- y is the actual output
- m is the number of observations on the training set

Cross-Validation Cross-validation is a technique that involves reserving a particular sample of a data set not used for training the model. Model is tested on this data set to evaluate its performance. Cross-validation can be used for model selection or hyper parameter tuning (see hyper parameter). There are various methods to perform cross-validation such as:

- K-fold cross-validation
- Leave one out cross-validation (LOOCV)
- Stratified K-fold cross-validation

Data Frame Data frame is a two-dimensional labeled data structure with columns of potentially different types defined in the context of use of R and Python.

Data Mining Data mining is a study of extracting useful information from data. Data mining is done for the purposes of market analysis, customer purchase pattern, fraud detection, predicting annual sales, etc.

Dataset A dataset (or data set) is a collection of data. A dataset is organized into some type of data structure.

Data Science Data science is a combination of data analysis, algorithm development, and technology in order to solve analytical problems. The main goal is to generate business values.

Data Transformation Data transformation is the process to convert data from one form to the other. This is usually done during preprocessing step.

Decision Boundary A decision boundary or decision surface is a hyper surface that partitions the underlying vector space into two or more sets, one for each class. How well the classifier works depends upon how closely the input patterns to be classified resemble the decision boundary.

Deep Learning Deep learning is associated with machine learning algorithms (artificial neural networks), which uses the concept of human brain to facilitate the modeling of some arbitrary function.

Descriptive Statistics Descriptive statistics comprises those values, which explains the spread and central tendency of data. For example, mean is a way to represent central tendency of the data, whereas range is a way to represent spread of the data.

Degree of Freedom It is the number of variables that have the choice of having more than one arbitrary value.

Dimension Dimension means how many features one has in data.

Dimension Reduction Dimension reduction refers to the process of converting a set of data having vast dimensions into data with fewer dimensions ensuring that it conveys similar information concisely and consistently. Dimensionality reduction is helpful in training a *model* using a bigger dataset. Also, in many cases, the *accuracy* of the model increases after the original dataset is transformed into a dimensionality-reduced dataset.

Dummy Variable Dummy Variable is another name for Boolean variable derived from some other variable or a combination of variables. An example of dummy variable is that it takes value 0 or 1. 0 means value is true (i.e., gender = male) and 1 means value is false (i.e., gender = female).

Ensemble Learning Ensemble learning is a problem of learning a strong classifier by combining multiple weak classifiers.

Ensemble Learning Algorithm An ensemble learning algorithm combines multiple weak classifiers to build a string classifier (the one with a higher accuracy than that of individual classifier).

Epoch An epoch is one pass through the training set by a machine learning algorithm.

Exploratory Data Analysis (EDA) EDA is a phase used for data science pipeline in which the focus is to understand insights of the data through visualization or by statistical analysis.

Evaluation Metrics The purpose of evaluation metric is to measure the quality of the ML model.

False Negatives Points that are actually true but are incorrectly predicted as false.

False Positives Points that are actually false but are incorrectly predicted as true.

Feature Also known as variable or attribute, is an observable quantity, used to describe an object (e.g., color, size, age, weight) and is used to create a model. It can be numerical or categorical.

Feature Selection Feature selection is the process of selecting relevant features from a data set that are required to explain the predictive power of an ML model. Note that it results in dropping irrelevant features.

Feature Vector A *feature vector* is a vector in which each dimension represents a certain *feature* of an *example*.

F-Score An evaluation metric that combines both precision and recall as a measure of effectiveness of classification.

Goodness of Fit The goodness of fit of a model describes how well it fits a given set of observations capturing the discrepancy between observed value and predicted value.

Gradient Descent Gradient descent is an iterative optimization technique for finding the minimum of a function. In ML algorithms, we use gradient descent to minimize the cost function. It finds out the best set of parameters.

Grid Search Grid search is a way of hyper parameter tuning. The process consists of training the model on all possible combinations of hyper parameter values and then selecting the best combination. The best combination of hyper parameters is the one that performs the best on the validation set (see hyper parameter).

Holdout Sample While working on the dataset, a small part of the dataset is not used for training the model instead, it is used to check the performance of the model. This part of the data set is called the holdout sample.

Hybrid Cloud When an organization maintains a local Private Cloud to perform on-premises computing, as well as uses servers in a Public Cloud for backup or overflow of computing tasks, is called a Hybrid Cloud.

Hyper Parameter Hyper parameter is parameter whose value is set before training an ML model. Different models require different hyper parameters and some require none. Hyper parameters should not be confused with the parameters of the model because the parameters are estimated from the data. They are tweaked to improve the performance of the model. How fast model can learn (learning rate) or complexity of a model, and number of hidden layers in a Neural Networks are some examples of hyper parameters.

Hyper Plane A *hyper plane* is a boundary that separates a space into two subspaces. For example, a line is a hyper plane in two dimensions and a plane is a hyper plane in three dimensions. In machine learning, a hyper plane is usually a boundary separating a high-dimensional space.

Hypothesis Hypothesis is a possible view or assertion of an analyst about the problem he or she is working upon. It may be true or may not be true.

Imputation Imputation is a technique used for handling missing values in the data.

Inferential Statistics In inferential statistics, one tries to hypothesize about the population by only looking at a sample of it.

Instance A data point, row, or sample in a dataset. Another term used is observation.

Iteration Iteration refers to the number of times an algorithm's parameters are updated while training a model on a data set.

Labeled Data The output of an observation in supervised learning has a "class" or "tag" associated with each of its observation.

Learning Algorithm A learning algorithm, or a machine learning algorithm, is an algorithm that can produce a model by analyzing a dataset.

Learning Rate The size of the update steps to take during optimization loops like Gradient Descent. During each iteration, the gradient descent algorithm multiplies the gradient by the learning rate. The resulting product is called the gradient step.

Log Loss Log Loss or Logistic Loss is one of the evaluation metrics used to find how good the model is. Lower the Log Loss better is the model. Log Loss is the logarithm of the product of all probabilities. Log Loss for two classes is defined as:

$$-(y \log(p) + (1 - y) \log(1 - p))$$

where y is the class label and p is the predicted probability.

Machine Learning Machine learning is at the intersection of subfield of computer science, mathematics, and statistics that focus on the design of systems that can learn from and make decisions and predictions based on data.

Maximum Likelihood Estimation It is a method for finding the values of parameters that maximize the likelihood. The resulting values are called maximum likelihood estimates (MLE).

ML-as-a-Service (MLaaS) ML-as-a-service is an array of services that provide machine learning tools as part of cloud computing services. This can include tools for data visualization, facial recognition, natural language processing, predictive analytics, and deep learning.

Model A mathematical representation of a real-world process. A predictive model is used to forecast a future outcome based on the past behavior. Models are trained using a data set.

Model Selection Model selection is the task of selecting a statistical model from a set of known models.

NaN NaN stands for "not a number." It is a numeric data type representing an undefined or unrepresentable value.

Natural Language Processing (NLP) Natural Language Process, or NLP for short, is a field of study focused on the interactions between human language and computers. NLP helps machines "read" text by simulating the human ability to understand language. It sits at the intersection of computer science, artificial intelligence, and computational linguistics.

Nominal Variable Nominal variables are categorical variables having two or more categories without any kind of order.

Normalization Normalization is the process of rescaling data so that they have the same scale typically in the interval [−1, +1] or [0, 1]. Normalization is used to avoid over fitting and improving computation speed. Normalization is used when the attributes in data have varying scales.

Noise Any irrelevant information or randomness in a dataset that obscures the underlying pattern.

Observation A data point, row, or sample in a dataset.

One Hot Encoding One Hot Encoding is done usually in the preprocessing phase. It is a technique that converts categorical variables to numerical value.

Ordinal Variable Ordinal variable are those variables that have discrete values but have total order involved.

Outlier An observation that deviates significantly from normal observations in the dataset.

Over Fitting Over fitting occurs when model learns the training data too well and incorporates details and noise specific to dataset. One can tell a model is over fitting when it performs great on training set, but poorly on test set (or new data).

Parameters Parameters are properties of training data learned by training a machine learning model. They are adjusted using optimization function and unique to each experiment.

Pattern Recognition Pattern recognition is the ability to detect arrangements of characteristics or data that yield information about a given system or data set. Pattern recognition is essential to many overlapping areas of IT, including big data analytics, biometric identification, security, and artificial intelligence (AI).

Precision Precision can be measured as of the total actual positive cases; how many positives were predicted correctly. It can be represented as:

$$\text{Precision} = \frac{\text{TP}}{\text{TP} + \text{FP}}$$

Predictor Variable Predictor variable is used to make prediction for dependent variable.

Private Cloud If an organization maintains its compute resources in-house and shares them with internal users is called a Private Cloud. Typically, these users belong to the organization that maintains the Private Cloud resources.

Public Cloud When Cloud Resources are located remotely and made available to the users, anytime and anywhere, on a commercial basis is called a Public Cloud.

***P*-Value** *P*-value is the probability of getting a result equal to or greater than the observed value when the null hypothesis is true.

Random Search Random search is a hyper parameter tuning technique in which the combinations of different hyper parameter values are first generated and then they are sampled randomly and used to train a model.

Range Range is the difference between the highest and the lowest value of the data. It is used to measure the spread of data.

Recall Recall is described as the measured of how many of the positive predictions were correct. It can be represented as:

$$\text{Recall} = \frac{\text{TP}}{\text{TP} + \text{FN}}$$

Recommendation Algorithms Algorithms that help machines suggest a choice based on its commonality with historical data.

Regression A prediction method whose output is a real number, i.e., a value that represents a quantity along a line. It is a supervised learning method. Example: predicting the stock price or the revenue of a company.

Regularization Regularization is a technique utilized to combat the over fitting problem. This is achieved by adding a complexity term to the loss function that gives a bigger loss for more complex models.

Reinforcement Learning Training a model to maximize a reward via iterative trial and error. *Reinforcement learning* is a subfield of machine learning where the machine perceives its environment's state as a vector of features. The machine can execute actions in every state and different actions bring different rewards and move the machine to another state. The goal of the reinforcement learning is to learn a policy that is the prescription of the optimal action to execute in each state. The action is optimal if it maximizes the average reward.

Residual Residual of a value is the difference between the observed value and the predicted value of the quantity of interest. Using the residual values, one can create residual plots that are useful for understanding the model.

Response Variable Response variable (or dependent variable) is that variable whose variation depends on other variables. (Predictors or independent variables).

Receiver Operating Characteristic (ROC) Curve A plot of the true positive rate against the false positive rate at all classification thresholds. This is used to evaluate the performance of a classification model at different classification thresholds.

Robotic Process Automation (RPA) RPA uses software with AI and ML capabilities to perform repetitive tasks earlier performed by humans.

Root Mean Squared Error (RMSE) RMSE is a measure of the differences between predicted value and observed values. It is the standard deviation of the residuals. The formula for RMSE is given by:

$$\text{RMSE} = \sqrt{\frac{\sum_{i=1}^{N}\left(y_i - \hat{y}_i\right)^2}{N}} \text{ where}$$

- $\hat{y}_i$ —Predicted value
- y_i—Actual observed value

- N—Total number of data points

Similarity Metric A similarity metric is a function that takes two feature vectors as input and returns a real number that indicates how these two feature vectors are "similar." Usually, the more two vectors are similar, the higher is the real number. Most often similarity metrics are used in clustering.

Skewness Skewness is a measure of symmetry. A data set is symmetric if it looks the same on the left and right of the center point.

Standardization Standardization (or Z-score normalization) is the process where the features are rescaled so that they'll have the properties of a standard normal distribution with mean equal to zero and standard deviation equal to 1. Z-score is calculated as follows:

$$Z = \frac{x - \mu}{\sigma}$$

where μ is the mean and σ is the standard deviation.

Standard Error A standard error is the standard deviation of the sampling distribution of a statistic. The standard error is a statistical term that measures the accuracy of which a sample represents a population.

Supervised Learning Training a model using a labeled dataset. The goal is to predict a label from a given set of predictors. Using the set of predictors, a function is generated which maps inputs to desired outputs. The goal is to approximate the mapping so well that predicted output to the new data is close to the desired output.

Target In statistics, it is called the dependent variable; it is the output of the model or value of the variable one wishes to predict.

Test Set A dataset with the same structure as training data, used to measure the performance of a model. How generalizable is model to unseen data?

Tokenization Tokenization is the process of splitting a text string into units called tokens. The tokens may be words or a group of words.

Training Set The *training set* is a collection of randomized *examples* used an input to the *learning algorithm* to create a *model.*

Training The process of creating model from the training data. The data is fed into the algorithm, which learns a representation for the problem, and produces a model that can be used for prediction.

True Negative These are the points that are actually false, and we have predicted them false.

True Positive These are the points that are actually true, and we have predicted them true.

***T*-Test** T-test is used to compare two data sets by comparing their means.

Type I Error False Positives. Consider a company optimizing hiring practices to reduce false positives in job offers. A type 1 error occurs when candidate seems good and they hire him, but he is bad. The decision to reject the null hypothesis could be incorrect.

Type II Error False Negatives. The candidate was great, but the company passed on him. The decision to retain the null hypothesis could be incorrect.

Under fitting Under fitting occurs when model over-generalizes and fails to incorporate relevant variations in the data that would give model more predictive power. One can tell a model is under fitting when it performs poorly on both training and test sets.

Unstructured Data Information that either does not have a predefined data model or is not organized in a predefined manner.

Unsupervised Learning Training a model to find patterns in an unlabeled dataset (e.g., clustering). In the type of learning, we do not have a target or outcome variable.

Validation Set A set of observations used during model training to provide feedback on how well the current parameters generalize beyond the training set. If training error decreases but validation error increases, model is likely over fitting, and one should pause training.

Variance The variance is an error from sensitivity to small fluctuations in the *training set.*

- **Low variance** suggests model is internally consistent, with predictions varying little from each other after every iteration.
- **High variance** (with low bias) suggests model may be over fitting and reading too deeply into the noise found in every training set.

Weak Classifier In ensemble learning, a weak classifier is usually one of a collection of low-accuracy classifier, which when combined by an ensemble learning algorithm, can produce a string classifier.

Z-Test *Z*-test determines to what extent a data point is away from the mean of the data set in standard deviation.

Index

A
Accuracy, 427
Activation Function, 427
Advanced Encryption Standard (AES), 296
Agile Scrum, 13–15, 33
AI Act, 31–33, 227, 306, 368–373, 382
AI deployment, 258, 263, 273, 282
AI integration, 251–252, 258–259
AI21 Jurassic, 195
AI labeling, 352–353
AI/ML, 2, 61, 74, 135, 202, 214, 253, 298, 317, 375, 406, 410
AI/ML-enabled medical device regulatory approval, 356, 363, 370
AI/ML-enabled medical devices, 3, 4, 19, 20, 23, 27–33, 37, 136, 194, 237, 243, 289–313, 350–383, 410, 424, 425
AI/ML lifecycle, 22, 33
AI model testing, 246
AI risk management, 227
AI software development, 382
Amazon Titan, 195, 283
Anomaly detection, 79, 80, 125, 133, 135, 174, 278, 284, 357
Any-to-any multimodal models, 207, 208
Artificial intelligence (AI), xxvii, 2, 63, 74, 114, 115, 172, 195, 214, 252, 302, 315, 354, 394, 419, 427, 433, 434
Artificial intelligence software development, 4, 20, 21, 32, 33
Autoencoder, 427
Autonomy, 224, 356, 359, 360, 368–370, 375

B
Baseline, 427
Bias, 22, 25, 26, 28, 32, 59, 60, 62–64, 80, 97, 98, 105, 136, 154–155, 177–179, 183, 199, 215–217, 221, 222, 227–232, 237, 240, 243, 246, 252, 253, 255–257, 307, 333, 350, 354, 355, 357, 360, 361, 364, 365, 374, 376–379, 381, 382, 403–405
Bidirectional Encoder Representations from Transformers (BERT), 195
Big data, 428
Binning, 428
BLA, 358
Blood flow, 386, 387
Bloom, 196

C
Cancer, 39, 42, 78, 206, 229, 385, 388, 391, 404, 406
Cardiovascular, 183, 385, 386, 404, 406, 411
Catastrophic forgetting, 199–200, 208
Central Processing Unit (CPU), xxvii
Change control, 11, 33, 291
ChatGPT, 308
Classification, 2, 6, 7, 11, 27, 28, 31, 34, 61, 76–78, 80–82, 86–89, 92–94, 96–101, 103–106, 108, 109, 112, 113, 115, 117–119, 122, 123, 125, 127, 128, 132, 133, 135–141, 145, 147–149, 160–161, 170, 180, 181, 186, 191, 192, 209, 222,

A. Pandey et al., *Artificial Intelligence in Medical Software*,
https://doi.org/10.1007/978-3-032-10808-1

223, 244, 251, 259, 278, 279, 298, 299, 302, 321, 342, 345, 356, 369, 374, 396–398, 413, 415, 416, 428, 429
Claude, 195
Claude 3 Haiku, 196
Claude 3 Opus, 196
Claude 3.5 Sonnet, 159, 196
Clinical evaluation, 7, 27–28, 243–246
Clinical research, 50, 52, 53, 56–59
Clinical testing, 414, 417, 419, 424, 425
Clinical validation, 32, 63, 135, 183, 193, 215, 230, 237–240, 242, 244, 246, 255, 359, 361–362, 381, 403
Cloud computing, 291, 299
Cloud multimodal, 269
Clustering, 78, 79, 111–114, 128, 133, 149–151, 174, 186, 297, 298, 342, 345, 398, 429
Cohere, 196
Comorbidity, 48, 49, 63, 402
Compliance, 4–7, 15, 17, 18, 23, 28, 30, 32, 33, 57, 58, 60, 63, 64, 66, 68, 136, 193, 214, 216, 217, 225, 227, 230, 244, 246, 252, 253, 255–260, 269–273, 276, 278, 282–284, 293, 302–306, 312, 350, 354, 355, 362, 364, 366, 369, 373, 381–383, 405
Computational costs, 181–183
Confidence interval (CI), 27, 260, 272, 274, 326–328, 330, 331, 415, 416, 422
Confusion matrix, 138–139, 147–149, 353, 378
Containerization, 260, 270, 272, 273, 275, 276, 282
Convergence, 113, 429
Convolutional neural networks (CNNs), xxvii
Cross validation, 62, 106, 147, 167–170, 174, 176–179, 181, 183, 217, 239, 241, 365
Cybersecurity, 2, 5, 10, 17–19, 23, 24, 28–30, 32, 255, 289–295, 301–305, 310–311, 351, 355, 357, 378, 381, 416, 420

D

Data annotation, 62, 355
Data collection approaches, 50
Data lifecycle, 63–67, 269, 344, 355
Data management, 2, 4, 22, 24, 37, 214, 276, 295, 297–300, 306, 311, 344, 351, 354, 364–365, 373, 417
Data mining, 430
Data preprocessing, 24, 66–70, 252, 273, 365
Data quality, 21, 24, 39, 40, 47, 57–59, 61, 63–69, 136, 151, 172, 195, 217, 239–243, 252–254, 258, 283, 364, 369, 374, 376–379, 403
Dataset splitting, 183
Decision tree (DT), 81, 99–103, 105, 106, 108, 109, 128
Deep learning (DL), xxvii, 114–127, 134, 168, 175–176, 184, 195, 203, 234, 239, 256, 279, 307, 316, 321, 322, 338, 339, 342, 345, 346, 350, 388, 395, 396, 398–399, 401, 402, 415, 430
De Novo (DEN), 6, 356, 358, 410, 412, 413, 425
Deployment, 4, 64, 74, 148, 197, 216, 255, 291, 315, 352, 390
Deployment strategies, 32, 256, 260, 263–268, 271, 282, 325–326
Design for interpretability, 232–234
Device classification, 369
DevOps, xxvii, 26, 253, 318, 329
Dice score, 185
Dimensionality reduction, 69, 79, 133, 149–151, 174, 184, 242, 398
Dimension reduction, 430
Distributed denial of service (DDOS), xxvii
Docker, 260, 270, 273, 276, 278, 317, 319, 340–341
Domain knowledge, 157, 165–167, 193, 220, 233, 242, 251–252
Dropout, 173, 180, 200, 242

E

Early stopping, 173, 175, 179–181, 183, 415
Edge deployment, 197, 255, 272
Electronic health records (EHRs), 37–40, 45–47, 49, 255, 258, 276, 284, 399
Encryption, 5, 17, 19, 198, 255, 260, 271, 283–285, 292, 296–297, 299, 300, 311, 357, 405
EU AI Act, 31–33, 63, 246, 260, 305–306, 312, 367–374
Evaluation, 7, 57, 82, 135, 199, 214, 257, 306, 332, 350, 389, 414
Evaluation metrics, 22, 23, 82, 145, 151, 162, 168, 170, 178, 193, 239, 255, 365
Explainability, 22, 28, 82, 135, 136, 165, 168, 217–227, 230, 232–234, 246, 252, 272, 289, 362, 378, 380–381

F
False negatives, 431, 436
False positives, 431, 435
FDA guidance, 11–13, 230, 238, 244–246, 290, 355
Federated learning (FL), 297, 388, 389, 405
Field programmable gate array (FPGA), xxvii
Fine-tuning, 127, 156, 158, 174, 192–195, 198–200, 208, 242, 275, 276, 309
510(k), 6, 135, 350, 356, 358, 410, 412, 413, 425
Food and Drug Administration (FDA), 2, 6, 12, 13, 17, 18, 22, 23, 27, 28, 33, 34, 57, 135, 136, 216, 231, 232, 243, 253, 259, 271, 290, 297, 300–302, 306, 307, 311, 347, 350, 354, 356, 357, 363, 365, 374, 381, 382, 406, 407, 409, 412
Foundation models (FMs), 183, 193, 195–199, 208
4M multimodal AI, 207

G
GE Healthcare, 410, 412, 425
General Data Protection Regulation (GDPR), 6, 17, 28–30, 198, 260, 269, 290, 312, 405
Generalizability, 25, 52, 59–61, 104, 136, 170, 177, 182, 183, 193, 215, 230, 237, 239–243, 254, 307, 354, 355, 364, 396, 401, 402
Generative AI (Gen AI), 32, 33, 125, 151–160, 183, 207, 283, 284, 311, 339, 342, 403, 405, 406
Generative pre-trained transformer (GPT), 152, 195, 308, 309
Glaucoma, 385, 391–403, 406
Good Clinical Practice (GCP), xxvii
Good machine learning practices (GMLP), 23–26, 135, 216, 219, 259–260, 306, 359
Google Cloud Platform, xxvii
Graphical processing units (GPU), xxvii, 117

H
Hallucination Index Report, 159–160, 183
Hallucination rate, 155
Hashing, 296
Health Canada, 216, 219, 311, 358–363, 381, 382
Healthcare, 5, 16–20, 28–30, 33, 38–51, 53, 60, 68, 78, 135, 154, 165, 182, 183, 192, 193, 200–205, 215, 216, 219, 220, 222–230, 232, 233, 236–240, 252–260, 262, 263, 269, 276, 277, 281, 283, 284, 298, 302, 307, 309, 310, 350, 353, 358, 359, 362, 364, 372, 377, 379, 383, 385, 389, 394–396, 402–407, 412
Health Insurance Portability and Accountability Act (HIPAA), 17, 28–30, 49, 50, 198, 254, 255, 257, 258, 260, 269–271, 283, 284, 290, 293, 295, 312, 405, 406, 408
Health level seven (HL7), 5, 17, 46, 283, 284
High performance computing (HPC), xxviii
Homomorphic encryption (HE), xxviii
Hugging Face, 196, 209
Human-AI interaction, 215
Human factors, 5, 18, 25, 33, 355, 356, 362, 379–381, 416, 420, 424, 425
Human-in-command (HIC), 32, 234–237, 246, 257–258, 378
Human-in-loop, 32, 234–237, 246
Human-in-the-loop, 25, 234–237, 254, 257–258, 378
Hybrid Cloud, 431
Hybrid Environment, 270
Hypothesis, 432

I
Idempotency, 325
IEC 62304, 2–4, 8–13, 18, 20, 21, 34, 352, 380
IEC 62366, 245
Indication for use, 4, 413
Information Security (InfoSec), xxviii
Infrastructure as a Service (IaaS), xxviii
Integration, 4–7, 10, 17, 19, 21, 23–25, 32, 46, 53, 63, 127, 135, 158, 159, 184, 198, 199, 201, 207, 224, 251–285, 318, 327, 331, 341, 345, 350, 366, 379, 381, 402, 404
Internet of Things (IoT), xxviii
Interpretability, 22, 23, 25, 32, 136, 145, 149, 151, 165, 167, 172, 206, 218–227, 232–234, 239, 242, 246, 252, 254, 306, 355, 396, 402
In Vitro Diagnostic Medical Devices Regulation (IVDR), 243–244, 259, 290, 302, 306, 369, 373
ISO 14971, 8, 354, 360, 374, 380, 381

K

K-means, 79, 111–114, 133, 174

L

Large language models (LLMs), xxviii, 125, 156–160, 171–174, 183, 199, 200, 226, 254, 255, 283, 284, 307–309, 311, 313, 404
Latency, 20, 158, 173, 197, 254, 255, 263, 270, 271, 274–276, 278, 279, 284, 299, 300, 343, 390, 405
Legacy devices, 310, 311
Legacy operating systems, 310
Linear regression, 77, 80, 82–86, 89–91, 132, 220, 232

M

Machine learning (ML), 24, 38, 74, 132, 195, 214, 251, 308, 315, 350, 387, 425
Machine learning integration, 251–285
Machine learning operations (MLOps), xxviii, 26–27, 33, 127, 272, 273, 282, 315–348
Manufacturer Disclosure Statement for Medical Device Security (MDS2), 309, 310, 312
Marketing submissions, 22, 307, 347, 350–358, 363, 382
Massively Multimodal Masked Modeling (4M), 207–208
MDCG guidance, 243
Medical Device Coordination Group (MDCG), 7, 28, 243
Medical Device Regulation (MDR), 2, 7, 17, 28, 29, 31, 243–246, 259, 260, 290, 302, 305, 306, 369, 373, 382, 383
Medical device software (MDSW), 2–20, 23, 27–34, 243, 244, 260, 347, 363
MLOps levels, 326–331
Mobile deployments, 270, 282
Model, 86, 92, 94, 430, 432
Model evaluation, 75, 82, 144, 148, 169, 176, 214, 215, 228, 332, 352, 397
Multi-clouds, 269, 274, 285, 389–391, 405–407
Multimodal, 32, 184, 191–210, 254, 269, 270, 274, 279, 284, 389, 391–396
Multimodal AI, 191–210, 254, 269–270
Multimodal machine learning models, 208, 269

N

Naïve Bayes (NB), 80, 92–96
National Institute of Standards and Technology (NIST), xxviii
Natural language processing (NLP), xxviii, 45, 114, 121, 122, 127, 132, 134, 152, 153, 160–162, 171, 183, 184, 307, 308, 399, 421, 433
Noise, 433
Non-clinical testing, 415–416, 420
Normalization, 433

O

Observational studies, 52
On-premise deployment, 273
Optical coherence tomography (OCT), 395–397, 401, 402, 413, 419–422, 424, 425
Outlier, 433
Overfitting, 25, 26, 59, 62, 69, 79–81, 91, 97, 103, 105, 106, 128, 135, 144–147, 165–168, 172–175, 177–183, 195, 199, 214, 216, 240, 241, 354, 357, 360, 365, 402, 433
Overtrust, 377, 378

P

Parameters, 433
Penetration, 291, 294, 301, 304, 312, 357
Performance, 6, 44, 74, 134, 191, 214, 251, 289, 315, 352, 396, 416
Performance characteristics, 260, 375, 418
Performance measures, 156, 253
Performance metrics, 32, 82, 131–186, 197, 214, 217, 224, 226, 234, 237, 242, 246, 253, 260, 263, 322, 353, 361, 363, 365, 378, 397, 421–422
Performance validation, 351, 355, 356
Personally Identifiable Information (PII), 290, 291, 297, 303, 308, 312, 321
Platform as a Service (PaaS), xxviii
Precision, 433
Predetermined change control plan (PCCP), 12, 23, 33, 135, 237, 243, 253, 260, 306, 347, 358, 360, 363–367, 381, 417–418, 425
Premarket approval (PMA), 6, 300, 356, 358, 363, 410, 412
Premarket clearance, 425

Privacy, 2, 5, 17, 18, 21, 22, 28–30, 32, 33, 38, 46, 59, 65–66, 68, 158, 172, 192, 198, 223–225, 238, 243, 253–258, 260, 269–271, 276, 285, 289–313, 333, 357, 378, 388, 403
Private Cloud, 431, 433
Prospective clinical validation, 238–239
Protected health information (PHI), 29, 30, 290, 291, 293, 294, 297–300, 302, 312
Public Cloud, 433
PyTorch, 167, 184, 209, 210, 271, 323, 342, 344

Q
Quality System Regulation (QSR), 2, 6, 366

R
Random clinical trial, 239, 240
Random forests, 77, 81, 104–108, 132, 135, 170, 178, 242
Real-world evidence (RWE), 27, 237, 243, 246
Recurrent neural networks (RNN), xxviii
Reference standards, 25, 60–62, 355, 361, 362, 364, 366, 422
Regression, 51, 76, 77, 80–85, 88–92, 94, 96, 99–101, 104–106, 108, 117, 118, 127, 128, 132, 133, 135, 136, 141–147, 149, 168, 170, 209, 233, 299, 328, 342, 345, 374, 434
Regularization, 97, 147, 167, 168, 173, 175, 178–183, 200, 233, 241, 365
Regulations, 2, 4–8, 17, 18, 28–31, 38, 46, 47, 56, 57, 193, 198, 216, 226, 244, 246, 253–255, 257–260, 269, 271, 290, 300–302, 311, 321, 347, 357, 358, 363, 364, 367, 368, 373, 382, 402, 405, 406
Regulatory frameworks, 2, 6–8, 23, 29, 33, 225, 237, 253, 255, 305, 306, 381, 382
Reinforcement learning (RL), 133, 134, 162–165, 171, 183, 360
Retrieval-Augmented Generation (RAG), 156–159, 183
Retrospective clinical validation, 237–239
Risk management, 5, 8, 11, 18, 23, 24, 33, 63, 64, 221, 292, 301, 302, 309, 310, 312, 350–383, 425
Risks, 5, 43, 105, 157, 195, 215, 252, 290, 294, 354, 386, 416
Robustness, 23, 25, 32, 62, 104, 146, 151, 163, 164, 174–176, 181, 193, 197, 206, 217–218, 241, 246, 247, 271, 307, 355, 361, 366, 374, 383, 402

S
Secure multi-party computing (SMPC), xxviii
Security, 5, 6, 10, 17, 19, 20, 22–25, 28, 30, 33, 38, 47, 59, 62, 66, 68, 158, 172, 198, 217, 253–258, 260, 268–273, 278, 289–313, 328, 343–345, 378, 388, 405, 408
Segmentation quality, 185
Sensitivity, 51, 63, 135, 137–138, 140, 145, 146, 177, 179, 185, 193, 214, 230, 260, 353, 365, 381, 397, 399, 415, 416, 418
Siemens Medical, 410, 412, 425
Sleep apnea, 413–418, 425
Software as a Medical Device (SaMD), 4, 6, 8, 13, 27, 230, 244–246, 306, 307, 312, 425
Software as a Service (SaaS), xxviii
Software Bill of Materials (SBOM), 310
Software development lifecycle (SDLC), 293
Software in Medical Devices (SiMD), 2–4, 6, 8, 13, 16, 19–20
Software safety class, 8, 9, 34
Specificity, 51, 135, 138, 140, 185, 193, 198, 214, 230, 260, 353, 365, 381, 397, 415, 416, 418
Stable diffusion, 196
Standardization, 435
Supervised, 64, 66, 76, 80, 92, 96, 99, 115, 128, 132–134, 136, 151, 167, 183, 360, 397
Supervised learning, 64, 76–80, 82, 115, 117, 128, 132, 133, 151, 171, 298, 396, 435
Support vector machines (SVMs), 77, 80, 81, 96–99, 101, 132, 149, 358, 360
Systems and Organization Controls 2 (SoC2), 302, 303, 305

T
TensorFlow, 167, 184, 209, 210, 271, 273, 284, 323, 342, 344, 346
Tensor processing unit (TPU), xxviii

Testing, 2, 4, 5, 10, 12, 21–25, 32, 33, 59, 61–63, 66, 79, 88, 94, 112, 145, 159, 166, 168, 169, 176, 177, 214–247, 253, 256, 260, 263–265, 270–273, 282, 283, 290, 293, 294, 300, 301, 304, 305, 315, 318, 322–324, 326–330, 350, 356, 357, 359, 361, 362, 364–366, 371, 378, 380, 381, 383, 392, 420, 424
TIR45, 13–16, 352
Training, 4, 22, 24, 25, 27, 28, 32, 57, 59, 61–63, 66, 68, 71, 75, 76, 80, 81, 86–88, 93, 94, 97–100, 102, 103, 105, 106, 108, 109, 112, 113, 115–121, 125–127, 131–186, 193, 194, 199, 200, 203, 207–209, 214, 217, 223, 224, 226–230, 232, 237, 241, 242, 253, 255, 258, 260, 271–274, 279, 280, 283, 285, 286, 304, 306, 308, 309, 311, 318–322, 327, 328, 330, 332, 337–339, 342, 350, 353, 354, 357–361, 364, 366, 373, 377–379, 388, 395, 402, 403, 415, 422, 424
Transfer learning, 164, 181, 183, 191–210, 242, 271, 365
Transparency, 7, 23, 28, 31, 32, 61, 63, 64, 135, 157, 172, 193, 199, 216, 218–227, 230, 232–234, 242, 246, 252, 253, 255, 271, 289, 290, 301, 307, 333, 350, 355, 358, 359, 362–363, 365–367, 369, 371, 378, 380–383, 396, 402–404
Trusted execution environment (TEE), xxviii
Tuning, 21, 22, 32, 59, 61, 62, 85, 96, 168–169, 171, 172, 174, 176–179, 181, 183, 199, 273, 309, 321–324, 346, 350, 352, 360, 361, 364

U

Underfitting, 145, 147, 168, 178, 183, 360, 436
Unsupervised, 76–78, 80, 112, 113, 115, 117, 125, 132–133, 149–151, 174–175, 183, 360, 397
Unsupervised learning, 64, 77–79, 111–115, 117, 118, 128, 132–134, 150, 151, 174, 175, 298, 397, 398, 436
Usability, 5, 25, 59, 155, 238, 275, 307, 350, 355, 356, 362, 366, 380–382
US FDA, 11–13, 21–26, 29, 33, 34, 219, 244–246, 260, 306–307, 312, 350–358, 363, 409–425

V

Validation, 5, 52, 106, 136, 193, 214, 252, 305, 318, 350, 402, 415
Variance, 80, 85, 91, 105, 106, 112, 135, 142, 144, 150, 163, 177–179, 183, 428, 436
Vision AI, 254–256
Vision processing unit (VPU), xxviii

W

Waterfall, 2, 13–16, 20, 21, 383

MIX
Papier aus verantwortungsvollen Quellen
Paper from responsible sources
FSC® C105338

If you have any concerns about our products,
you can contact us on
ProductSafety@springernature.com

In case Publisher is established outside the EU,
the EU authorized representative is:
Springer Nature Customer Service Center GmbH
Europaplatz 3, 69115 Heidelberg, Germany

Printed by Libri Plureos GmbH
in Hamburg, Germany